History of Vacuum Science and Technology

A Special Volume Commemorating the 30th Anniversary of the American Vacuum Society, 1953–1983

Edited by

Theodore E. Madey
National Bureau of Standards
Gaithersburg, MD 20899

William C. Brown
General Electric Company
Neutron Devices Department
St. Petersburg, FL 33733

Published for the American Vacuum Society by the American Institute of Physics

Library of Congress Catalog Card Number: 84-70451
International Standard Book Number: 0-88318-437-0

Published by the American Institute of Physics, Inc.,
335 East 45 Street, New York, New York 10017

Printed in the United States of America

Foreword

The present volume commemorates the 30th Anniversary celebration of the American Vacuum Society (AVS), an association of 5000 scientists, engineers, and technologists dedicated to the production, characterization, and use of vacuum in modern science and technology. The evolution of this monograph and its contents are outlined in the following paragraphs.

It has been standard practice that during the National AVS Symposium each year the Program Chairman for each Technical Division supplies the session titles and proposed invited speakers to next year's National Program Chairman for use in developing the next national program. At the 29th Annual AVS Symposium held in Baltimore, Maryland in early November 1982, it was proposed by the 1983 Program Chairman of the Vacuum Technology Division that since the 1983 Symposium would coincide with the 30th Anniversary of the AVS, a commemorative session on the History of Vacuum Science and Technology be presented as a separate session. Also at the Baltimore Symposium, plans were discussed for borrowing the Magdeburg Hemispheres from the Deutsches Museum in Munich, Germany to be put on display at the 1983 Symposium. Both proposals were received enthusiastically and supported by the 1983 Program Chairman and Committee as well as by the National AVS Board of Directors.

After these endorsements, action immediately began in selecting subjects for papers to be presented in the session on the History of Vacuum Science and Technology. The plan to obtain the Magdeburg Hemispheres was expanded to include an exhibit of early vacuum equipment obtained from many locations in this country and overseas. Invitations to present the historic papers were sent to distinguished persons in their particular fields of Vacuum Science and Technology. The positive responses to these invitations were extremely gratifying and this part of the program was very quickly formulated. Concurrently, commitments for the borrowing of early pieces of vacuum equipment were progressing at a rapid pace. Some administrative obstructions were encountered; however, these were overcome in very short order. Cooperation of all involved was extraordinary. In addition to the oral presentations and the History of Vacuum exhibit, an actual reenactment of the Magdeburg Hemispheres experiment was planned as a part of the 30th anniversary commemoration.

With all plans in order, the historical session and exhibit were presented at the 30th Anniversary of the AVS in Boston, Nov. 1–3, 1983. During the oral presentations of the papers on the history of Vacuum Science and Technology, standing room only was available at each and every presentation. In the exhibit hall, where the early vacuum equipment and explanations were on display, excellent attendance was noted at all times. The success of the display and oral session was so recognized that a suggestion was made to publish a monograph including all the information presented, providing a single document for reference purposes. This proposed monograph was expanded somewhat to include reprints of some early historical papers. The AVS Board of Directors approved the necessary finances for publication, and with much work and cooperation by many people this monograph is now a reality.

This monograph is divided into three sections: (1) texts of the oral presentations at the History of Vacuum Science and Technology session in Boston (reprinted from the *Journal of Vacuum Science and Technology*), (2) photographs and explanations describing the History of Vacuum exhibit as well as a description of the reenactment of the Magdeburg Hemispheres experiment, and (3) a collection of early papers in the field of Vacuum Science and Technology.

Many people are to be acknowledged for their work in making this entire program a success. At the risk of omitting some, a listing of the people to be thanked is given on a separate page.

W. C. Brown

T. E. Madey

CONTENTS

Acknowledgments

Many people who contributed to the success of the AVS 30th Anniversary activities and to this volume deserve special mention. The History of Vacuum Science session at the November 1983 AVS Symposium and the accompanying papers in the present volume were organized and coordinated by W. C. Brown, with the enthusiastic support of Len Brillson and assistance by Paul Johansen, Colin Alexander, Rey Whetten, Ray Berg, and Len Beavis. The History session was chaired by Luther Preuss, who introduced the speakers: J. M. Lafferty, T. E. Madey, J. H. Singleton, M. H. Hablanian, C. B. Duke, P. A. Redhead, and J. P. Hobson.

The History of Vacuum Science exhibit, coordinated by T. E. Madey, was made possible by the continual support and assistance of Lawrence Bell, Director of Exhibits at Boston's Museum of Science. Exhibit arrangements were ably handled by Ed Greeley, with assistance from John Sullivan and L. Elder. John Sullivan also worked with Lawrence Bell to arrange the Magdeburg Hemispheres reenactment, using hemispheres fabricated by C. Hemeon and M. Hablanian and draft horses provided by the New England Draft Horse Association. Thanks are due to a number of persons and institutions who provided information, lent exhibits, or assisted in other ways: D. Menzel, T. U. München; A. Brachner, Deutsches Museum, München; S. Weart, American Institute of Physics; M. W. Yeates, MIT Museum; R. W. Hoffman, Case Western Reserve University; J. A. Freeman, Balzers; R. Sherman, Smithsonian Institution, Washington; W. Weinstein, National Bureau of Standards; R. Swinburne, H. Hagstrum, AT&T/Bell Laboratories; J. Maddocks, J. Vossen, RCA Laboratories; A. O. Nier, U. Minnesota; L. Arnold, J. Helmer, R. Powell, Varian; J. Singleton, Westinghouse Research Laboratories; G. Wise, General Electric Research Laboratories; R. Weber, Physical Electronics Industries; C. Alexander, Technical Enterprises, Inc.; T. P. Shaughnessy, AVS New England Combined Chapter; D. Meyer, ITT; L. Johnson, National Bureau of Standards.

The AVS Headquarters staff—M. Churchill, N. Hammond, and M. Schlissel—provided support and guidance in many aspects of the projects. This volume was published by the American Institute of Physics on behalf of the AVS, with much effort and cheerful cooperation by Larry Feinberg, Kathy Strum, Maria Garruppo, and Kenneth Dreyhaupt of the AIP, and by G. Lucovsky, the editor-in-chief of the *Journal of Vacuum Science and Technology.*

Finally, thanks are also due to the following individuals or institutions who gave permission for us to reprint articles, photographs, etc. They include Roger Sherman of the Smithsonian Institution National Museum of American History, who translated the von Guericke work and lent many photographs reproduced herein; the publishers of *Scientific American* for permission to reprint two articles; The American Institute of Physics; Professor Robert K. de Kosky; Taylor and Francis, Ltd.; Pergamon Press; Dr. R. K. Gehrenbeck; Professor Daniel Alpert; Dr. P. A. Redhead; and Mrs. James C. Lawrence.

1

AVS 30th Anniversary Session on the History of Vacuum Science and Technology

The following seven papers are the texts of talks presented in the special 30th Anniversary session at the 30th National AVS Symposium in Boston, Nov. 1, 1983. In these papers, J. M. Lafferty outlines the history of the American Vacuum Society and the International Union for Vacuum Science, Technique and Applications; T. E. Madey reviews some early applications of vacuum science; M. H. Hablanian presents a short history of the development of vacuum pumps; J. H. Singleton concentrates on the development of valves, connectors, and traps for vacuum systems; P. A. Redhead recounts the development of measurement methods for vacuum pressures; and C. B. Duke describes the extensive and profound changes which have occurred recently in surface science, and indicates the role of the AVS in promoting and facilitating the development of this exciting field. Finally, J. P. Hobson closes with his thoughts on the future of vacuum technology.

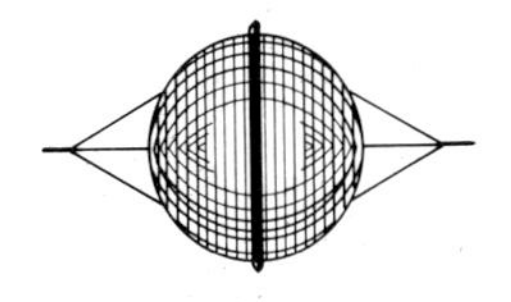

Shown here are speakers at the special 30th anniversary session on the History of Vacuum Science and Technology organized by W. C. Brown and held at the 30th National AVS Symposium in Boston, Nov. 1, 1983. From left to right: J. M. Lafferty, T. E. Madey, M. H. Hablanian, J. H. Singleton, P. A. Redhead, C. B. Duke, J. P. Hobson, and L. E. Preuss (Session Moderator). The written texts of their presentations appear on the following pages.

(Photo by J. Lyn Provo)

History of the American Vacuum Society and the International Union for Vacuum Science, Technique, and Applications

J. M. Lafferty

Corporate Research and Development, General Electric Company, Schenectady, New York 12301

(Received 29 August 1983; accepted 1 November 1983)

On the 30th anniversary of the American Vacuum Society, the Vacuum Technology Division presented a session on the History of Vacuum Science and Technology. The author was asked to present a paper on the history of the AVS and the IUVSTA for this session. Topics selected for discussion include the formation, growth, organizational structure, and strategy of operation of the AVS. Other subjects covered are publications, scholarships, and the author's own personal experience in setting up two major awards for the Society. For a more comprehensive and detailed history of the AVS, the reader should consult "The First Twenty Years of the American Vacuum Society" [J. Vac. Sci. Technol. **10**, 833 (1978)] and "The American Vacuum Society—1973 to 1983" [J. Vac. Sci. Technol. A **1**, 1351 (1983)]. Other countries have also organized their own vacuum societies. Many of these, some 22 in number, including the AVS, are members of the IUVSTA, an international confederation of national vacuum organizations. The formation and early history of this organization is described, including its new scientific divisions patterned after the AVS. Various objectives and activities of the Union are discussed.

PACS numbers: 01.10.Hx, 01.65. + g, 01.60. + q

I. INTRODUCTION

Vacuum research appears to have had its beginning around the middle of the seventeenth century with the discovery of the vacuum pump by the great German scientist and technologist, Otto von Guericke. Some time in the 1650's he succeeded in developing a vacuum pump with an exhaust arrangement that separated a definite volume of air from the vessel to be exhausted and gave it up to the atmosphere—the principle used in most mechanical pumps today. With this pump, von Guericke was able to exhaust quite large vessels and study the properties of vacua. He appears to have been much impressed with the magnitude of the force exerted by atmospheric pressure; many of his experiments were arranged to demonstrate this force. His most notable and highly publicized was the Magdeburg hemispheres experiment (Fig. 1), in which he put two small hemispheres together to form a sphere and evacuated it by means of his vacuum pump. Sixteen strong horses were unable to overcome the immense force of the atmosphere and pull the hemisphere apart.

From the beginning, vacua have been produced for a purpose. Scientists needed a vacuum environment in which to do research—for example, to increase the mean free path of an electron, ion or neutral-particle beam in the residual gas of some devices, or to provide an environment free of chemically active gases for surface or deposition studies. Industry's first need for vacuum was for the production of light bulbs, x-ray, and radio tubes. Whatever the reason, in the early days, vacuum technology was generally practiced by scientists and engineers who lacked training in the field, and whose main pursuits were research and technology in other fields. Under such circumstances, it was not too surprising that vacuum technology came to be known as black magic with string, Glyptal, sealing wax, and all the rest.[1] It was clear that professionalism was lacking in the practice of vacuum science and technology and that there was a need for a technical society devoted exclusively to this field.

II. COMMITTEE ON VACUUM TECHNIQUES

The person who apparently first recognized the need for a professional vacuum society, or at least did something about it, was Frederick A. McNally, a young scientist with the Jarrel–Ash Company. Through his efforts, an organizational meeting was held in New York on 18 June 1953, to discuss the formation of a permanent organization that could bring together both the theoretical and practical knowledge of the many different fields using vacuum as a production or research tool. This meeting resulted in the formal organization of the "Committee on Vacuum Techniques" less than a week later, and its formal incorporation in Massachusetts on 19 October 1953. Since its first symposium the following June in Asbury Park (New Jersey), with 35 papers and 295 registrants, the group has convened an annual symposium in every succeeding year.

III. THE AMERICAN VACUUM SOCIETY

In 1957, the CVT changed its name to the American Vacuum Society, Inc. The first biennial equipment exhibit was held in conjunction with its 1961 symposium. Since 1965, an equipment exhibit has been a part of every symposium. These exhibits have proved to be of great interest to the registrants, and of considerable financial help to the AVS. More than 1300 attended the 30th National Symposium in Boston in 1983, and over 400 papers were presented there.

The AVS had continued to grow (Fig. 2) and prosper, reaching a peak of over 3200 members in 1968. Its membership then declined to less than 2200 by 1982 following government cutbacks in the space and other R & D programs. During the past seven years, however, the AVS has started to grow again at an almost constant rate of about 435 members per year. Its current membership exceeds 5000.

Originally published in J. Vac. Sci. Tech. A **2**(2), 104–109 (1984).

FIG. 1. In von Guerike's experiment demonstrating the immense force of the atmosphere, 16 horses were unable to separate the evacuated Magdeburg hemispheres.

Over the years, the AVS has developed a close association with the American Institute of Physics. In April 1963, it became an affiliate of AIP, and a full member society in 1976. The AIP has been publishing the *Journal of Vacuum Science and Technology* for the AVS since its inception in 1965. The Institute also has managed the equipment exhibits at the national symposia since 1965. In December 1967, AVS moved its offices from Boston to the AIP building in New York City, and in May 1968, a full-time Executive Secretary, Nancy Hammond, was engaged by AIP for the Society. In 1979, the AVS office expanded to include word-processing equipment and another full-time employee. In 1981, a Meetings Manager, Marion Churchill, was employed to manage the nontechnical aspects of the National Symposia and the short course operations. The modern large-scale facilities of the AIP for publishing and for society membership processing have been very beneficial to the AVS.

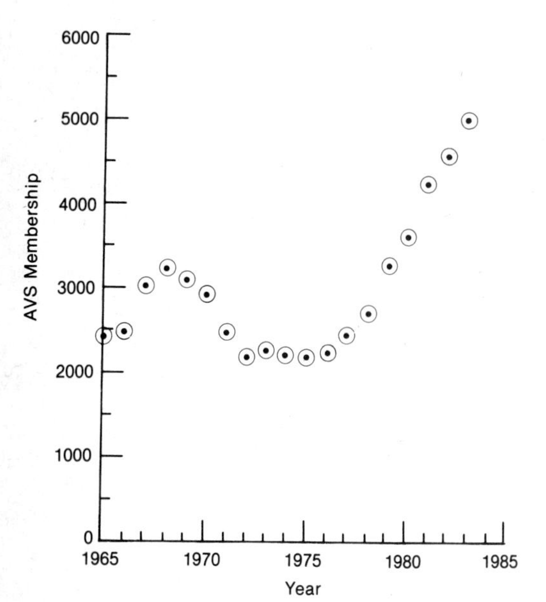

FIG. 2. American Vacuum Society membership since 1965.

IV. AVS DIVISIONS AND CHAPTERS

The AVS is governed by an elected Board of Directors consisting of a President (Fig. 3), President-Elect, Secretary-Clerk, Treasurer, and six Directors. Three directors are elected each year and they serve a two-year term.

If, in the 1960's, the AVS Board of Directors had insisted on limiting their interests to the production and measurement of vacuum, it is doubtful whether the Society would be in existence today. At best, I suspect, it would have a greatly diminished level of activity. Instead, the Board formed new divisions to provide a home and give support to those new

American Vacuum Society Presidents, 1953-83

1953-54 J.B. Merrill · 1954-55 R.A. Koehler · 1955-56 J.R. Bowman · 1956-57 A. Guthrie · 1957-58 M.W. Welch · 1958-59 A.J. Gale

1959-60 W.G. Matheson · 1960-61 B.B. Dayton · 1961-62 L.E. Preuss · 1962-63 D.J. Santeler · 1963-64 C.H. Bachman · 1964-65 G.H. Bancroft

1965-66 H.A. Steinherz · 1966-67 H.W. Schleuning · 1967-68 P.A. Redhead · 1968-69 J.M. Lafferty · 1969-70 W.J. Lange · 1970-71 R.F. Bunshah

1971-72 D.G. Bills · 1973 M.H. Francombe · 1974 D.M. Hoffman · 1975 L.W. Hull · 1976 N.R. Whetten · 1977 R.W. Hoffman

1978 L.C. Beavis · 1979 C.B. Duke · 1980 J.L. Vossen · 1981 T.E. Madey · 1982 J.A. Thornton · 1983 J.R. Arthur

FIG. 3. Presidents of the AVS, 1953–83.

disciplines that had benefited most from improved vacuum technology. These included a Vacuum Metallurgy Division in 1961, a Thin Film Division in 1964, a Surface Science Division in 1968, and Electronic Materials and Processes Division in 1979 and a Fusion Technology Division in 1980.

To keep the tail from wagging the dog that started all of this, the vacuum technologists, in self-defense, formed a home for themselves by setting up their own Vacuum Technology Division in 1970.

The divisions provide technical guidance to the Society. They play a strong hand in the content of the National Symposia and they sponsor or co-sponsor topical symposia related to their own special fields of expertise. New divisions will undoubtedly be formed in the future as new scientific and technological areas develop that are related to vacuum.

It will be noted that even though the AVS has now become a multidisciplinary society, there is a synergistic interaction between the divisions that extends beyond the fact that they all use vacuum as an important tool in practicing their trade. This has given the Society strength and tenacity.

Local Chapters are a vital part of the AVS structure and have contributed substantially to its growth. They are geographical in nature and serve to bring important technical information to local communities on a continuing basis through the year between the annual National Vacuum Symposia. This is done through regional symposia, courses, newsletters, dinner meetings, etc. The first Chapter, Pacific Northwest, was organized in the Seattle area in 1962. Since then, the chapters have grown to 21 in number throughout the United States and in Canada. The latest chapter formed was Texas in 1981.

V. AVS SERVICES AND AWARDS

In addition to its annual National Vacuum Symposia and the publication of its own journal, the AVS has been very active in promoting vacuum education by publishing monographs and teaching courses at both the apprentice and advanced theory levels. Its Short Course program alone has become a major operation with a cash flow of nearly one-half million dollars annually, thanks to the pioneering efforts of Vivienne Harwood. The Society publishes a Membership Services Booklet, a biennial Membership Directory and a bimonthly newsletter. It has an active Standards Committee, which has published many of its own standards and helped in developing national and international vacuum standards.The AVS also has established grants, graduate scholarships, and major awards in vacuum science and the related fields covered by its divisions.

Since I was intimately involved in setting up two of these major awards, perhaps the readers will allow me to digress for a moment to describe my own personal experiences on this subject. In 1969, it occurred to me that the AVS should establish an award to recognize and encourage outstanding theoretical and experimental research in vacuum science and related fields of interest to its divisions. I had in mind that the award should consist of a gold medal, certificate, and a monetary prize and that it would be called the *Saul Dushman Award* in honor of the grand old man who in 1949 published the classic book, *Scientific Foundations of Vacuum Technique*. As it turned out, financial support for the award was not forthcoming from his institution and I turned to the AIRES Institute through Luther Preuss, a close friend of the M. W. Welch family. This approach was more successful. The Institute contributed funds for ten $1000 awards and later contributed an additional $9000.

M. W. Welch had long been a close friend of the AVS, participating in its formation, and later serving as its president in 1957–58. To commemorate his pioneering efforts in the founding and support of the Society, it was decided to call the award *The Medard W. Welch Award*. I accordingly laid out a design for a gold medal (Fig. 4) with a profile of Medard W. Welsh on one side and the classic scene of the Magdeburg hemispheres (Fig. 1) on the reverse side. Luther Preuss engaged the Weyhing Brothers in Detroit, Michigan to fabricate the dies for the metal. The first medal struck was given to M. W. Welch in appreciation for his founding the award.

The monies from the AIRES Institute were placed in a separate interest-bearing account and the interest was used to defray the cost of the metals. In 1978, the AVS Board of Directors adopted a policy of paying the travel costs of the award recipient to attend the National Vacuum Symposium at which the presentation was to be made. In memory of M. W. Welch, who died in 1980, the Board established an additional endowment to ensure that the gold medal associated with this award could be given in perpetuity.

In 1972, when I was chairman of the Awards, Grants, and Scholarships Committee, I had the pleasure of presenting the Welch Award to Kenneth C. D. Hickman (Fig. 5) for his outstanding research and development related to vapor pumps and their working fluids and for his invention of the self-fractionating vapor pump. Hickman had always been a very active participant in the AVS technical programs and had a keen interest in the Society. It was not long after he received the Welch Award that Hickman approached me on the subject of establishing another major AVS award. He did not want it to detract from the Welch Award and had in mind that it should recognize and encourage outstanding *single* discoveries or inventions in fields of interest to the AVS. He also stipulated that the award should be conferred

FIG. 4. The gold medal given with *The Medard W. Welch Award*.

FIG. 5. A photograph taken at Chicago in 1972 on the occasion of the presentation of *The Medard W. Welch Award* to Kenneth C. D. Hickman by James M. Lafferty, Chairman of the Awards, Grants, and Scholarship Committee. Mr. Welch, founder of the Award, is in the center of the photo and Dr. Hickman is on the right.

FIG. 6. The first Gaede–Langmuir plaque given to Kenneth C. D. Hickman by the American Vacuum Society in appreciation of his founding *The Gaede–Langmuir Prize.*

only when an outstanding candidate appeared and in no case more than once every two years. He also had in mind that the award should be called a *prize* to distinguish it from the Welch *Award.* In addition to a $2000 cash prize, there should be a plaque with the citation for the award inscribed on it that could be displayed rather than a gold medal to be tucked away. We had many long discussions on what the prize would be called. He finally decided on *The Gaede–Langmuir Prize* because he felt that these two men were eminently successful in exemplifying the type of accomplishments that should be recognized in awarding the new prize. Hickman decided that the plaque should be made of antique silver and there should be a bust of Gaede and Langmuir facing each other from opposite sides. The details were left up to me. The biggest problem I had was finding a photograph of Professor Wolfgang Gaede for the sculptor at Medallic Arts Company whom I had engaged to design the plaque. This was finally found with the help of my good friend, Professor Dr. Günter Ecker, at the University of Bochum in Germany. The first plaque (Fig. 6) produced was given to Hickman in appreciation of his founding *The Gaede–Langmuir Prize.* The prize was first awarded in 1978 to Professor Pierre V. Auger.

The Board of Directors also provides the recipients travel expenses to receive the prize.

Hickman had stipulated that the founder of *The Gaede–Langmuir Prize* should remain anonymous during his lifetime. After a brief illness, he died on November 3, 1979.[2] In consultation with Mrs. Eleanor Hickman, it was agreed that it would be appropriate to reveal the founder of *The Gaede–Langmuir Prize* on the occasion of the 30th anniversary of the American Vacuum Society.

VI. THE INTERNATIONAL UNION FOR VACUUM SCIENCE, TECHNIQUE, AND APPLICATIONS

Other countries have also organized their own vacuum societies or national committees on vacuum. Many of these, some 22 in number (including the AVS), are members of the International Union for Vacuum Science, Technique, and Applications.

The Union had its origin in June 1958 in Namur, Belgium under the leadership of Professor Emil Thomas at a meeting called the "Premier Congres International pour L'Etude des Techniques du Vide." At this meeting, the International Organization for Vacuum Science and Technology (IOVST) was structured with Professor Thomas as president. The IOVST accepted an invitation from the AVS to hold its second International Congress in Washington, D.C. in combination with the 8th National Vacuum Symposium of the Society in 1961. An equipment exhibit was also held for the first time during this meeeting. The exhibit proved to be very successful from both a technical and financial point of view for both organizations.

On December 8, 1962 in Brussels, Belgium the IOVST was dissolved and a new constitution was set up with the formation of the International Union for Vacuum Science, Technique, and Applications (IUVSTA). All the assets and records of the IOVST were transferred to the new organizatin and Medard W. Welch, who played a leading role in its formation, was chosen to be its first president. At this time, vacuum groups from 13 countries were members of the Union. The essential difference between the IOVST and the new IUVSTA is that the old organization was based upon individual members in the various countries; whereas now each country is represented by a councilor. The councilor presently representing the AVS is L. C. Beavis.

In its present form, the IUVSTA is an international confederation of national vacuum organizations and excludes private membership. It was organized to promote vacuum science and technology on an international level. This includes promotion of education and research and the establishment of international standards for nomenclature, measuring techniques, and fittings on vacuum equipment. The

IUVSTA encourages the establishment of national vacuum committees (societies) in countries where they still do not exist and helps coordinate the activities of the national committees.

The IUVSTA organizes an International Vacuum Congress every three years. Since 1971, this Congress has been held in conjunction with the International Conference on Solid Surfaces which was initiated by the AVS Surface Science Division. The latest Congress was held at Madrid, Spain in September 1983 with nearly 1000 in attendance. Two earlier Congresses were held jointly with National Vacuum Symposia of the AVS in 1961 and 1971, and a third one is planned for 1986 in Baltimore.

The IUVSTA publishes a quarterly News Bulletin that is sent to its member National Committees and others to inform them of the activities of the Executive Council and the Divisions. It also contains a calendar of events and other articles of international interest. Starting in 1984, J. L. Provo, editor of the AVS Newsletter, will be the new editor of the IUVSTA News Bulletin. The AVS has agreed to handle the publication and mailing of the Bulletin at cost from their office in New York, taking advantage of the large-scale automated publication facilities of the American Institute of Physics.

The interests of the IUVSTA encompass not only vacuum *per se*, but those disciplines that use vacuum as an important tool. Like the AVS, the Union has established divisions in surface science, thin films, vacuum science, electronic materials and processing, and fusion. These divisions will now play an active role in the organization of the technical program of the triennial congresses and conferences. This was done for the first time at the Madrid meeting. All of the divisions worked together with the Spanish Congress organizers in an International Advisory Committee. They had an important role in the selection of scientific papers, in the selection of invited speakers and of moderators for the different sessions at the meeting which were divided on the basis of the Union's divisional structure.

The IUVSTA serves as a "clearing house" for the organization of international congresses and conferences on vacuum and related fields by granting sponsorship for these meetings. The formalities of the sponsorship requests are handled by the Secretary-General,[3] but divisional approval is also required. The purpose of the sponsorship is to avoid date conflicts and coordinate international meetings in fields of interest to the Union.

VII. THE M. W. WELCH INTERNATIONAL SCHOLARSHIP

In 1965, M. W. Welch made a proposal to the AVS Board of Directors for the establishment of a $5000 international scholarship to be used for postgraduate study in vacuum science and technology. Under the terms of the proposal, the AVS would be custodian of the funds, advancing $5000 each year to the IUVSTA on the acceptance of a satisfactory protocol describing how it would select the scholars. A protocol was accepted in 1966 and the first scholar was chosen in 1968. A scholar has been selected each year since then and the Welch family has continued to advance money annually to support the scholarships. The amount of the scholarship was increased to $7000 in 1974. On the establishment of an Executive Secretariat by the IUVSTA in London in 1971, all administrative chores associated with the Welch Scholarship were transferred to that office. On the demise of the Secretariat in 1980, for economic reasons, the administrative work was returned to the AVS and is now managed by a Scholarship Administrator, J. P. Hobson. Under the present protocol, revised in 1980, the Scholarship Administrator is responsible to the Welch Foundation consisting of the Director of the Scientific and Technical Directorate of the IUVSTA who serves as *ex officio* chairman and four "IUVSTA Welch Scholarship Trustees" elected by the Executive Council of the IUVSTA.

VIII. CONCLUSION

The unprecedented growth of the AVS during the extended period of the past seven years in a period of economic stagnation can only mean that the Society must be doing something right. In my judgment, what it is doing is to provide a needed service to the scientific and engineering community in selected areas related to vacuum, some of which are growing much more rapidly than the national economy.

The IUVSTA should prosper also now that they have their finances under control and continue to develop the technical related activities of their new Divisions.

ACKNOWLEDGMENT

The author wishes to express his appreciation to Nancy Hammond for her help in obtaining the photographs for Fig. 3.

[1]J. M. Lafferty, Phys. Today **34**, 211 (1981).
[2]J. M. Lafferty, J. Vac. Sci. Technol. **16**, 2117 (1979).
[3]H. Jahrreiss, IUVSTA Information Booklet, 2nd ed. (1981).

Early applications of vacuum, from Aristotle to Langmuir

Theodore E. Madey

National Bureau of Standards, Washington, D.C. 20234

(Received 15 September 1983; accepted 11 October 1983)

Highlights of the development of vacuum science and technology from ancient times to the early 20th century are reviewed. The view of the Greek philosophers that vacuum was an impossibility hampered understanding of the basic principles of vacuum until the mid-17th century. Verifiable vacua were first produced in Italy by Berti and Torricelli; von Guericke's dramatic experiments vividly demonstrated atmospheric pressure. Pistonlike "air pumps" were widely used in England and the European continent through the 18th and early 19th centuries to produce and characterize the properties of vacuum (lack of sound transmission, inability to support life, gas discharges, etc.). The Industrial Revolution was made possible through the genius of Newcomen, who designed huge atmospheric engines (based on condensation of steam to form a vacuum beneath a piston, which was then driven by the pressure of the atmosphere). A system of "atmospheric railways" propelled by vacuum pistons was built in England in the mid-19th century. Serious scientific developments of the 19th century which necessitated vacuum included Crookes' and Faraday's gaseous discharge measurements, the first sputtering experiments by Grove, the isolation of the rare gases by Ramsay, the standards work of Miller, the discovery of the electron by Thomson, and of x rays by Röntgen. The development of the incandescent light by Edison provided a background for the remarkable achievements of Langmuir in vacuum and surface science at the dawn of the 20th century. An Appendix is included which lists museums containing vacuum-related exhibits.

PACS numbers: 01.65. + g, 07.30. – t

I. INTRODUCTION

In the following pages, we present a brief description of the early history of vacuum science and technology, from ancient times to the early 20th century. Emphasis is placed on the development of *concepts* relating to the production and use of vacuum in science and engineering; other particles in this volume address the history of the pumps, gauges, etc., needed to generate and characterize vacuum through the ages.

The expression "nature abhors a vacuum" is certain to bring a smile to the lips of any vacuum scientist who hears it. Yet it summarizes the ancient view, originating with Aristotle and developed by his followers, that vacuum is a logical impossibility.[1] Aristotle defined space in terms of the extension of the body which occupies it; if "vacuum" means a space which contains nothing but is capable of containing something, there can be no vacuum! In general, the belief that vacuum was impossible was almost universally held for 1800 years, until the end of the 16th century. Even in the 17th century, such a respected thinker as Descartes wrote forcefully in denying the possibility of a void.[1] Such phenomena as the limited height to which water could be sucked by a pump, and the ineffectiveness of a siphon above a certain height, were attributed to imperfections in the pump or mysterious forces in the siphon. The concept that air had weight and exerted pressure was slow to be grasped.

It was against this tradition of many centuries, in which scientific concepts had been based more on the musings of philosophers than on the results of well-conceived experiments, that revolutionary strides were made in the science of vacuum by the great experimentalists of the 17th century.

II. THE FIRST VACUUM EXPERIMENTS: THE 17TH CENTURY

Although Torricelli is widely regarded as the inventor of the barometer, Middleton's[2] careful research has revealed that the first experiment in which a vacuum was known to be produced was performed by a young Italian, Gasparo Berti, around 1640. It appears that he was motivated by the siphon problem to devise an experiment to study the production of vacuum. His apparatus was basically a water barometer, and is shown in Fig. 1; it consisted of a long lead tube AB which could be filled with water from above through cock G with the lower cock R immersed in water and closed. Upon closing cocks G and D and opening cock R, the water in the initially filled tube dropped to a height L, leaving the space above the water evacuated. Unfortunately, a tiny bell M contained in the upper vessel was heard to ring clearly (presumably because sound was transmitted via the support) so that skeptics refused to believe that a vacuum was present. Thus, despite Berti's clever experimental design and execution, the correct explanation of this brilliant experiment was not agreed upon at the time.

Evangelista Torricelli, who may have been influenced by reports of Berti's experiment, had a very clear concept of air pressure. He designed an experiment in 1644 not simply to produce a vacuum, but to make an instrument to show changes in air pressure (the actual experiment may have been performed first by Viviani, to whom Torricelli had confided his thoughts.)[3] A sketch of the Torricellian apparatus using mercury-filled glass tubes as the first Hg barometers is given in Fig. 2. The experiment demonstrated in a convincing fashion that the space left above the mercury, when a Hg-filled

Originally published in J. Vac. Sci. Tech. A **2**(2), 110–117 (1984).

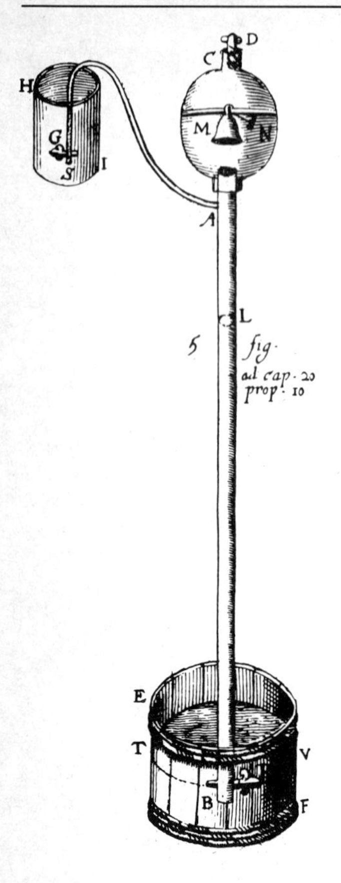

FIG. 1. Diagram of the apparatus which Berti used to produce vacuum for the first time around 1640 (see the text). The lead tube AB was about 35 ft long. The entire tube and reservoir N were filled with water, after which all cocks (G, D, B) were closed. When cock B was opened, the water level dropped to the level L, leaving a void in the space above. (Ref. 1)

tube is inverted in a Hg-filled vessel, is a void: The Hg level was independent of the volume above it, and it could be completely filled by water admitted from below. Even so, intelligent Aristotelians did not believe a vacuum was present, concluding that the apparently empty part of the tube is filled with a tenuous substance or spirit "breathed out" of the mercury, and that an attraction between this substance and the mercury supports the latter in the tube.[3]

In an experiment in 1648 which Middleton[4] calls "one of the great moments in the history of ideas," Pascal's brother-in-law Perier carried a barometer and about 4 kg of mercury up a mountain, the Puy de Dome, which was estimated to be about 1000 m high. When the barometer was set up at the top, the height of the mercury column was more than 7.5 cm shorter than the height of a similar column left at the bottom of the mountain, and watched continuously by a monk throughout the time of the climb. This experiment, together with the experiment of a "vacuum within a vacuum" (i.e., one barometer inside the evacuated space of another), were crucial in establishing that a vacuum could be produced, and that air pressure was a reality.

Many demonstrations of the properties of vacuum were performed during the mid-17th century (expansion of a small air-filled bladder in vacuum, the inability of vacuum to support combustion or to sustain life, and the lack of sound-transmission), but the mechanism for transmission of light through vacuum remained an unsolved puzzle until the work of Maxwell in the 19th century.

III. THE INVENTION OF THE AIR PUMP: OTTO VON GUERICKE

Otto von Guericke made a substantial contribution to 17th century science by his invention of the air pump (i.e., vacuum pump) which has been called "one of the four greatest technical inventions of the century, the others being the telescope, the microscope, and the pendulum clock."[5] In an early experiment sometime around 1640, he attached what was basically a water pump to a wooden cask filled with water in order to see whether an empty space would remain in the cask upon pumping out the water.[6,7] When several strong men working the pump were able to pump out the water (Fig. 3), a noise was heard as air rushed in through the pores of the wood. In a later experiment, Guericke had a large copper sphere fabricated, which could be attached to the pump; he decided to omit the water and pump out the air directly. When it appeared that all the air was nearly pumped out, the sphere was suddenly crushed with a loud noise: Guericke recognized atmospheric pressure as the

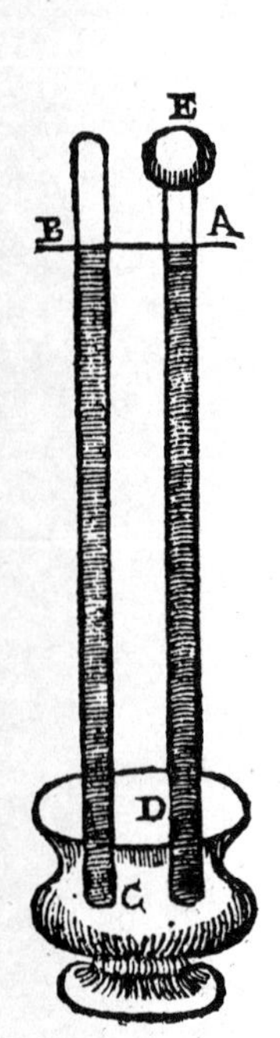

FIG. 2. Schematics of the mercury barometers reportedly made by Torricelli in 1644 (Ref. 1).

FIG. 3. Woodcut depicting the experiment designed by Otto von Guericke to pump the water from a sealed barrel, in an attempt to produce a vacuum. Several strong men were required to operate the pump in this experiment performed around 1640.

cause, and ascribed the weakness of the sphere to its departures from sphericity. The problem was solved after a new and more perfect sphere was constructed.

Guericke then proceeded to design and fabricate a special piston-type machine for generating a vacuum: the first air pump. This later underwent various transformations, as he improved the seals and valves and diminished the dead space in subsequent piston-type pumps. A long period of experimentation followed in which he studied the properties of vacuum. Among other things, his observations that the energy associated with equalization of pressure caused stones and nuts to be blown about led him to conclude that winds and storms could only be caused by differences in atmospheric pressure.[8]

Guericke's most famous experiments involved the famous "Magdeburg hemispheres," the smaller pair of which, after evacuation, could only with difficulty be separated by a team of eight horses on either side, whereupon a loud report resulted; the simple opening of a stopcock caused them to fall apart themselves. A larger pair was calculated by Guericke to require two teams of 24 horses! In a dramatic display for potential 17th century funding agents, Guericke performed these experiments in Regensburg at the Reichstag in 1654, first to a small group, and later by request to the Emperor and the princes.[8]

News of Guericke's experiments spread throughout Europe, and people began to repeat them in many places.[6] An enthusiastic experimenter was Robert Boyle, one of the founders of the Royal Society in England. Boyle made a number of improvements to Guericke's air pump, and was also responsible for developing the law relating gas pressure and volume at constant temperature (Boyle's law). Middleton[9] credits Boyle with the first use of the word "barometer."

IV. DEVELOPMENTS IN THE 18TH AND EARLY 19TH CENTURIES

Studies of electrical discharges have been interlinked with developments in vacuum science for over three centuries. They had their beginnings in 1676 when Jean Picard, a French astronomer, carried a barometer into a dark room and noted that light was produced as the mercury bounced up and down.[10] Some of the earliest systematic experiments on what we now call electroluminescence or triboluminescence were performed by Hauksbee between 1705 and 1711; he attributed the luminous effects to friction and connected them with electrification.[11] The explanation of triboluminescence was not understood until this century, when even now the details have not been fully worked out.[12]

Nollet in Paris generated an electric discharge in an exhausted egg-shaped vessel (the "electric egg") in 1740. Davy and later Faraday (~1838) in England worked more extensively with discharge tubes. The name of Faraday (the "Father of Electricity") has become associated with a dark space near the cathode during discharge. However, little understanding of any of these phenomena was available until the experiments of Crookes, Thomson, and others in the late 19th century (cf. Sec. VI, below).

In the century after Torricelli, the barometer was developed as a measuring instrument. Many of the modifications and improvements to the basic design were aimed at improving the sensitivity, i.e., magnifying the scale. Much attention was devoted to multiple liquid barometers, although other schemes were used as well.[13] Also in the 17th century, a number of investigators addressed the issue of the improvement and maintenance of the vacuum in barometers. Boiling the mercury in the tube was a common practice by 1740, and de Luc reported in 1772 that it was as important to outgas the tube as to get air out of the mercury itself.[13] This seems to be the first example of a vacuum bakeout!

The inability of vacuum to support life was a subject of drawing room entertainment, as illustrated in the famous 1763 painting "Experiments with the Air Pump" by Joseph Wright of Derby.[14] In a dramatic setting, Wright's picture depicts the responses of a family group to a scientist's experiments using a live dove inside an evacuated vessel.

Improvements in air pumps were attributed to Hauksbee and Nollet in the 18th century,[15] but the basic design was still of the Guericke type. The best such mechanical pump of the age was built by Newman of the United Kingdom and exhibited at the Great Exhibition in London in 1851.[16] In a competition against other pumps from France, Denmark, and the United Kingdom, Newman's pump was the clear winner: "In the experiments which were tried, the reading of the barometer at the time being 30.08 in., the gauge of the air pump stood at 30.06 in." Prior to the development of the mercury Geissler, Töpler, and Sprengel pumps in the mid 19th century,[11,15] the Torricellian method maintained its supremacy as a means of producing high vacuum: the experimental chamber was simply the evacuated reservoir at the top of a mercury barometer. Of course, the early workers were unaware of the presence of mercury vapor, and often assumed their vacuum was "perfect."

V. LARGE-SCALE VACUUM TECHNOLOGY IN THE 18TH AND 19TH CENTURIES

While scientists were uncovering the basic principles of vacuum science and philosophers were debating the existence of the void, clever engineers—practical men—set about to apply the newly found principles. It is no exaggeration to state that the Industrial Revolution was made possible by the genius of Thomas Newcomen, who designed the first practical "atmospheric pumping engines." The basis of Newcomen's design, first applied in 1712, was the following.[17] He was unable to use steam under pressure since no boiler existed, but he was aware of the pressure of the atmosphere. If an upright cylinder with a sliding piston was filled with steam which was then condensed, a vacuum would be formed and the pressure of the atmosphere would drive the piston to the bottom of the cylinder with considerable force. By attaching the piston to a great pivoted beam, Newcomen designed a series of "beam engines" used extensively in England to pump water from deep mine shafts, to pump domestic water supplies, and to supply water for industrial water wheels in time of drought. These engines predated rotary steam engines by 70 years, and ushered in the Industrial Revolution.

A fascinating but little-known episode in the history of vacuum technology concerns the system of "atmospheric railways" constructed in Ireland, England, and on the European continent in the mid-19th century.[18] In the 1830's when steam locomotives were rather unreliable, dirty, noisy, heavy in relation to their power, and not able to face steep gradients, some imaginative engineers conceived a plan to build clean, silent, light, cheap trains powered by the force of the atmosphere pushing a piston laid between the rails!

One such "atmospheric" system was built in 1846 by Brunel[19,20] on the South Devon coast of England (cf. Fig. 4 for the principles of the system). A continuous line of cast-iron tube was fixed centrally between the rails; the tube varied from 15 in. in diameter on the level to 22 in. on steep gradients. Along the top of the tube was a slot several inches wide closed by the "longitudinal valve," a continuous flap of leather strengthened with iron framing and hinged along one edge, the other edge closing on the opposite side of the slot; the valve was rendered air tight by a sealing composition of soap and cod oil (Fig. 5 is a schematic of a similar valve).[18]

Within the tube, a close-fitting piston was propelled by the pressure of the external atmosphere on its rear surface. The air in the tube in front of the piston was exhausted by huge air pumps operated by stationary engines ($\sim$80 hp steam engines) in pump houses spaced at three-mile intervals along the track. The typical working vacuum in advance of the piston was $\sim$16 in. of mercury.[18]

Motion of the train was obtained using a support which extended downward at an angle from the underside of the first railway coach, and which connected to a frame forming the rear end of the piston. This support was admitted to the iron tube by raising of the valve along its hinge, the valve being closed behind the bar by rollers which pressed the flap down into the sealing composition. The highest speed recorded with these trains was an average of 64 mph over 4 miles (maximum 68 mph), an impressive figure, indeed!

The atmospheric railways suffered a very short lifetime, however.[18,19] Factors such as accidents in starting the trains, the lack of control by the engineer on board the train, damage to the pistons, and breakdowns in the pumping stations contributed to their demise. The fatal flaw, however, was due to the inefficiency of the continuous valve under the combined effects of climatic variations, chemical reactions which rotted the leather, and the eating of the sealing composition by rats. Stephenson[22] had concluded already in 1844 that "the requisities of a large traffic cannot be attained by so inflexible a system as the atmospheric, in which the efficient operation of the whole depends so completely upon the perfect performance of each individual section of the machinery." The last atmospheric system to close was a 5.5 mile line near Paris in 1860. In the U. S., pneumatics methods were used to propel the first subway, a short-lived demonstration system one block in length, built in New York City in 1870.[23]

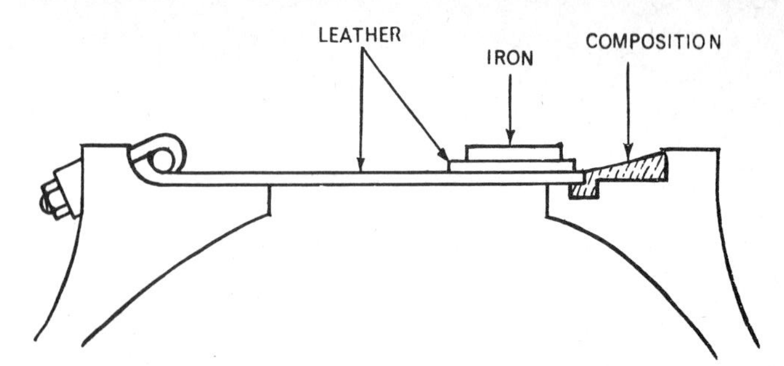

FIG. 5. Drawing made by the engineer Samuda in 1846 of the double leather valve used to seal the vacuum traction tube on one of the atmospheric railways in Britain, the Croydon line. The sealing composition on the Croydon line was a mixture of beeswax and tallow; Brunel (Ref. 19) used a more viscous compound of soap and cod oil to seal a similar valve on the South Devon line. (Ref. 18, p. 129).

An application of vacuum technology which is still in use today in various parts of the world involves the use of pneumatic tubes to propel mail and messages.[1] By 1886, the British Post Office had installed in London 34 $\frac{1}{2}$ miles of tubes, in which carriers were propelled either by compressed air at 12 psi or atmospheric pressure at 6 $\frac{1}{2}$ psi. By 1897, New York, Boston, and Philadelphia had similar systems. Many department stores in the United States used pneumatic tubes for cash control until a few years ago.

FIG. 4. A close-up drawing of a vacuum traction tube used to propel an "Atmospheric Railway." (a) is the piston inside the tube, (b) is the apparatus connecting the piston to the leading vehicle of the train, (c) is one of the wheels which raised the sliding valve (d), and (e) is the wheel that closed the sliding valve after passage of the connecting apparatus (b). This drawing is from the Illustrated London News of 1845, and is reproduced in Ref. 21.

VI. VACUUM SCIENCE AND THE BIRTH OF MODERN PHYSICS AND CHEMISTRY: THE LATE 19TH CENTURY

The science and technology of vacuum played a key role in many of the experiments which provided the basis for modern physics and chemistry. Such far-reaching results as the discovery of the electron and of x rays, the isolation and identification of the rare gases, explanation of gaseous discharges, and advances in cryogenics were made possible only because of measurements made using vacuum methods. A striking demonstration of the impact of vacuum techniques at the birth of modern science is the fact that 5 out of the first 12 Nobel prizes[24] awarded in physics and chemistry in the period 1901–1906 were based on vacuum measurements made in the late 1800's (Röntgen—x rays, Rayleigh—discovery of Argon, Ramsay—discovery of rare gases, Lenard—research on cathode rays, Thomson—electrical conductivity of gases).

By the latter part of the 19th century, the invention of the Töpler pump (1862) and the Sprengel pump (1865) together with the McLeod gauge (1874) had made possible the routine generation and characterization of vacuum at the 10^{-3} to 10^{-4} Torr level ($\sim 10^{-1}$ to 10^{-2} Pa). Most of the experiments of that era were performed in glass vessels using Pt sealed through glass for electrical feedthroughs.[11]

In the following paragraphs, we briefly summarize some of the vacuum-related scientific and technological advances of the late 19th century.

A. Gaseous discharges

It seems certain that the beauty of electrical discharges in gases was responsible, in large part, for the enthusiasm with which they were studied, and for providing the motivation for scientists to improve the vacuum in their experimental tubes.[11] Crookes began a long series of experiments on discharges around 1879. He observed the "dark space" seen near the cathode, now associated with his name; he showed that beams of "cathode rays" carried sufficient energy to cause heating, and to set his radiometers into motion; he observed fluorescence of glass by cathode rays; he demonstrated that their paths could be bent by magnets, and identified them as negative particles.[25] Elster and Geitel showed, in experiments beginning in 1880, that heating wires in vacuum or in gases leads to emission of negative particles at a rate which increases rapidly with temperature.[25,26] Wehnelt[25] soon showed that the emission of negative particles by metallic oxides, notably those of calcium and barium, is far greater than that of any known metal: the first oxide cathodes! The later use of Wehnelt cathodes and storage batteries (rather than induction coils) increased stability in gaseous discharges and and allowed studies of the various dark spaces and striated discharges.

B. Photoelectric effect

Hertz's discovery in 1887, that ultraviolet light incident on a spark gap facilitates the passage of a spark, led to a series of investigations on the effects of UV and ordinary light on electrified bodies. Elster and Geitel's work enabled them to arrange the metals in the order of the "facility with which negative electrification is discharged by light": rubidium and potassium were most effective, with carbon and mercury least effective. Lenard discovered that the energy of the emitted negative charges was independent of the intensity of the incident light, although the yield increased with intensity. The explanation of these surprising observations remained a mystery until the photoelectric theory of Einstein in 1905, which correlated photon energy, surface work function, and energy of the emitted electron.[27]

C. Discovery of the electron

J. J. Thomson[28,29] synthesized a number of diverse results in 1898 when he demonstrated that cathode rays in a discharge all had the same value of charge to mass e/m irrespective of their source, and that they were all due to the same particle—the electron (discharge initiated by induction coil, negative particles photoemitted from surfaces, and thermal emission of negative particles from hot wires). He further recognized that the mass of the electron was over 1000 times smaller than that of the hydrogen atom. His observations and conclusions settled a controversy over the nature of the cathode rays which had raged for years.

D. Discovery of x rays

W. K. Röntgen discovered in 1895 that when a discharge tube is highly exhausted (i.e., typically $\lesssim 10^{-2}$ Torr) so that the glass walls fluoresce,[30] a highly penetrating radiation is produced capable of passing through air, flesh, and even thin sheets of metal. He performed many experiments to characterize the properties of these "x rays".

E. Sputtering

W. R. Grove[31] in 1852 first observed evidence for sputtering of a cathode and transfer of material to a polished plate, during a gaseous discharge (Fig. 6 is a diagram of his apparatus). Fifty years later, it was realized that "positive rays"—ions bombarding the cathode—were responsible for the sputtering phenomenon.[11] J. J. Thomson was the first to recognize that bombardment of a cathode by positive ions also produced secondary electrons.[25]

F. Chemistry

The discovery and isolation of the rare gases[32] by William Ramsay and Lord Rayleigh in the 1890's and the identification of their place in the periodic table of elements, was facilitated by the use of Töpler pumps by these investigators.

G. Cryogenics and cryopumping

Air was first liquified in 1885, but a suitable method for storage of cryogenic fluids was not available until Dewar stored liquified air in a vacuum vessel in 1892.[33] He found that the total effect of high vacuum plus "silvering" the walls of the vacuum vessel reduced the influx of heat by a factor of 30 times! At about the same time, Dewar recognized the remarkable power of cryogenically cooled charcoal for absorbing gases, and that it could be used as a vacuum pump. The vacua obtained using cooled charcoal were so high—

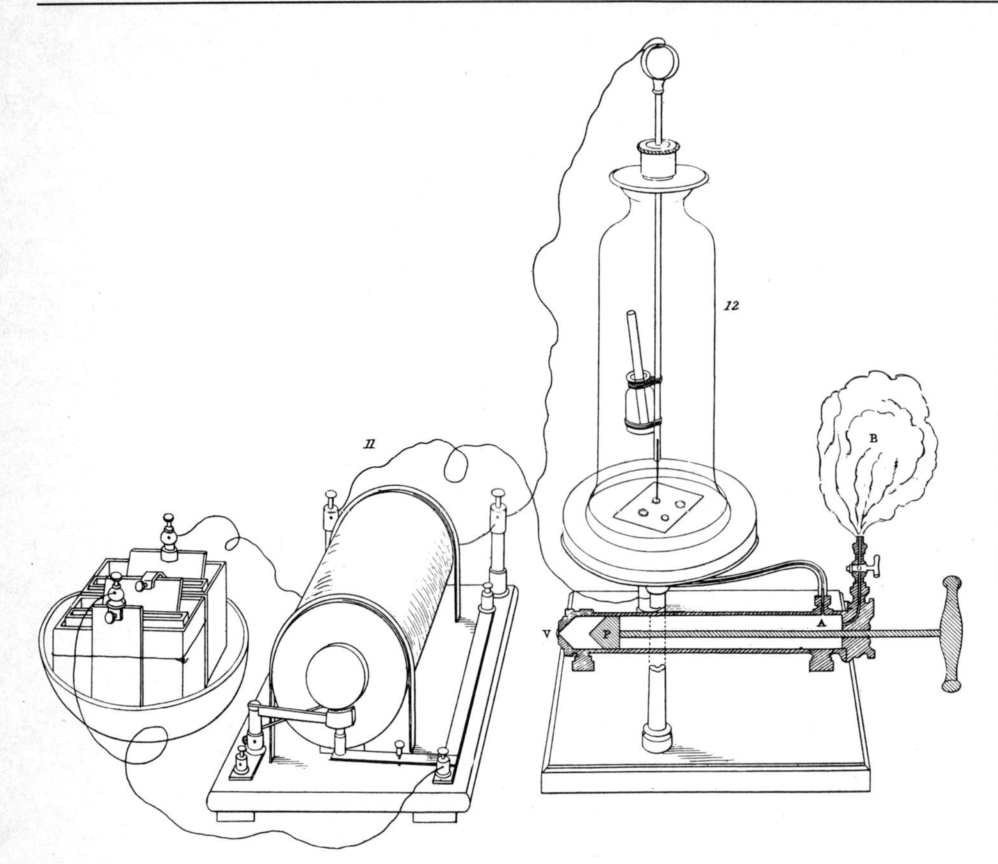

FIG. 6. Drawing of apparatus used by Grove for his first sputtering studies. (Ref. 31). II is the coil apparatus used for generating a high voltage; the contact breaker is shown in front. 12 is the vacuum vessel containing an electrode and a polished silver plate, on which deposits were observed. The air pump used to evacuate the vessel is (A), and a bladder (B) contained a gas sample used for studies of discharges in different gases.

better than 10^{-5} Torr—that it was difficult to record the pressures with a McLeod gauge.

H. Metrology

In the accurate measurement of mass using an analytical balance, the object to be weighed and the weights are rarely of the same density; hence, they are subject to different buoyancy effects, and the apparent weight of the object will differ from the true weight. Different surface areas of object and weights can also lead to differential adsorption of moisture. Measurement in vacuum is an obvious way of eliminating both problems, although manipulation is more difficult and the case of the balance must be vacuum tight and capable of withstanding considerable pressure. A vacuum balance made to the design of W. H. Miller has $1\frac{3}{4}$ in. thick glass plates at the front and back of the case. The instrument was used from 1872 to about 1892 for accurately comparing copies of weights with the primary standards in England.[34]

I. Incandescent lighting

In the successful research which led to the development of the carbon filament lamp in 1879, Edison used one of the first Sprengel pumps in America, borrowed from Princeton University.[35] His patent states: "I have discovered that even a cotton thread, properly carbonized and placed in sealed glass bulbs, exhausted to one-millionth of an atmosphere, offers from 100–500 Ω resistance to the passage of current and that it is absolutely stable at very high temperature." This was the first practical and economic electric light for universal domestic use.

VII. INTO THE 20TH CENTURY: LANGMUIR

In the early years of the 20th century, great advances were made in both the science and technology of vacuum. The development of the kinetic theory of gases, the invention of rotary mechanical pumps and mercury diffusion pumps, the development of thermal conductivity gauges and the ionization gauge, the development of traps and bakeout procedures all occurred within a relatively short time, and have been described in detail by Dushman in 1922.[36] At the same time, major studies were made in the understanding of gaseous discharges, x rays, thermionic emission, photoelectron emission, vaporization of solids, and surface chemistry. Although many scientists and engineers contributed to these developments, one man stands out for his contributions to the science, techniques, and applications of vacuum: Irving Langmuir.[37]

Langmuir spent most of his professional career at the General Electric Research Lab in Schenectady, New York, beginning in 1909. His work was motivated at the outset by a desire to understand the influence of vacuum on the life and efficiency of light bulbs, and on why they turned black in use. By 1920 he had performed pioneering experiments in measurements of vapor pressures, chemical reactions at low pressures, dissociation of gases, adsorption of gases on surfaces, blackening of tungsten lamps, thermionic emission, and atomic structure.[38] He had also developed both glass and metal diffusion pumps (which he called "condensation pumps"), and two sensitive molecular viscosity gauges.

Some of his most far-reaching studies were in the areas of surface chemistry and catalysis.[37] He revolutionized the thinking in these areas by suggesting that heterogeneous catalytic processes occurred not in "thick films" of molecules condensed on surfaces (as was widely believed at the time) but in a single layer of gas molecules held on a solid surface by the same forces that held the atoms of the solid together. In his 1915 study of the catalytic oxidation of CO on Pt,[38] he concluded that the oxygen and CO layers are never more than one molecule thick, and that the unsaturated character of the carbon atoms leads to chemical interaction ("chemisorption") of CO on the Pt surface. The pregnant scientific problems addressed by Langmuir are still being studied today by surface scientists! No one exemplifies the complete 20th century vacuum scientist better than he.

VIII. CONCLUSIONS

The techniques and understanding of vacuum processes have paralleled and influenced major developments in physics, chemistry, and engineering for over 300 years. Vacuum technology of the 18th century provided a technical basis for the huge engines which made possible the Industrial Revolution; the highly sophisticated vacuum science and technology of the late 20th century have given birth to a new Revolution—the Microelectronics Revolution—with its far-reaching impact on the lives of all. The author hopes that this brief description of the concepts of vacuum, and the evolution of those concepts, has provided the reader with insights into some of the developments which impacted on those revolutions.

ACKNOWLEDGMENTS

The author acknowledges with pleasure the generous assitance of several colleagues who called my attention to key references: P. A. Redhead, G. K. Wehner, M. Hablanian, and W. Weinstein. European colleagues whose cooperation and advice are appreciated include A. Brachner (München), A. Legon and A. Bellion (London), T. Bryan (Swindon, United Kingdom).

APPENDIX

During the preparation of this paper, the author has visited or been advised of a number of museums which contain exhibits of interest to the vacuum scientist. Although the list is not intended to be exhaustive, some interesting exhibits are on display at the following:

I. United States

(a) The Smithsonian Museum of American History, Washington, D. C. Pneumatics exhibit, several 18th and 19th century piston-type air pumps, guinea-and-feather tube. Also, the storage rooms (open by special permission only) contain a wealth of old vacuum apparatus, historic vacuum tubes, etc. (b) National Bureau of Standards Museum, Gaithersburg, Maryland. Original 1917 Stimson diffusion pump; early mercury manometer used for standards work. (c) The Wythe House, Williamsburg, Virginia. Eighteenth century air pump. (d) The Franklin Court Museum, Philadelphia, Pennsylvania. Air pump used by Benjamin Franklin around 1779. (e) MIT Museum, Boston, Massachusetts. Ninteenth century air pump used at MIT.

II. United Kingdom

(a) The Science Museum, Exhibition Road, London. This museum contains the most extensive display of vacuum apparatus in the English-speaking world, including replicas of the Magdeburg hemispheres and von Guericke's air pump, Boyle's air pump, and a wealth of elegant 18th and 19th century air pumps and vacuum apparatus (bell-in-vacuum, bladder in vacuum, etc.). Original apparatus fabricated and used by Dewar, Crookes, Ramsey, and Miller (vacuum balance) are on display, along with a Newcomen beam engine.

(b) The Faraday Museum, Royal Institution of Chemistry, London. A complete collection of Faraday memorabilia, including his air pump, "electric egg" used for studies of gaseous discharges, and pump used for liquefaction of gases.

(c) University College, London, Dept. of Chemistry. Original tubes made by Ramsay to illustrate gaseous discharge in rare gases (can be seen by special appointment only).

(d) The Whipple Museum of the History of Science, Free School Lane, Cambridge. A remarkable collection of historic scientific apparatus of all types, much of which is not catalogued. Collection includes Newman's air pump from the 1851 Great Exhibition, several English and continental air pumps, a display of Crookes tubes, early glass x-ray tubes.

(e) The Cavendish Museum, Cavendish Laboratory, Cambridge. An outstanding collection of historic vacuum tubes and apparatus used by Cambridge physicists: J. J. Thomson's tube for the electron e/m measurements, x-ray tubes used by Bragg, early mass spectrometers, tubes used to identify "positive rays," etc. A visit here is a profound experience.

(f) The Great Western Railway Museum, Swindon. The "Brunel room" describes Brunel's development of an atmospheric railway on the South Devon line, and contains the only three-dimensional relic of the railway: a section of 22 in. diam pipe.

(g) Museum of the History of Science, Broad St., Oxford. 18th and 19th century air pumps and vacuum apparatus, Dewar flasks, Edison light bulb, Moseley's x-ray tubes.

III. Germany

(a) The Deutsches Museum, München. The original "Magdeburg hemispheres" and other historic vacuum apparatus.

[1]W. E. K. Middleton *The History of the Barometer* (Johns Hopkins University, Baltimore, 1964), pp. 3–5.
[2]Reference 1, p. 10ff.
[3]Reference 1, p. 19ff.
[4]Reference 1, p. 51.
[5]Reference 1, p. 61.
[6]A. Wolf, *A History of Science, Technology, and Philosphy in the 16th and 17th Centuries* (Macmillan, New York, 1935), p. 99ff.
[7]A. Brachner, *Kultur and Technik* (Zeitschrift des Deutschen Museum, München, 1979).
[8]P. Lenard, *Great Men of Science* (Macmillan, New York, 1933), p. 54ff.

[9]Reference 1, p. 71.
[10]Reference 1, p. 356.
[11]G. W. C. Kaye, *High Vacua* (Longmans, Green, and Co., London, 1927).
[12]J. Walker, Scientific American, July 1982, p. 146.
[13]Reference 1, pp. 87 and 244.
[14]Exhibited at the Tate Gallery, London.
[15]E. N. da C. Andrade, in *Advances in Vacuum Science and Technology, Proceedings of 1st International Congress on Vacuum Techniques*, edited by E. Thomas (Pergamon, New York, 1960), p. 14.
[16]J. A. Bennett, *Science at the Great Exhibition* (Whipple Museum, Cambridge, 1983).
[17]T. E. Crowley, *Beam Engines* (Shire, Aylesbury, Bucks, United Kingdom, 1982).
[18]C. Hadfield, *Atmospheric Railways* (David and Charles, Newton Abbot, 1967).
[19]L. T. C. Rolt, *Isambard Kingdom Brunel* (Penguin, Middlesex, England, 1980).
[20]Great Western Railway Museum, Swindon, United Kingdom (unpublished information).
[21]G. [illegible]. Allen *The Illustrated History of Railways in Britain* (Marshall Cavendish, London 1982), p. 37.
[22]R. Stephenson, Report on the Atmospheric Railway System, London, 1844.
[23]L. K. Edwards, Sci. Am. **213** (8), 30 (1965).
[24]*Encyclopedia Brittanica* (Benton, Chicago, 1973).
[25]*Encyclopedia Brittanica* 11th ed. (Cambridge University, Cambridge, 1911), Vol. 6, p. 864ff.
[26]O. W. Richardson, *The Emission of Electricity from Hot Bodies* (Longmans, Green, and Co., London, 1916).
[27]H. S. Allen, *Photoelectricity* (Longmans, Green, and Co., London, 1913).
[28]J. J. Thomson, Philos. Mag. **44**, 293 (1897).
[29]J. J. Thomson, *Conduction of Electricity Through Gases* (Cambridge University, Cambridge, 1906).
[30]Reference 25,Vol. 23, p. 694.
[31]W. R. Grove, Philos. Trans. Faraday Soc. **1852**, 87.
[32]W. Ramsay and J. N. Collie, Proc. Roy. Soc. London **60**, 206 (1896); Ref. 25, Vol. 2, p. 475ff; Vol. 13, p. 233ff.
[33]Reference 25, Vol. 16, p. 744ff.
[34]J. T. Stock, *Development of the Chemical Balance*, A Science Museum Survey (Her Majesty's Stationery Office, London, 1969).
[35]M. Josephson, *Edison* (McGraw–Hill, New York, 1959), pp. 198 and 221.
[36]S. Dushman, *Production and Measurement of High Vacuum* (General Electric Review, Schenectady, New York, 1922).
[37]G. L. Gaines, Jr., and G. Wise, *Heterogeneous Catalysis—Selected American Histories*, ACS Symposium Series No. 225 (American Chemical Society, Washington, D. C. 1983).
[38]*The Collected Works of Irving Langmuir* edited by C. G. Suits, and H. E. Way (Pergamon, New York, 1961).

Comments on the history of vacuum pumps

M. H. Hablanian

Varian Associates, Lexington, Massachusetts 02173

(Received 31 August 1983; accepted 24 October 1983)

A brief outline of the development of methods for pumping of fluids is presented, emphasizing parallel progress of conceptual understanding and technological achievement. Early experiments with vacuum pumping in the 17th and 18th centuries are reviewed, but the main discussion is reserved for the developments made during the last 100 years when vacuum technology became an industrial practice. Accelerated progress in the development of new pumping methods and in the design of vacuum pumps was made in response to requirements of industrial applications. These include incandescent lamps, vacuum tubes, vacuum distillation, vacuum instruments such as mass spectrometers, electron microscopes, particle accelerators and cathode ray tubes, nuclear energy, vacuum metallurgy, space simulation, thin film deposition for microelectronics, and surface analysis. Each of the above led to some progress in vacuum pump technology ranging from mechanical pumps and diffusion pumps to turbopumps, cryopumps, and ion-getter pumps. A brief history of the development of high vacuum pumps in the United States, written by B. B. Dayton, is included in the Appendix.

PACS numbers: 07.30.Cy, 01.65. + g

I. BEFORE GUERICKE

Most writers on the history of vacuum pumps attribute the first invention to Otto von Guericke,[1] approximately in 1650, following Torricelli's experiments with vacuum in mercury-filled tubes in 1644 (or 1643). However, others mention Galileo[2] and his attempts to measure the force of vacuum formation using cylinders, pistons, and weights (1640).

The initial attempts to create vacuum were made by pumping water out of filled containers, using a variety of well-known water pumps. Water pumps go back at least to ancient Alexandria and were used in ancient Roman mines.[3] Many 16th century mining engineers probably produced 250 Torr vacuum without knowing it when they tried to pump water from above. Galileo reported in his book (Leyden, 1638) that he first heard about the 33 or 34 ft limit from a mine worker.

This discovery must have been before 1630 because in that year G. B. Baliani reported to Galileo the failure of a water pipe siphon at Genoa where water would run down the hill on both sides of the summit even though no leaks were discovered.[4] Galileo thought this was due to the break in the "rope of water," associating the phenomenon with his experiments with breaking wires and ropes in tension.

Baliani appears to have disagreed with that and suggested the possibility of vacuum above the water column.

Galileo attempted to measure the force needed to create vacuum by using a cylinder open at the bottom and suspending weights or sand in the bucket attached to the piston. This apparatus was made of wood and it was difficult to seal despite the use of water. The idea of pumping air directly did not occur until Guericke's third pump version. However, interestingly enough, air pumps for pipe organs existed in ancient Greece,[2] as shown in Fig. 1.

The thought that the force of atmospheric air had something to do with the action of water pumps had been recorded as early as 1615 in the writings of Isaac Beekman[5] (1558–1637). According to de Waard, Beekman was a man of genius who did not publish, and is therefore not well known. In his thesis (University of Caen, 1618) he did not hestitate to state "Aqua suctu sublata non attrahitur vi vacui sed ab aere incumbente in locum vacuum impellitur" which could be translated as "water raised by suction is not pulled by the

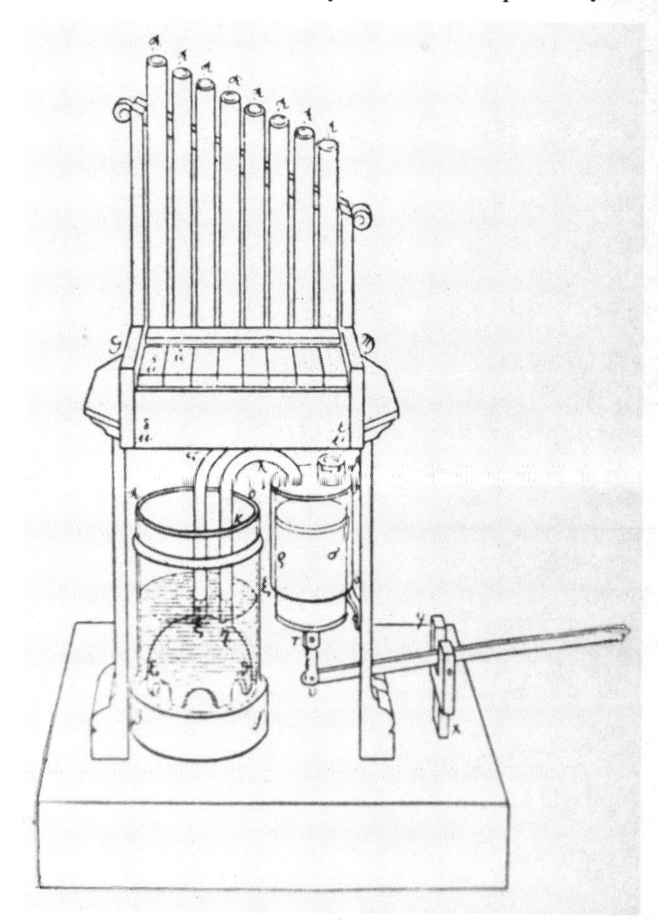

FIG. 1. An organ from ancient Greece [from Bernal (Ref. 2)]. Piston pump is on the right with an inlet valve at the top. Accumulator under water on the left.

Originally published in J. Vac. Sci. Tech. A 2(2), 118–125 (1984).

force of vacuum but is driven into the empty space by the still air." Beekmans's journals[6] have many sketches of water pumps and siphons as well as a pear-shaped bulb filled with water dipped with the open end into a pool of water. It is not clear whether Beekman's observation was completely original. In those days, they did not give careful references to sources (at least Descartes accused Beekman of this[4]).

The ancient "fear of vacuum" was beginning to be questioned, perhaps as early as 1600. Kepler (1603) realized the limited extent of the atmosphere (estimated 4 kM from astronomical studies of bending of light). Did he suspect vacuum or ether beyond?

The rejection of the possibility of vacuum by ancient philosophers[7] was mainly based on the premise that, if vacuum were possible, the absence of resistance would permit infinite speed. In retrospect, it is difficult today to appreciate the confusion. Notable in the ancient writings is the absence of discussion on the degree of vacuum. They appear to have thought of air as a given substance and did not seem to realize that gases can have variations in density (the kinetic theory of gases was developed in the 18th century and was, and is still, called "theory").

Guericke's interest in vacuum appears to have been a reaction to Descartes *Principia* which was published in 1644. He decided that the question of existence of vacuum should be answered by experiment rather than by philosophical speculation.[8] However, Torricelli's experiments were already known at that time (1643). Torricelli and his associates had the brilliant idea of using mercury instead of water and glass tubes instead of wood, clay, or metal pipes. But what preceded his discoveries? The sequence of events and thoughts is not easy to follow,[5] but de Waard suggests that the stimulation came from Gasparo Berti's experiments in Rome in 1640, as reported by a German Jesuit priest, Kaspar Schott (Gasparis Schotti) in his *Technica Curiosa* published in 1664 in Nuremberg.[9] Berti's apparatus involved a modified "pump" attached to the front of his house, as shown in Schott's book (Fig. 2). Berti's experiments provoked long discussion. Torricelli's more practical, and of course more complete, experience was the consequence.

Schott's book describes in great length Guericke's Magdeburg hemispheres and has a chapter on pneumatic experiments in France, Italy, and other countries. Schott had written an earlier book in 1657[1] in which he shows Guericke's early pumps.

Guericke was apparently first to realize that air can be pumped directly by appropriately designed pumps, just like water, and his third generation pump was built for that specific purpose. His pumps are well described in E. Andrade's article (Fig. 3). It is essentially similar to the water pump and has manual valves. It is more carefully built to make a tighter seal around the cylinder and tighter valves. In principle, the only difference between such pumps used for creating vacuum rather than compressing air (or lifting water) is that the working stroke is pulling instead of pushing, with corresponding valve sequences. The inlet is still at low pressure and the discharge at the higher pressure. The basic difference is that a sealed container is attached to the inlet instead of the discharge (compare with Fig. 1).

FIG. 2. Berti's experimental setup, from Schott's *Technica Curiosa*, courtesy MIT rare book collection.

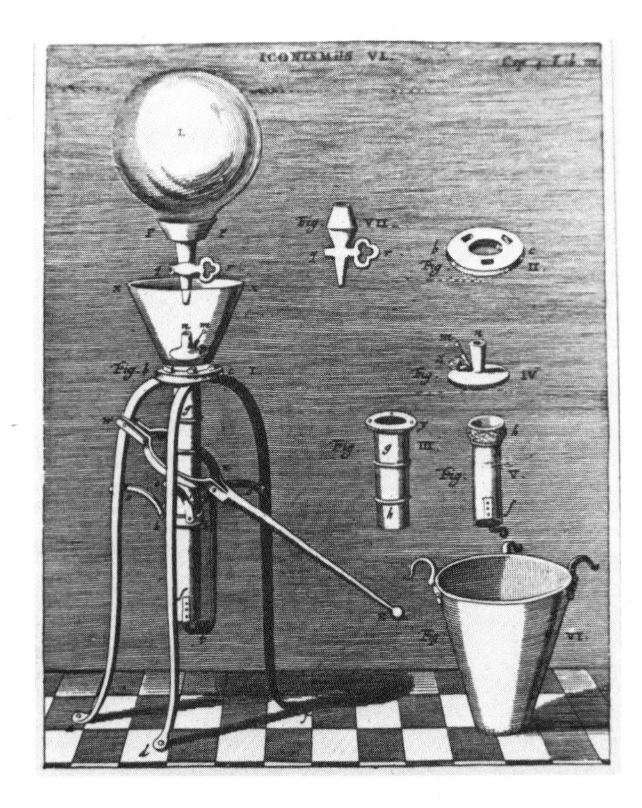

FIG. 3. Guericke's transportable pump used in demonstrations in Magdeburg and Berlin (1663, Ref. 8).

II. FROM GUERICKE TO 1900

A. Mechanical pumps

In general, the history of the development can be traced as follows: first, modification of existing water pumps with pistons and valves, which lasted to the end of the 19th century, then a return to a more mechanically primitive concept of liquid mercury "piston" pumps, later the establishment of rotary mechanical pumps, followed by adaptations of vapor jet pumps, turbomachinery and finally, pumps based on ionization, chemical combination, and cryogenic adsorption.

For a long time, vacuum pumps were not even called vacuum pumps. Guericke called them syringes[11]; Boyle,[12] pneumatic machines; later the term "air pumps" was established. The use of the word "pump" for the device, instead of the rarefied gas compressor (which it is), is due to the association with water.

The publication of Schott's books and, later, Guericke's own book (1672) spread the new knowledge in Europe, and better pumps were built by others. Notable was Boyle, who published his own designs in 1660. Engineering improvements by Hooke and Hauksbee, among others followed. Pumps of that general type were in use until almost the 20th century, when an Englishman, H. A. Fleuss, introduced a more or less industrial version of a piston pump (Fig. 4) which remained in use until the development of the incandescent lamp industry sparked the early advances in vacuum technology at the turn of the century.

Before the 20th century, vacuum pumps were used for laboratory work and were often made by local instrument makers.[10]

Excellent reviews of history and descriptions of early pumps have been published in technical literature associated with the high vacuum technology. Notable is E. N. Andrade's article.[1] Three early textbooks with a wealth of information of historical interest are by L. Dunoyer,[13] S. Dushman,[14] and R. Jaeckel.[15] Additional information can be obtained from two more recent reviews by M. Dunkel[16] and J. M. Lafferty.[11]

A historically important development was the Sprengel's pump (1865) shown in Fig. 5, which was based on the continuous application of Torricelli's method of obtaining vacuum.

Sprengel's pump, with improvements and modifications, was used by Edison in 1879 for evacuation of the first incandescent lamps.[17]

The evacuation was started with a Geissler pump[1] and finished by a Sprengel pump. Edison and his associates designed a unique new Sprengel–Geissler pump which was used until 1896. A person continuously lifted and lowered mercury bottles, connected to the pump by flexible tubes, to trap and eject volumes of air. Good vacuum was obtained when the recompressed air volume turned into small bubbles. A solid mercury column when the bubbles became invisible was an indication of high vacuum.

The first pump that was adaptable for using the emerging electric motors was introduced by W. Kaufman in 1905. It consisted of a "Torricelli" tube twisted into a helix and a return line for mercury, so that the device resembled a tubular Archimedes screw in which captured volumes of air between slugs of mercury would transfer from the inlet to the discharge side. A much more practical design was developed by Gaede in the same year.

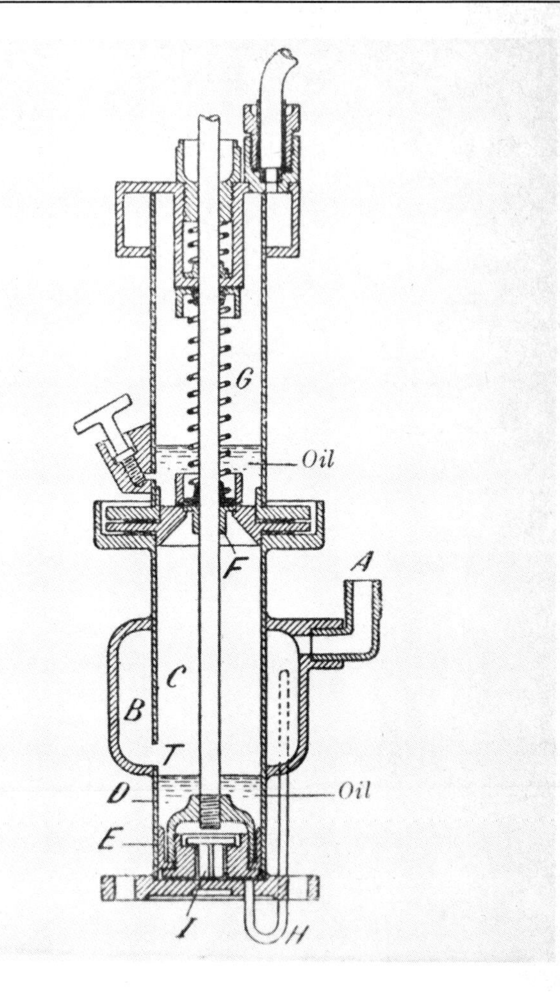

FIG. 4. Fleuss pump called "The Geryk" (Ref. 13). Inlet is at A and T. Discharge valve is at F.

Gaede demonstrated his rotary mercury vacuum pump in 1905 at the 77th meeting of German Scientists (and Physicians) in Meran[18] (Fig. 6). This demonstration was a starting point of the close cooperation between the Leybold Company and Prof. Gaede (University of Freiburg).

This pump is representative of attempts to reach low pressure by mechanical pumps alone. It was used in series with water jet pumps, and produced 7×10^{-5} Torr at 20 rpm. It was ten times faster in pumping a 6 l volume than the Sprengel-type pumps used then. In principle, it was essentially a rotary liquid piston device. Such pumps enjoyed a brief period of popularity until a line of oil sealed rotary mechanical pumps was developed (before 1910, according to the Pfeiffer Co. bulletin).

Mechanical pumps existed in Edison's time, but, interestingly enough, the vapor pressure of the oils used in those days was higher than that of mercury. However, by 1910, mechanical vacuum pumps, more or less similar to present designs, were fully developed. Of course, the basic designs are very old (at least 1588, see Fig. 7), and they were adapted for vacuum pumping using tighter mechanical tolerances and better oils and seals.

For successful industrial or scientific laboratory device

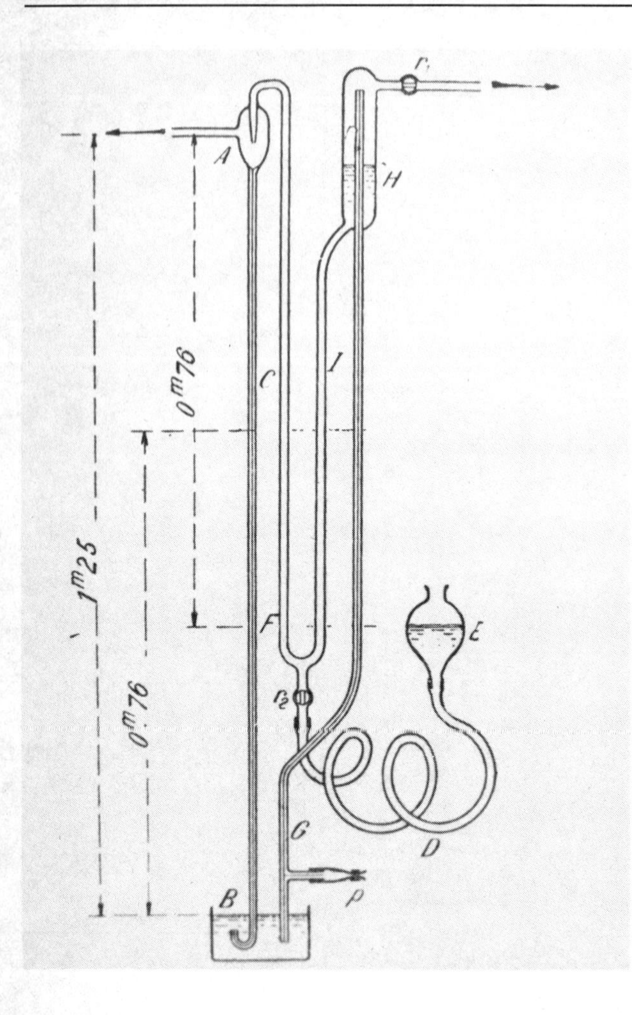

FIG. 5. Sprengel pump (Ref. 13). Inlet is at A. Mercury droplets, acting as pistons, fall in the tube AB. The right-hand side is a mercury level restoring system.

development, there must often be the confluence of a pressing need, available technology, an inventive idea, and some scientific understanding of underlying principles. Many early inventive ideas were impractical because of the lack of available technology.[19] Thus, Guericke struggled with the problem of fitting a piston tightly into a cylinder, just as before him Galileo struggled[2] to seal his cylinder when he tried to measure the force of formation of vacuum, just as 100 years later James Watt struggled to produce a tight fit between the piston and the cylinder of his steam engine. Similarly, Gaede was ahead of his time with his molecular pump.

An interesting pattern can also be traced regarding the attempts to reach a better degree of vacuum without any apparent regard to pumping speed. Until the 20th century, the pumping speed of most vacuum pumps appears to have been less than 1 l/s!

Writing in 1922, Dunoyer[13] gives pumping speeds for many early pumps. Geissler and Toepler pumps, which Edison used for his incandescent lamps (1880), were essentially arrangements for continuous repetition of Torricelli's upside-down mercury-filled tube; pumping speed—an astonishing 4 cc/s. Early rotary mechanical pumps produced 2 l/s, another mechanical pump 120 cc/s, Gaede's rotary mercury pump 0.15 l/s maximum, Gaede's first molecular pump 1.4 l/s, Holweck's molecular pump 2.5 l/s, first Gaede diffusion pump maximum 80 cc/s, Langmuir diffusion pump, less than 4 l/s, Crawford's diffusion pump 1.3 l/s.

Only in the 1930s did larger pumps appear producing pumping speeds of a few hundred l/s. Characteristic of the time is a description of a "high capacity" vacuum installation in 1929 with twelve glass and three steel Langmuir "condensation" pumps (used in parallel) and producing a total pumping speed of 40 l/s at 5×10^{-3} Torr.[20] At that pressure, the pumps were probably overloaded.

III. THE MODERN PERIOD

A. The diffusion pump

In retrospect, the advantage of the diffusion pumps for the creation of high vacuum, compared to mechanical pumps, is that they can produce much higher pumping speed for the same size, weight, and cost. This was not realized at the beginning. The first idea of the diffusion pump appears to have been somewhat fallacious, the fallacy already alluded to by Dunoyer.[13]

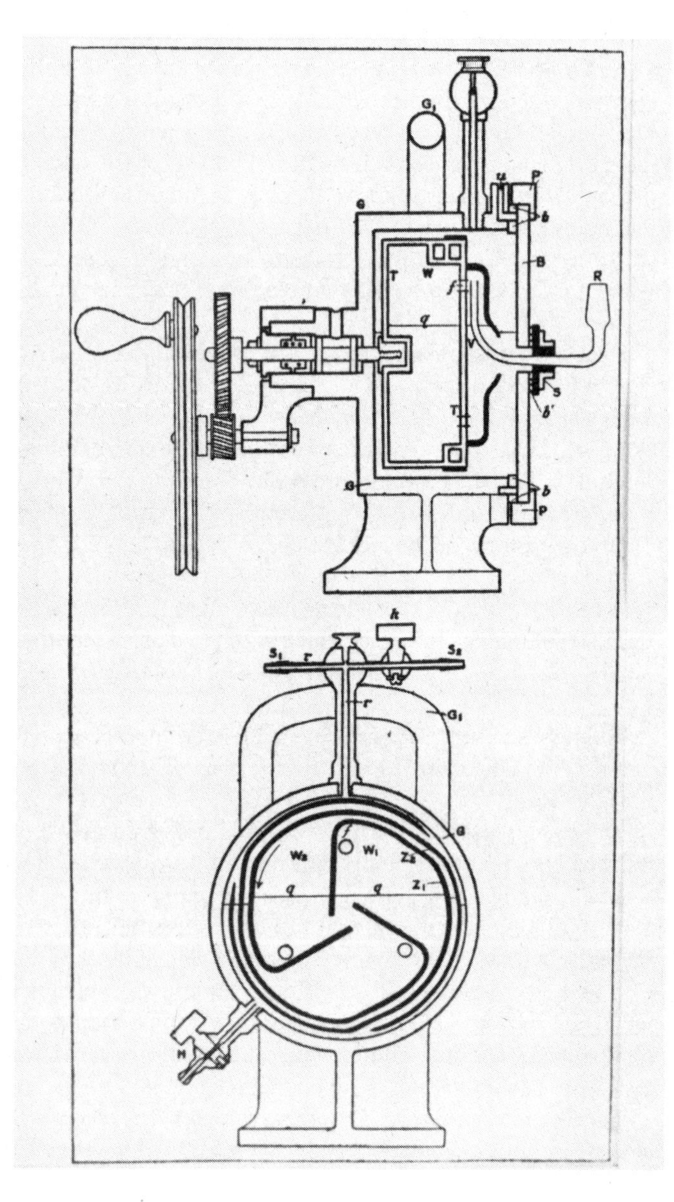

FIG. 6. Gaede's mercury rotary pump (Ref. 12). The inlet is at R. Foreline connection is the horizontal tube at the top.

FIG. 7. Vane pump after Rumelli (1588). Hagley Museum.

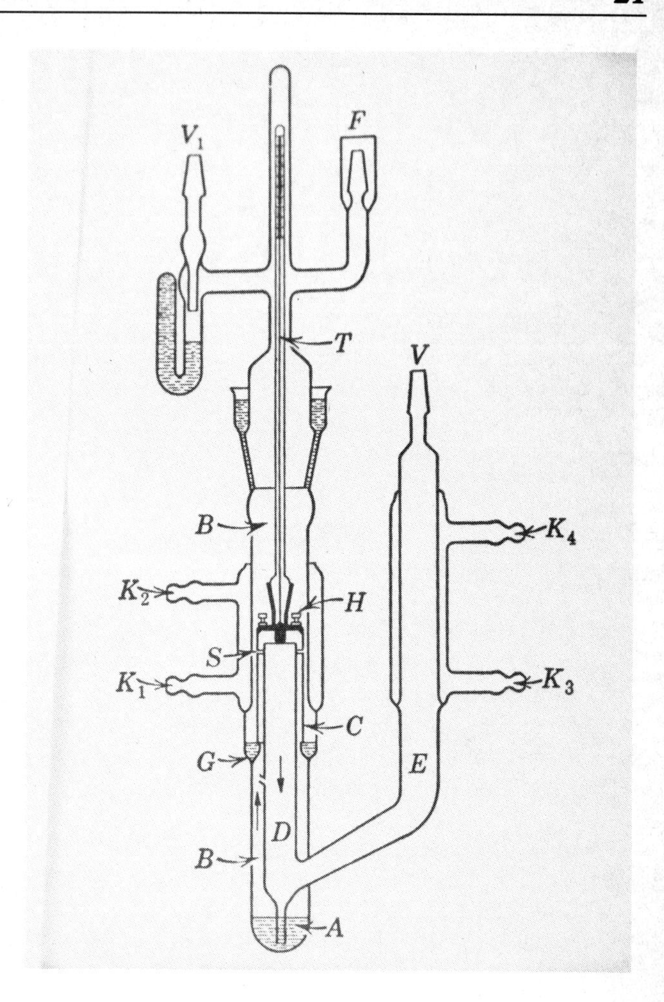

FIG. 8. Gaede's very early diffusion pump design (Ref. 13). The inlet is at F. The slot for pumping is at S. Foreline connection is at V.

Gaede, thinking in terms of kinetic theory, correctly deduced that gases can be pumped by the molecular drag effect only to be defeated by mechanical difficulties. In the case of diffusion pumps, he assumed the pumping action was due to the classical picture of the diffusion of air into a "cloud" of mercury vapor. Actually, if the mercury cloud were immobile, no pumping action would result. This confusion kept diffusion pump speeds at very low levels and the full realization of their capabilities was achieved only 20–25 years later. A very early pump by Gaede is shown in Fig. 8.

Langmuir followed with a better design (1916) but he overemphasized the condensation aspects of the pumping mechanism. In fact, condensation is basically not required for the pumping action but only for the separation of pumped gas and the pumping fluid.

The first correct design from the point of view of fluid mechanics was made by W. W. Crawford[21] in 1917, but the maximum pumping speed was only about 2 l/s for a 7 cm diam pump (roughly 50 times lower than modern designs). Such pumps were continuously improved during the next 20 years. The diffusion pump technical literature contains at least 300 items and it is beyond the scope of this paper to trace their development.

Low vapor pressure oils were introduced by C. R. Burch[14] in 1928. In 1937, C. M. Van Atta[14] reported a design with 200 l/s for an 18 cm diam pump, about five times lower than present pumps. Large pumps were designed in the early 1940's for large gas separation plants and later for vacuum metallurgy, industrial coating, and space simulation chambers.

Modern designs were made near 1960 by CVC Products, National Research Corporation (now Varian), and Edwards Company in England, while in Europe the emphasis was shifting toward molecular pumps.

B. "Molecular pumps"

Dushman[14] and most other writers indicate 1913 as the year of invention of the molecular pump by Gaede. However, the first mention of Gaede's pump occurs in 1912 in The Engineering Index Annual (Transactions of the German Physical Society). Dunkle[16] notes that the pump was not successful because in 1912 the art of making high speed rotating machinery had not yet been developed (Fig. 9).

It took 45 to 50 years before molecular pumps were made as an industrial device of adequate reliability,[22] the pioneering work being done by the Pfeiffer Company. Other attempts by Holweck and Siegbahn[22] were hardly more successful. They utilized, respectively, a cylinder and a disk as the molecular drag surface. One advantage of the bladed, multistaged, axial-flow, compressor-type vacuum pump is that each rotor surface is exposed only to a particular vacuum level. In cylindrical designs, the same surface passes from the high pressure to the low pressure side (see Fig. 9).

The idea of using the axial design seems to have occurred

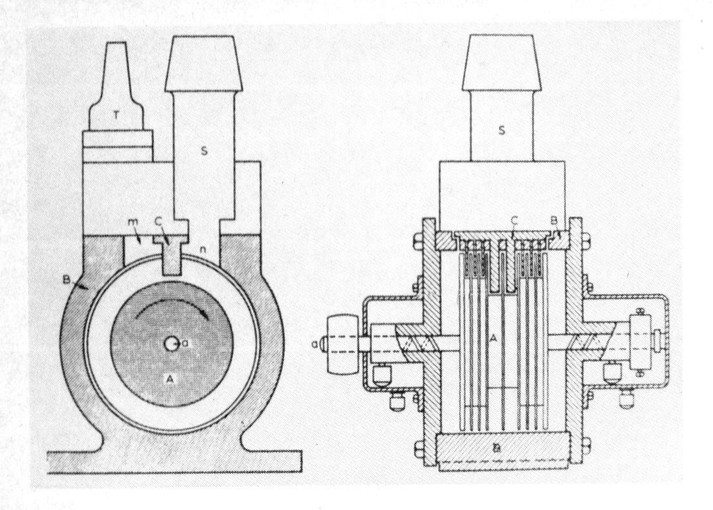

FIG. 9. Gaede's molecular drag pump (Ref. 13).

simultaneously to at least two persons, H. A. Steinherz[23] and W. Becker[22], the former attempting simply to utilize an existing compressor device as a vacuum pump. A. Shapiro[24] and his associates analyzed the open blade design, which was found to be entirely practical and efficient. The open blade designs are generally adapted now by most manufacturers.

C. Ion-getter and sputter-ion pumps

Many phenomena utilized today for high and ultrahigh vacuum pumping were known, and basically understood, for a long time.

These include outgassing, adsorption, gettering, sputtering, and cryogenic temperatures. They were not used earlier because the technology was not ready and the need was not pressing.

Vacuum of a few microns (10^{-3} Torr) was required in Edison's original electric lamps. In 1881, it took 5 hours.[25] To handle the outgassing of water vapor, bulbs were heated to about 300 °C during evacuation. Gettering by phosphoric anhydride was introduced from the beginning. Sprengel pumps are associated with the initial successes of the lamp industry. Improvements in pumping speed reduced evacuation time to 30 min. This process was used until 1896 when Malignani's (Udine, Italy) chemical exhaust processing was adopted. This consisted of placing inside the lamp red phosphorus, which would vaporize when the filament was heated. The evacuation time was reduced to 1 min. The effect of phosphorus is partly chemical and partly due to adsorption.[25]

The clean up of gases by an electrical discharge has been observed for over a century.[11] Gettering for lamp making is at least that old. T. M. Penning mentions sputtering as a method for reducing gas pressure at least as early as 1939.[26] The mention of ionic pumping appears at least as early as 1953,[27] associated with the initial development of ultrahigh vacuum techniques.

R. Herb, late in the 1940's or in the early 1950's, became determined to eliminate vapors in vacuum systems. One of his associates suggested titanium for gettering.

A container was pumped only by a forepump and a titanium sponge evaporator. It was successful but had a low pressure limit, which we suspected to be argon. They then built an ionizer to ionize residual gas and drive ions into the wall.

Such "Evapor-ion" pumps were rather difficult to operate and maintain (for example, Fig. 10). Sputter-ion pumps developed by L. D. Hall and associates at Varian were simpler and achieved universal acceptance a few years later.[28]

The first ion pumps were single cell appendage pumps[29] developed for maintaining high vacuum in large, high-power, microwave tubes. They resembled Penning gauges with the cathode made of titanium. Multicell designs (Fig. 11) and triode pumps for better argon pumping quickly followed.[30] Further details about ion-getter and sputter-ion pump development are given, for example, by G. F. Weston.[31]

Later, orbitron pumps, also introduced by Herb and his associates, enjoyed brief popularity but were withdrawn mainly due to problems with mechanical reliability.

Most modern ultrahigh vacuum instruments would not be possible without the development of sputter-ion pumps. Just as Leybold Company and W. Gaede were associated with the developments in the high vacuum technology at the turn

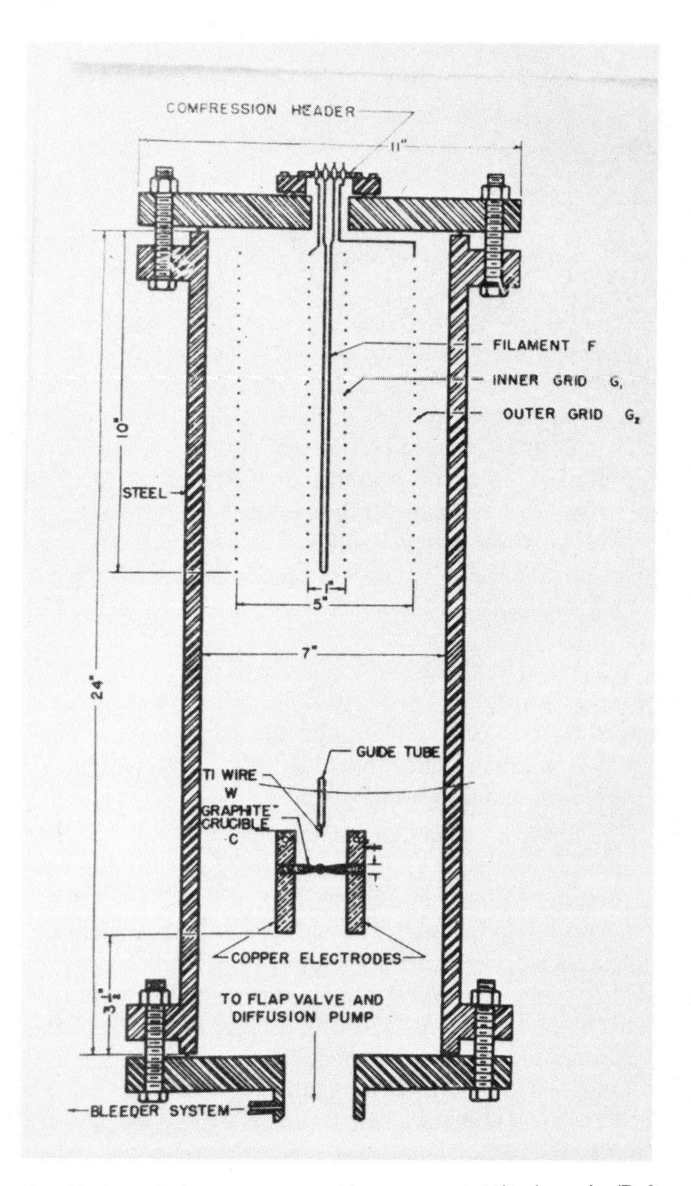

FIG. 10. An early ion-getter pump with an evaporated titanium wire (Ref. 27).

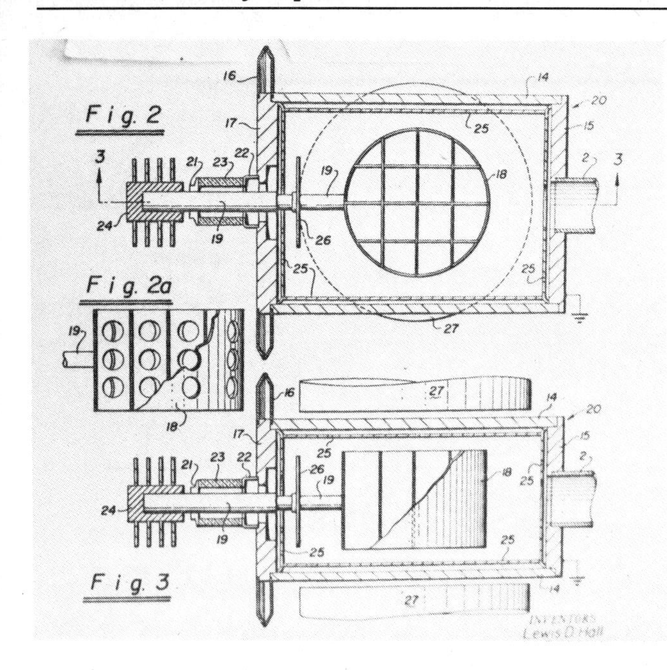

FIG. 11. A figure from the patent describing the sputter-ion pump (Ref. 30).

of the century, a group of engineers and scientists at Varian Associates pioneered industrial utilization of sputter-ion pumps and other ultrahigh vacuum pumping devices, such as sorption pumps, as well as valves and flanges.

D. Cryopumping

Cryopumping effects with and without high surface sorption materials were understood at least 100 years ago. Early in the 20th century, one of the methods for obtaining vacuum was to use activated charcoal cooled with liquid air.

With the invention of diffusion pumps, liquid nitrogen traps were generally used. Although the intended use of the traps was for suppression of the mercury vapor, they did pump water vapor remaining in the chamber after evacuation.

During the late 1950's and 1960's, cryogenic shrouds were often used in the construction of space simulation chambers. Occasionally, even liquid helium was used for enhanced cryopumping in early extreme high vacuum chambers.

The industrial art of cryopumping had to wait until the development of the reliable, closed-loop, gaseous helium cryorefrigerators.[32] These refrigerators were developed for other than high vacuum uses, but have recently found wide acceptance, the industrial introduction having been pioneered by CTI. However, their development is so recent that they hardly qualify for a discussion of history.

ACKNOWLEDGMENTS

Several people responded generously to requests for historical information: Dr. G. Wise of General Electric Co. (Schenectady), Dr. Bächler of Leybold Co., K. Seibel of Pfeiffer Co., L. Herbert and N. Gross of Varian Associates (Palo Alto), K. Johnson of Varian Central Library, R. Herb of National Electrostatics Co., J. Singleton of Westinghouse Co. (Pittsburgh), Z. Dobrowolski of Kinney Vacuum Co., the personnel of MIT central library archives, and finally, B. Dayton, whose contribution appears in the Appendix in a shortened version.

[1]E. N. Da and C. Andrade, *Advances in Vacuum Science and Technology*, edited by E. Thomas (Pergamon, New York, 1960).
[2]J. D. Bernal, *The Extension of Man*, (MIT, Cambridge, 1972).
[3]J. K. Finch, *The Story of Engineering* (Doubleday, Garden City, 1960).
[4]S. Drake, *Galileo at Work* (University of Chicago, Chicago, 1978).
[5]G. Sarton, ISIS, XXVI, I (French), p. 212. C. de Waard, *L'experience Barometrique, Etude Historique* (Thouars, Deux-Sevres, 1936).
[6]Isaac Beekman, *Journals*, edited by C. de Waard (Martinus Nijhoft, The Hague, 1939).
[7]G. Sarton, *A History of Science*, (Harvard University, Cambridge, 1952).
[8]A. Brachner, *Kultur & Technik, 3, 2,* 1979; German Museum, Munich (Thiemig, Munich, 1979).
[9]G. Schott, *Technica Curiosa* (Nuremburge, 1664).
[10]R. T. Gunther, *Early Science in Oxford*, Vol. 1 (Oxford University, Oxford, 1923), pp. 250, 16.
[11]J. M. Lafferty, Phys. Today, **34**, 211 (1981).
[12]Encyclopedia Brittanica, 1947 edition.
[13]L. Dunoyer, *Vacuum Practice* (Bell, London, 1926).
[14]S. Dushman, *Scientific Foundations of Vacuum Technology* (Wiley, New York, 1949).
[15]R. Jaeckel, *Kleinste Drucke* (Springer, Berlin, 1950).
[16]M. Dunkel, Vak.-Tech. **24**, 133 (1975).
[17]R. Conot, *A Streak of Luck* (Seaview, New York, 1979).
[18]W. Gaede, Trans. German Phys. Soc. **1905**, 287 (1905).
[19]F. Klemm, *A History of Western Technology* (MIT, Cambridge, 1984).
[20]W. Gaede and W. H. Keeson, Z. Instrumentenkd. **49**, 298 (1929).
[21]W. W. Crawford, Phys. Rev. **10**, 557 (1917).
[22]W. Becker, *Advances in Vacuum Science and Technology*, edited by E. Thomas (Pergamon, New York, 1960).
[23]M. H. Hablanian, *Advances in Vacuum Science and Technology*, edited by B. Thomas (Pergamon, New York, 1960).
[24]C. H. Kruger and A. H. Shapiro, *1960 Transactions of the 7th National Symposium, American Vacuum Society* (Pergamon, New York, 1961).
[25]J. W. Howell and H. Schroeder, *History of the Incandescent Lamp* (The Magna Company, Schenectady, New York, 1927), p. 123.
[26]T. M. Penning, US Patent No. 2 146 025.
[27]R. H. David and A. S. Davatia, Rev. Sci. Instrum. **25**, 1193 (1954).
[28]L. D. Hall, Rev. Sci. Instrum. **29**, 367 (1958).
[29]W. F. Westendorp, US Patent No. 2 755 014, filed April 1953.
[30]L. D. Hall, US Patent No. 2 993 638, filed July 1957.
[31]G. F. Weston, Vacuum **28**, 209 (1978).
[32]W. E. Gifford and H. O. McMahon, Adv. Cryog. Eng. **5**, 354 (1959).

APPENDIX: A BRIEF HISTORY OF THE DEVELOPMENT OF DIFFUSION PUMPS IN THE UNITED STATES (By B. B. Dayton)

In 1916, Irving Langmuir at the General Electric Co. made improvements on the mercury diffusion pump invented by W. Gaede in 1913. Langmuir showed that the slit through which the gas was pumped in the Gaede pump could be widened if the mercury vapor was directed away from the slit by a suitable nozzle. The pumping speed was thereby greatly increased. In 1916 Langmuir applied for a patent on the inverted or "mushroom cap" nozzle design.

In 1917, W. Crawford improved the Langmuir pump by using an expanding conical nozzle and a well designed condenser and entrance chamber. He also developed a multiple stage inverted nozzle mercury pump and obtained a patent in 1921. In 1918, O. E. Buckley applied for a patent on the concentric nozzle multistage mercury pump.

In 1922, Gaede in Germany developed a highly effective four stage steel mercury pump.

Mercury vapor pumps became widely used in the electronic tube industry in the U. S. from 1920 to 1940, but they required the use of refrigerated traps to keep the vapor out of the tubes (except for mercury rectifier tubes).

In 1928, C. R. Burch at Metropolitan–Vickers in England discovered that by replacing the mercury in Gaede's diffusion pump with low vapor pressure petroleum fractions, a good vacuum could be produced without a liquid air trap. These new oils were produced by molecular distillation. K. Mees, of Eastman Kodak Laboratories, brought the work of Burch to the attention of K. C. D. Hickman who had been working with low vapor pressure synthetic organic esters in place of mercury in special manometers for measuring the pressure in an apparatus for the drying of photographic film where mercury vapor was harmful. In 1929, Hickman found that these esters gave good results when used as the pump fluid in small glass diffusion pumps.

During the period from 1929 to 1937, the factors involved in designing diffusion pumps for use with the new synthetic oils were studied by Hickman and his co-workers in the Eastman Kodak Company resulting in the development of multiboiler "self-fractionating" oil diffusion pumps which extended the lowest pressure attainable without cold traps from 10^{-5} Torr to about 5×10^{-8} Torr. In 1935, J. Bearden introduced the use of immersion heater coils of nichrome wire for oil diffusion pumps. In 1937, N. Embree, a co-worker of Hickman, applied for a patent on the streamlined inverted nozzle design which greatly improved the pumping speed. Also in 1937, L. Malter of the Radio Corporation of America applied for a patent on the concentric vapor conduit tube system for multistage self-fractionating oil diffusion pumps, and Hickman applied for a patent on a horizontal metal fractionating oil diffusion pump.

Beginning in 1928, experiments in vacuum distillation of plasticizers for motion picture film at Kodak resulted in the development in late 1930 of a falling-film molecular still. Kodak and General Mills then organized Distillation Products, Inc. to produce vitamin concentrates and investigate other commercial applications of high vacuum molecular distillation.

Visitors to Kodak expressed interest in purchasing the diffusion pumps and vacuum gauges which Hickman and his co-workers had developed mainly for use on the stills. As a result of the demand for vacuum equipment based on the oil diffusion pump, Distillation Products, Inc. formed a vacuum equipment group to produce the stills and pumping accessories to manufacture the Hickman glass fractionating pumps, metal pumps, pump fluids, and accessories.

In 1940, Morse formed a new company, National Research Corporation, which soon obtained contracts for lens-coating equipment from the Navy and pumps for vacuum furnaces to produce magnesium for the Defense Plants Corporation.

In 1941, in response to a problem in handling the high gas load from 32 in. diam centrifugal stills, the gap in the range between diffusion pumps and steam ejectors was filled by Hickman's invention of an oil–vapor ejector pump. This was developed into a series of "booster pumps" which found application in production of wartime magnesium and in the dehydration of penicillin, blood plasma, and foods. A series of three-stage high-speed vertical all-metal oil diffusion pumps was developed in the period from 1940 to 1943 beginning with a 4 in. diam pump in 1940 for the stills and lens coaters followed by a 6 in. pump in 1941 for larger stills and coaters, a 14 in. pump in 1943 for cyclotrons and finally, 20 and 30 in. vertical pumps in 1943 for use on the Calutron electromagnetic isotope separation units of the Manhattan Project.

In 1943, Westinghouse Electric Corp. had organized a small vacuum group and began making a 20 in. oil diffusion pump patterned after a design developed by Normand and Thornton at the University of California Radiation Laboratory. Some of these were used at Oak Ridge during the Manhattan Project.

In 1944, M. E. Johnson of DPI developed a series of small all-metal two-stage fractionating pumps with two vertical stacks in either an air cooled model with fins or a water cooled model with copper cooling coils. These so-called VMF pumps found application on automatic rotary exhaust and sealing machines for production of electronic tubes.

In 1945, DPI began to manufacture vertical metal fractionating pumps after obtaining rights under the Malter patent owned by RCA. A 10 in. model, the MCF-1400, was the first pump in 1945 and was followed in succession by a 20 in. and a 32 in. model. The larger pumps found use on synchrocyclotrons. In 1946, this line of MCF pumps was extended downward in size to a 6 in. pump and later to a 4 in. and 2 in. size. By this time, the National Research Corporation had also developed a complete line of metal oil diffusion pumps.

Bibliography

1915 W. Gaede, Ann. Phys. **46**, 357–392.
1916 I. Langmuir, J. Franklin Inst. **182**, 719.
1916 I. Langmuir, Gen. El. Rev. **19**, 1060.
1917 W. W. Crawford, Phys. Rev. **10**, 557–563.
1923 W. Gaede, Z. Tech. Phys. **4**, 337–369.
1928 C. R. Burch, Nature **122**, 729.
1930 K. C. D. Hickman and C. R. Sanford, Rev. Sci. Instrum. **1**, 140–163.
1931 J. A. Becker and E. K. Jaycox, Rev. Sci. Instrum. **2**, 774.
1935 R. M. Zabel, Rev. Sci. Instrum. **5**, 54–55.
1935 J. E. Henderson, Rev. Sci. Instrum. **6**, 66–67.
1935 H. E. Edwards, Rev. Sci. Instrum. **6**, 145–146.
1935 M. J. Copley, O. C. Simpson, H. M. Tenney, and T. E. Phipps, Rev. Sci. Instrum. **6**, 265–267.
1935 J. A. Bearden, Rev. Sci. Instrum. **6**, 276.
1936 K. C. D. Hickman, J. Franklin Inst. **221**, 215–235, 383–402.
1936 I. Amdur, Rev. Sci. Instrum. **7**, 395–396.
1937 J. Holtsmark, W. Ramm, and S. Westin, Rev. Sci. Instrum. **8**, 90–91.
1937 A. E. Lockenvitz, Rev. Sci. Instrum. **8**, 322–323.
1937 C. M. Van Atta and L. C. Van Atta, Phys. Rev. **51**, 377.
1938 L. Malter and N. Marcuvitz, Rev. Sci. Instrum. **9**, 92–95.
1939 R. S. Morse, Electronics, November 1939.
1940 K. C. D. Hickman, J. Appl. Phys. **11**, 303–313.
1940 R. S. Morse, Rev. Sci. Instrum. **11**, 277–281.
1944 K. C. D. Hickman, Chem. Rev. **34**, 51–106.
1946 K. C. D. Hickman, *The Molecular Still in America* (Chemical Products, 1946), pp. 1–6.
1948 C. R. Burch, *The First Oil Condensation Pump* (Chemistry and Industry No. 6, London, 1949), pp. 87–88.

The development of valves, connectors, and traps for vacuum systems during the 20th century

J. H. Singleton

Westinghouse R & D Center, Pittsburgh, Pennsylvania 15235

(Received 8 September 1983; accepted 11 October 1983)

The components used in vacuum systems have undergone profound change during the 20th century. The development has been driven both by the needs of research and of commercial manufacturing. It is, however, remarkable that the sealing wax and string era persisted through the 1940's even in large research establishments. The present wide range of reliable commercial components became available only with the demand for ultrahigh vacuum capability, and with the development of helium leak detection techniques which permit rapid assessment of vacuum integrity. This paper provides a historical perspective of the evolution of: (a) sealing techniques for feedthroughs and for connections between components, (b) valves for isolation and for control of gas flow, and (c) traps to minimize the backstreaming of pumping fluids.

PACS numbers: 01.65. + g, 07.30. − t

I. INTRODUCTION

The intent of this brief review is to describe the historical development of some of the key hardware components of vacuum systems and to identify the major driving forces behind such developments. It is not possible to catalogue or to uniquely assign the credit for every advance in technique, but only to describe what components were in general use during selected time intervals. Major incentive for improvement came with specific changes in technology, e.g., with the development of the diffusion pump, but of equal significance was the impact of commercial requirements, such as the manufacture of incandescent lamps.

II. THE STATE OF VACUUM TECHNOLOGY BEFORE 1900

As a point of departure, it is useful to review the general state of vacuum technology at the end of the 19th century. The most common techniques for exhausting a system were mercury displacement pumps, such as those of Toepler (1862)[1] and Sprengel (1865).[2] These pumps were capable of achieving pressures in the 10^{-3} to 10^{-4} Torr range, when used with a refrigerated trap, but they were painfully slow; in the early production of Edison's incandescent lamps, the exhaust time was 5 h.[3] Vacuum systems, such as one used by Townsend in the study of electrical conductivity of gases,[4] were constructed of glass and of metals such as brass, aluminum, and zinc. Seals were made using commonly available materials such as Bankers Sealing Wax, "elastic glue", and rubber. Lubricated glass or brass stopcocks, or mercury-filled U-tubes were used to isolate parts of the system. However, the importance of heating a vacuum system, to minimize subsequent outgassing, was quite clearly understood. For example, the incandescent lamps of Swan (1878)[3] were flamed during exhaust. Such lamps, and the early Roentgen x-ray tubes,[5] used an all-glass envelope with platinum leads sealed directly through the glass and they were tipped off by fusing the glass after evacuation. Because the mercury displacement pumps did not effectively remove water, tubes or trays of phosphorus pentoxide were used to chemically trap the water, a procedure which continued in common use in some laboratories for the next 50 years.

III. DEVELOPMENTS BETWEEN 1900 AND 1920

In the first two decades of the 20th century, vacuum technology advanced rapidly. The trapping of gases, by adsorption on charcoal at room temperature, had first been reported by Tait and Dewar in 1875,[6] but the discovery that this effect was enormously increased by cooling the charcoal in liquid air[7] resulted in its application for exhausting vacuum chambers. Gaede developed the first modern rotary oil pump in 1907[8] and his mercury diffusion pump in 1915[9] In 1916, Langmuir[10] described the "condensation pump", a substantial improvement on the Gaede device, and the prototype of all modern diffusion pumps. It was in 1911 that Langmuir joined the research laboratories of the General Electric Company in Schenectady and commenced his long series of studies, initially concerned with the incandescent lamp, which form much of the basis of our understanding of vacuum technology.[11]

In 1900, the only bakeable metal to glass seal which could be used for electrical connections into a vacuum system used platinum and soft glass. Because of the enormous expansion in the production of incandescent lamps, much effort was expended to develop a cheaper and reliable substitute for platinum. The first successful replacement was developed by Eldred in 1911.[12] It utilized a wire made with a nickel–iron alloy core, having a thermal expansion similar to that of soft glass, coated with intermediate layers of copper and nickel, and finally with a fused sheath of platinum to facilitate sealing to glass. This rather complex seal was followed in 1913 by the development of Dumet by Fink,[13] consisting of a similar alloy core brazed in a thick outer sheath of copper. This wire was easier and cheaper to manufacture, and the copper could be sealed directly into soft glass. With the further development of a borax coating on the wire, to promote a better glass to copper bond, by Van Keuren (1913),[14] a highly reliable electrical feedthrough was obtained, providing both a

Originally published in J. Vac. Sci. Tech. A **2**(2), 126–131 (1984).

matched expansion joint and a good glass to metal surface bond. Dumet is still the dominant metal to glass seal used in incandescent lamp production. In 1914, Sand[15] developed a reliable technique for sealing molybdenum (or tungsten) through quartz by using lead, which adhered well to both metal and glass. This seal was used extensively in the manufacture of high-power vacuum tubes. In 1915, Ezechiel patented the tungsten to hard glass seal,[16] which permitted considerably larger conductors to be used than was possible with Dumet. Clearly, by 1920 the glass to metal sealing techniques were well developed. It is also interesting to note that, as early as 1916, the diffusive bonding of optically flat glass surfaces, at temperatures well below the melting point, had been demonstrated by Parker and Dallady.[17]

As was noted earlier, the importance of baking vacuum systems was already clearly understood in 1900. The extensive studies of Langmuir on the degassing of glass, as related to lamp production, and by many other workers[18] further defined the quantities of the principal gases which were evolved by different types of glasses. These studies emphasized the importance of thorough baking as a means of reducing the subsequent outgassing of glass systems. The importance of excluding greased stopcocks or seals, in order to attain the lowest ultimate pressure, was also clearly defined by Gaede[19] and Langmuir.[20] However, there was apparently no development of valves or demountable seals, which could tolerate bakeout. The mercury cutoff was virtually the only technique used, and that, of course, always implied the presence of mercury in the system, except where a refrigerated trap could be introduced. Stock and his co-workers, in their classic studies of the boron hydrides,[21] made extensive use of mercury cutoffs and described a number of useful modifications, e.g., in the use of float valves to reduce the danger of mercury ejection when employed with high gas pressures.[22] Another interesting design for a cutoff was based upon the use of a porcelain frit which was impervious to mercury but permeable by gas.[22] Two separate porcelain surfaces were immersed under mercury, attached respectively to a gas reservoir and to the vacuum system. By moving the surfaces into contact, the gas flow into the system could readily be controlled. This principle was used many times in the following years as a means of control of gas flow, and as recently as 1950 an improved version of a porcelain rod leak was described by Hagstrum and Weinhart.[23] Some valves were designed with the capability of moderate bakeout, such as a valve constructed of platinum and platinum–iridium described in 1913 by Bodenstein,[24] but they were primarily used in chemical reaction studies and not capable of providing an adequate closed conductance for high vacuum use.

With the development of the diffusion pump in 1916,[10] it became common practice to use a liquid air trap to exclude mercury vapor from the system. Such traps were commonly of the standard reentrant glass pattern and were small enough to ensure multiple collisions of gases with the walls and hence efficient trapping.

IV. DEVELOPMENTS FROM 1920 TO 1940

In these two decades, the use of vacuum processing expanded enormously into many new areas, e.g., vacuum metallurgy. The major innovation in pumping technique was the development of the oil diffusion pump, which permitted adequate ultimate pressures to be achieved while using an ambient temperature trap. This development was made possible by the isolation of low vapor pressure fractions from natural petroleum products by Burch (1928)[25] using molecular distillation, and by the subsequent synthesis of pure organic compounds of high stability.[26] Dynamic measurements at low pressures were now possible using ionization gauges of proven reliability, e.g., the design of Dushman and Found.[27] However, such gauges could only measure as low as the 10^{-8} Torr region, and the vacuum techniques available in 1920 were already capable of reaching this limit. Thus, the gauge was of limited value as a diagnostic tool in attempts to achieve even lower ultimate pressures. Pressures lower than could be registered by the ion gauge were undoubtedly obtained by workers such as Anderson[28] and Nottingham[29] in measurements of the work function of various metals, and this was inferred from the stability achieved in the measurements themselves. The lack of a quantitative measure of pressure was probably the major limitation in the rate of development of ultrahigh vacuum techniques at this time. The achievement of these ultimate pressures required rigorous processing which concluded with tipping off the experimental tube; there was virtually no capability for manipulation of gases. Consequently, such techniques were rarely applied to the study of subjects such as adsorption and catalysis when surface contamination is a dominant concern, and much of the experimental work in these areas was pursued under conditions of uncertain cleanliness.

The development of glass to metal seals continued throughout this period. In 1923, Housekeeper, at the Western Electric Corporation, perfected a radically new sealing technique in which the mismatch in thermal expansion between metal and glass was accommodated by the seal geometry which utilized a feathered edge on the metal at the glass interface.[30] These seals were primarily used for copper, but are still widely used today for both copper and stainless steel. Kaye[31] showed examples of 3 in. seals of soft glass to a nickel–iron alloy, and of 2 in. seals of soft glass to a chromium–iron alloy, produced prior to 1927 for commercial use. By 1934, Kovar[32] and Fernico[33], the important alloys of iron, nickel, and cobalt, were developed for use in sealing to hard glass. The reliability of these seals displaced such procedures as the metallization of glass to permit soldering to metal.[34] However, special applications, such as attaching optical windows to a system, often continued to utilize less sophisticated techniques, since it was impractical to develop matched-expansion alloys except for large-scale applications. For example, Palmer describes the use of a silver chloride seal to attach fluorite windows to glass,[35] and Strong[36] describes the use of waxes for this purpose. It is interesting to note that at this time the techniques for sealing glass to metal were far advanced, yet the materials generally available for system construction were not always satisfactory. Kaye[37] recommended the use of tinning or stove enameling to rescue faulty brass or iron castings, and Strong[38] suggested the routine use of several layers of Glyptal[a)] varnish to seal the outside of the finished vacuum system. Such techniques are not entirely unknown today, but these references serve to remind us that some of the convenience in routine vacuum practice today stems from the high quality of the materials of construction.

During this period between 1920 and 1940 there were still no commercially available demountable seals or valves specifically developed for vacuum use. The procedures of Anderson[28] and Nottingham,[29] for achieving very low ultimate pressures, used neither demountable seals nor valves. Their glass systems were thoroughly baked and tipped off from the pumps before a getter was activated to complete the cleanup and to maintain the vacuum. Such systems had little or no capability for research requiring control of gas atmospheres and the techniques may be considered an extension of the procedures used for the production of radio tubes. Sealing waxes were still commonly used to make semipermanent joints,[36] although the Apiezon series of greases and waves, initially developed by Burch,[25] provided a more uniform product of controlled vapor pressure and melting point than had previously been available in the mixtures of natural products. The use of low melting alloys to produce seals was also well known,[39] as was the use of soft metal gaskets. Strong[40] illustrates a lead gasket used in a corner seal to provide a highly reliable demountable joint. Extensive use of such fuse-wire gaskets for demountable flanges, and for metal to porcelain seals, was made in the 1937 van de Graaff generator built at MIT[41] but it should be noted that the greatest reliability was obtained after coating the gaskets with Glyptal!

Metal vacuum valves were made by reworking standard service valves, e.g., by replacing the seat with a rubber gasket, and the stem packing glands with rubber seals, string soaked in an Apiezon grease, or a metal bellows.[42] The use of Apiezon grease for glass stopcocks in unbaked systems provided some improvement in reliability, although the difficulty in removing dissolved air from the grease film, and the unavoidable permeation through the grease, remained a serious limitation to the attainable ultimate pressure of the systems.

Interesting developments occurred in traps used for diffusion pumps. In 1925, Hughes and Pointdexter[43] described the use of alkali-metal traps for mercury pumps. Such traps used a few grams of sodium or potassium deposited along the inside surface of a pumping line. The rate of formation of the mercury amalgam along the walls of the relatively narrow tubes used (~ 1 cm diam) was rapid enough to provide a trapping efficiency equivalent to that of a refrigerated trap. The life of the trap was predicted to be several years. This was the time necessary to form an equimolecular sodium amalgam, at which point it was estimated that the vapor pressure of mercury would reach $\sim 3 \times 10^{-8}$ Torr. Surprisingly, such traps were found to remain effective after several exposures to the atmosphere. Major problems were reported to be the attack of the glass by the alkali metal, and the difficulty in depositing acceptable pure metal films.

The major advantage obtained by the introduction of the oil diffusion pump was the ability to operate with only a simple ambient temperature baffle, resulting in reduced maintenance and also permitting a higher conductance, as compared to a refrigerated trap. In order to provide more efficient trapping at ambient temperatures, Becker and Jaycox[44] described the use of a charcoal trap which was baked to 400 °C prior to turning on the diffusion pump. Ultimate system pressures as low as 2×10^{-8} Torr were achieved. It was claimed that baking such a trap, even after substantial use, did not release the trapped oil, but resulted in decomposition. This claim was based upon the discovery that the electron emission from a oxide-coated cathode did not decrease during the baking, in marked contrast to emission on exposure to oil vapors. Clearly, these ambient temperature traps anticipated those later described by Alpert[45] and Biondi,[46] although it is to be emphasized that the charcoal traps were often used as a fully packed manifold and had very low conductance.

V. DEVELOPMENTS BETWEEN 1940 AND 1960

This period saw an enormous growth in the use of vacuum equipment. During the early years, the growth was mainly driven by the military requirements of World War II. In the United States, the Manhattan Project resulted in the aggressive development of very large vacuum systems[48,49] In addition, the techniques for producing complex sealed-off vacuum devices were developed to satisfy the requirements of radar and other electronic tubes. However, the most important single development in this period was the invention of the Bayard–Alpert ionization gauge[50] in 1950. This gauge extended the range of reliable pressure measurement to the low 10^{-10} Torr region, and provided the impetus for development of an ultrahigh vacuum technology which could be readily applied to both vacuum and gas physics research. Although many laboratories participated in this development, the group working with Alpert was particularly active. A fascinating account of the developments is given by Alpert.[51]

The major improvement in sealing techniques during this period was the production of successful metal to ceramic seals using a process of metallizing the ceramic surface followed by brazing to metal. These seals were used commercially in Germany during World War II and extensively developed following the war,[52] primarily for vacuum tubes. When such seals became available for general vacuum, some survived only for one or two bakeout cycles before developing leaks, but such problems declined over the years with improved design and such seals have now become the basic type of electrical feedthrough for vacuum use. An interesting technique for joining ceramic to metal was described by Martin and Tunis[53] in 1957 in which a metal cylinder was forced over the end of a ceramic tube giving a ram seal which was suitable for ultrahigh vacuum applications. A discussion of the sealing forces involved is included in an important analysis of seals between dissimilar materials, including glasses, ceramics, and metals.[54] This analysis provides an understanding of the critical parameters and gives procedures for the design of practical seals.

The development of ultrahigh vacuum technology required both flanges and valves with bakeout capability to at least 400 °C. This requirement automatically excluded the use of organic materials such as rubber, and of low melting or high vapor pressure metals. Although the use of wire gaskets such as aluminum, compressed between flat flanges, was well known, such seals were not of high reliability. Greater success was achieved by compressing copper gaskets between accurately located knife edges of a harder metal, such as the seal described by Pattee.[55] A more easily machined joint, which sheared the copper gasket to form a step seal, was described by Lange and Alpert.[56] Still better reliability was achieved using a gold wire gasket compressed to form a corner seal[51] in a geometry quite similar to that shown by Strong for a lead wire seal.[40] The survival of such seals after 40 to 50 cycles to 450 °C was reported by Grove.[57] It is probable that the copper and gold seals described above were not designed using the principle of a captured gasket geometry, such as that discussed by Bills and Allen in their development of a bakeable valve,[58] or by Wheeler[59] in his discussion of metal gasket seals. The reliability of the gold seals was undoubtedly attributable to the fact that some interdiffusion between the gold gasket and the stainless steel flange occurs during baking, resulting in a substantial bond. A properly sealed gold-gasketed joint must be separated by force after bakeout and the bonding presumably counteracts the reduced compression caused by stretching of the flange bolts during baking.

In 1940, metal valves still largely used rubber seals. Rebuilding of standard valves remained common practice,[47] although some of the large valves specifically designed and manufactured for the Manhattan Project include interesting features, such as the provision to replace shaft seals without breaking vacuum.[47] The O-ring seal was developed during the war years, for use in aircraft hydraulic systems, and with the ready availability of precise O-rings in various synthetic rubbers they were soon widely adopted for vacuum use.

In 1951, Alpert described the first all-metal valve designed for bakeable ultrahigh vacuum systems.[60] The valve used a highly polished conical Kovar nose in a flexible Kovar diaphragm and a copper body and was assembled by brazing in a hydrogen atmosphere to avoid contamination with flux. A differential screw mechanism was used to precisely control the motion of the nose, which could be forced into the 1/2 in. diam. opening in the copper with a force of 5–10 tons, forming the valve seat on the first closure and providing a closed conductance of $\sim 10^{-10}$ l/s. Subsequent refinements by Buritz and Carmichael, in Alpert's group, provided an all-Monel valve with a somewhat better closed conductance, Bills and Allen[58] described a major improvement in the sealing geometry by using a pure silver insert which was constrained against flow by the harder surfaces of Monel. This design permitted much higher loading along the seal area and achieved closed conductances in the range of 10^{-14} l/s. Another sealing principle utilized a knife-edge seal in a copper gasket; such a seal was described by Conner *et al.*,[61] for a large gate valve, the required high sealing pressure being developed by pneumatic pressure.

The early ultrahigh vacuum systems used an oil diffusion pump and a refrigerated reentrant glass trap,[51] and the variation in the liquid nitrogen level was often accompanied by a corresponding pressure fluctuation. With the intent of minimizing such fluctuations, a corrugated copper foil was inserted in the trap[45] to ensure a more uniform temperature. Such traps were baked at 400 °C along with the rest of the system, and inevitably the trap remained unfilled often enough that it was recognized that the trapping at ambient temperatures was as effective as that at liquid nitrogen temperatures, at least to the limit of measurement of the Bayard–Alpert gauge. Such traps typically provided protection for 3 weeks. With the objective of increasing the useful life, Biondi[46] substituted artificial zeolite or activated alumina for the copper, reasoning that the larger surface area would increase the adsorption capacity. Initially, such traps were shown to have an effective life of at least 75 days. Subsequent experience suggests that protection for at least one year is obtained in the small glass systems described by Alpert.[51]

With the extension of ultrahigh vacuum techniques to large systems, the standard trap designs for large diffusion pumps were found to be relatively ineffective. One reason was that some oil molecules could traverse the trap with one or even no contact with a cold surface, which was often inefficiently cooled. The second problem was that it was frequently possible for oil to diffuse along relatively warm surfaces through the traps. Efficient traps which avoided these problems were described by Ullman[62] and Milleron.[63]

VI. DEVELOPMENT SINCE 1960

The recent development of vacuum technology has been dominated by the continued need for reliable techniques for achieving ultraclean systems for research, for commercial processing—especially in the semiconductor industry, and for sophisticated analytical systems for surface studies. This has resulted in the easy availability of a wide range of demountable flanged fittings, such as electrical and liquid feedthroughs, mechanical manipulators, and optical windows from which complex systems can be readily assembled. Because of the overriding requirement of cleanliness, the quality of materials and of the assembly techniques has been greatly improved. Vacuum or hydrogen brazing is widely used for the construction of many components, paricularly those with alumina feedthroughs. The flexibility and reliability of the tungsten–inert–gas arc welding process, first developed in the 1950s,[64] has made it the premier technique for joining many different types of metal. This process has also made possible the manufacture of the high quality nested bellows so frequently used in manipulators and in valves. The recognition of the captured seal geometry[59] has resulted in the almost universal acceptance of the copper seal for demountable flanges; it is perhaps fortunate that such a dominant position has been developed in these seals since this has ensured a high degree of interchangeability between the equipment of many different manufacturers.

Valves designed for vacuum service are now readily available and the application of innovative engineering has resulted in greatly improved performance and reliability, e.g. in the development of large gate valves.[65,66] The use of Viton rubber gaskets[67] and polyimide seals[68,69] has permitted some modest bakeout capabilities in valves using such seals. In the area of mechanical motion in vacuum systems, both metal bellows and magnetic coupling have been widely used. The development of a magnetically confined "ferrofluid" of low vapor pressure, to seal rotating shafts through the vacuum wall,[70] has provided an unexpected solution to a difficult problem, particularly where high speed or torque is essential. Unfortunately, such seals cannot be baked adequately for the most demanding vacuum applications.

Refrigerated traps for diffusion pumps, which conform to the requirements described in earlier experimental traps,[62,63] have become commercially available, and, partially as a consequence of the functional requirements, such traps frequently have long retention times for each filling of refrigerant.

VII. CONCLUSIONS

This study of the development of a limited range of vacuum system hardware components reveals large differences in the rate at which they were developed. The sealing technology, specifically for glass to metal, was crucial to the reliability of both incandescent lamps and electron tubes and therefore was soon developed to a commercially available product. However, it is surprising to note that reliable large vacuum valves did not appear on the commercial market until the late 1940s. The process of upgrading a general service valve for vacuum use, by winding a new stem gasket from string soaked in an Apiezon grease, seems remote and probably an exception, until it is found to have been common practice even in one of the larger research establishments.[71]

The quality control of the materials used in the construction of vacuum systems components has improved over the years because of the universally increasing demands of modern technology in general. A similar improvement in fabrication techniques, such as welding and brazing has also occurred. It is not unreasonable to conclude that the high quality of components which are available today is possible because of such improvements, combined with the rigorous demands of ultrahigh vacuum technology. Such components must be free from real or virtual leaks and from oil or similar contaminants or they will not perform satisfactorily. The availability of sensitive helium leak detection techniques has permitted greatly improved quality control. Finally, as might be inferred from previous history, the large market for such vacuum components has stimulated the required effort to make them commercially available.

ACKNOWLEDGMENTS

The organization of this review has proved more enjoyable than anticipated in that it has provided the incentive to reexamine much of the earlier work on which we must always base future developments. It has afforded the opportunity to discuss some of the more recent past history with people closely associated with the field, particularly with M. W. Biondi and W. J. Lange, to whom thanks are due.

[a] The tradename for an alkyd or glycol-phthalate base resin manufactured by the General Electric Company. Glyptal was beloved by several generations of vacuum users and was used in several different colors, including a clear varnish by those who preferred to forget about their leaks.

[1] See E. Hagen, Wied. Ann. **12**, 425 (1881).

[2] H. J. P. Sprengel, Pogg. Ann. **129**, 564 (1865).

[3] J. W. Howell and H. Schroeder, *The History of the Incandescent Lamp* (The Maqua Company, Schenectady, New York, 1927), p. 38.

[4] J. S. Townsend, Phil. Mag. **1**, 198 (1901).

[5] W. C. Roentgen, Sitzungsber. Phys.–Med. Ges. 132 (1895).

[6] P. G. Tait and J. Dewar, Proc. R. Soc. Edinburgh **8**, 628 (1875).

[7] J. Dewar, Proc. R. Soc. London Ser. A **74**, 122 (1904).

[8] W. Gaede, reported by O. E. Meyer, Verh. Deutsch. Phys. Ges. **10**, 753 (1907).

[9] W. Gaede, Ann. Phys. **46**, 357 (1915).

[10] I. Langmuir, Phys. Rev. **8**, 48 (1916).

[11] *The Collected Works of Irving Langmuir* (Pergamon, London, 1960).

[12] B. E. Eldred, U. S. Patent 1 140 136 (1913).

[13] C. G. Fink, see Ref. 3, p. 160.

[14] W. L. Van Keuren, see Ref. 3, p. 161.

[15] H. J. S. Sand, Proc. Phys. Soc. **26**, 127 (1914).

[16] W. Ezechiel, U. S. Patent 1 154 081 (1915).

[17] R. G. Parker and A. J. Dallady, Trans. Faraday Soc. **12**, 305 (1916).

[18] See, for example, S. Dushman, *Vacuum Technique* (Wiley, New York, 1949), p. 520 *et. seq.*

[19] W. Gaede, Ann. Phys. **41**, 366 (1913).

[20] I. Langmuir, J. Am. Chem. Soc. **35**, 171 (1913).

[21] A. Stock, *Hydrides of Boron and Silicon* (Cornell University, Ithaca, 1933).

[22] A. Stock, Z. Elektrochem. **23**, 33 (1917).

[23] H. D. Hagstrum and H. W. Weinhart, Rev. Sci. Instrum. **21**, 394 (1950).

[24] M. Bodenstein, Z. Phys. Chem. **85**, 297 (1913).

[25] C. R. Burch, Proc. R. Soc. London Ser. A **123**, 271 (1929).

[26] K. C. D. Hickman, J. Franklin Inst. **221**, 215 (1936).

[27] S. Dushman and C. G. Found, Phys. Rev. **17**, 7 (1921).

[28] P. A. Anderson, Phys. Rev. **34**, 129 (1929).

[29] W. B. Nottingham, J. Appl. Phys. **8**, 762 (1937).

[30] W. G. Housekeeper, J. Am. Inst. Elect. Eng. **42**, 954 (1923).

[31] G. W. C. Kaye, *High Vacua* (Longmans Green, London, 1927).

[32] H. Scott, J. Franklin Inst. **220**, 733 (1935).

[33] A. W. Hull and E. E. Burger, Physics **5**, 384 (1934).

[34] E. C. McKelvy and C. S. Taylor, J. Am. Chem. Soc. **42**, 1364 (1920).

[35] F. Palmer, Phys. Rev. **45**, 557 (1934).

[36] J. Strong, *Procedures in Experimental Physics* (Prentice–Hall, Englewood Cliffs, NJ, 1938), p. 129.

[37] Reference 31, p. 67.

[38] Reference 36, p. 134.

[39] Reference 31, p. 68.

[40] Reference 36, p. 128.

[41] L. C. VanAtta, D. L. Northrop, R. J. van de Graaf, and G. M. VanAtta, Rev. Sci. Instrum. **12**, 534 (1941).

[42] Reference 36, pp. 132–3.

[43] A. L. Hughes and F. E. Pointdexter, Philos. Mag. **50**, 432 (1925).

[44] J. A. Becker and E. K. Jaycox, Rev. Sci. Instrum. **2**, 773 (1931).

[45] D. Alpert, Rev. Sci. Instrum. **24**, 1004 (1953).

[46] M. A. Biondi, Rev. Sci. Instrum. **30**, 831 (1959).

[47] See, for example, A. Guthrie and R. K. Wakerling, *Vacuum Equipment and Techniques* (McGraw-Hill , New York, 1949).

[48] H. A. Thomas, T. W. Williams, and J. A. Hipple, Rev. Sci. Instrum. **17**, 368 (1946).

[49] A. O. Nier, C. M. Stevens, J. A. Hustrulid, and T. A. Abbot, J. Appl. Phys. **18**, 30 (1947).

[50] R. T. Bayard and D. Alpert, Rev. Sci. Instrum. **21**, 571 (1950).

[51] D. Alpert, *Production and Measurement of Ultrahigh Vacuum, Handbuch der Physik* (Springer, Berlin, 1958), Vol. 12, p. 39.

[52] W. H. Kohl, *Materials Technology for Electron Tubes* (Reinhold, New York, 1951), p. 403 *et. seq*

[53] I. E. Martin and A. C. Tunis, R. C. A. Engineer **3**, 3 (1957).

[54] G. Lewin and R. Mark, *Transactions of the 5th National Symposium on Vacuum Technology* (Pergamon, London, 1959), p. 44.

[55] H. H. Pattee, Jr., Rev. Sci. Instrum. **25**, 1132 (1954).

[56] W. J. Lange and D. Alpert, Rev. Sci. Instrum. **28**, 726 (1957).

[57] D. J. Grove, *Transactions of the 5th National Symposium on Vacuum Technology* (Pergamon , London, 1959), p. 9.

[58] D. G. Bills and F. G. Allen, Rev. Sci. Instrum. **26**, 654 (1955).

[59] W. R. Wheeler, *Transactions of the 9th National Symposium on Vacuum Technology* (McMillan, New York, 1962), p. 159.

[60] D. Alpert, Rev. Sci. Instrum. **22**, 536 (1951).

[61] R. J. Conner, R. S. Buritz, and T. von Zweck, *Transactions of the 8th National Symposium on Vacuum Technology* (Pergamon, London, 1962),

p. 151.
[62]J. R. Ullman, *Transactions of the 4th National Symposium on Vacuum Technology* (Pergamon, London, 1958), p. 95.
[63]N. Milleron, *Transactions of the 5thNational Symposium on Vacuum Technology* (Pergamon, London, 1959), p. 140.
[64]P. T. Houldcroft, *Welding Processes* (Cambridge Univ., Cambridge, 1967), p. 61.
[65]W. R. Wheeler, J. Vac. Sci. Technol. **13**, 503 (1976).
[66]S. Schertler, U. S. Patent 3 368 792.
[67]R. R. Addiss, Jr., L. Pensak, and N. J. Scott, *Transactions of the 7th National Symposium on Vacuum Technology* (Pergamon, London, 1961), p. 39.
[68]B. R. F. Kendall and M. F. Zabielski, J. Vac. Sci. Technol. **3**, 114 (1966).
[69]M. F. Zabielski and P. R. Blaszuk, J. Vac. Sci. Technol. **13**, 644 (1976).
[70]K. Raj. Industrial Research and Development, **21** 115 (1979).
[71]Unpublished Westinghouse reports from 1943.

The measurement of vacuum pressures

P. A. Redhead

Division of Physics, National Research Council, Ottawa, K1A OR6, Canada

(Received 9 September 1983; accepted 2 November 1983)

A historical review of the science and technology of pressure measurement, in the vacuum range, is given from the first measurement of a subatmospheric pressure in the 1650's until 1945. The modern period (from 1945 to the present) is reviewed with emphasis on the elucidation of the processes limiting the performance of various vacuum gauges (both total and partial pressure) and the practical solutions to these limitations. A comparison of the performance of modern vacuum gauges is presented.

PACS numbers: 07.30.Dz, 01.65. + g

I. HISTORICAL SKETCH: 1643–1954

The first production of a vacuum, as we all know so well, occurred in 1643 when Torricelli performed his famous experiment with a mercury filled tube.[1] Historical research has cast doubt on this common knowledge and suggests that although Toricelli undoubtedly conceived the experiment, it was actually performed by Viviani, not Torricelli, in 1644, not 1643.[2] Leaving that controversy for others to pursue, there seems to be no doubt that Descartes invented the barometer by affixing a paper scale to a Torricellian tube, as he indicated in a letter to Mersenne dated December 13, 1647.[3] Périer (Pascal's brother-in-law) then performed his celebrated experiment on the Puy de Dôme near Clermont–Farrand on September 16, 1674, where for the first time a barometer was set up at various heights on the mountain and the variation of atmosphere pressure with altitude established.[4] Figure 1 shows the earliest water barometer constructed by Otto von Guericke in about 1654. The top portion of the barometer was made of glass in which a manikin, mounted on a cork, floated on the water surface and pointed out the pressure on a scale.[5]

The date of the first vacuum measurement, i.e., a measurement of subatmospheric pressure, is less certain. The invention of the air pump by von Guericke made the production of a near vacuum feasible in much larger volumes than was possible with the Torricellian tube. Robert Boyle modified von Guericke's air pump and used it to establish many properties of vacuum and realized that the barometer made an excellent vacuum gauge. Figure 2 is a diagram of Boyle's apparatus (this diagram is from a later experiment) where he used a Torricellian barometer whose mercury reservoir was contained within a glass vessel evacuated by Boyle's air pump (Boyle called his pump a "pneumatic engine," it was known on the continent as the "machina Boyleana"). Boyle was able to achieve a minimum pressure of about $\frac{1}{4}$ in. Hg (6 Torr). The date of this experiment is not recorded but it must have been before 1660, when Boyle's first book on his pneumatic experiments was published.[6]

The mercury column manometer developed by Boyle, and its derivatives, were the only form of vacuum gauge available for the following two centuries. The next major advance was by McLeod[7] in 1874 who realized that the volume of a gas could be compressed by a known ratio, using a mercury column, to a higher pressure which could be easily measured, and that the original low pressure could be calculated by invoking Boyle's law. The McLeod gauge, shown in its original form in Fig. 3, is still in use for absolute pressure measurements in the range 1–10^{-6} Torr.[8]

Two decades after McLeod's invention Sutherland[9] proposed a vacuum gauge based on a measurement of the viscos-

FIG. 1. Water barometer of von Guericke (Ref. 5), circa 1654.

Originally published in J. Vac. Sci. Tech. A **2**(2), 132–138 (1984).

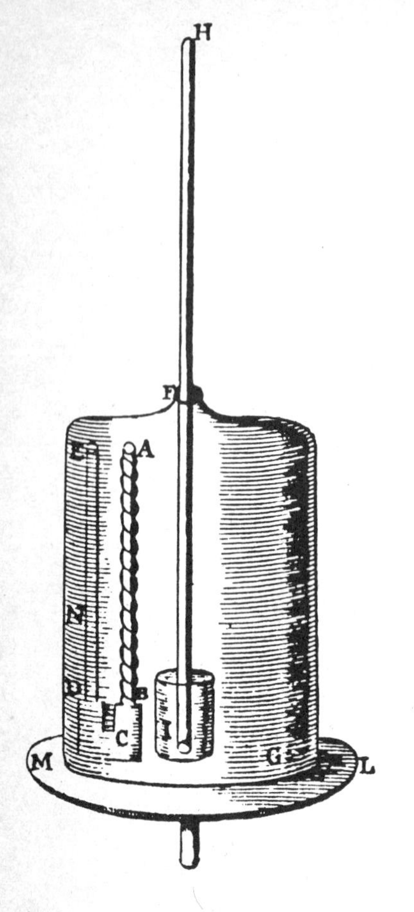

FIG. 2. Diagram of Boyle's "mercurial gauge, to discover the degrees both of rarefied and condensed air" (R. Boyle, The Excellency of Theology Compared with Natural Philosophy, London, 1665).

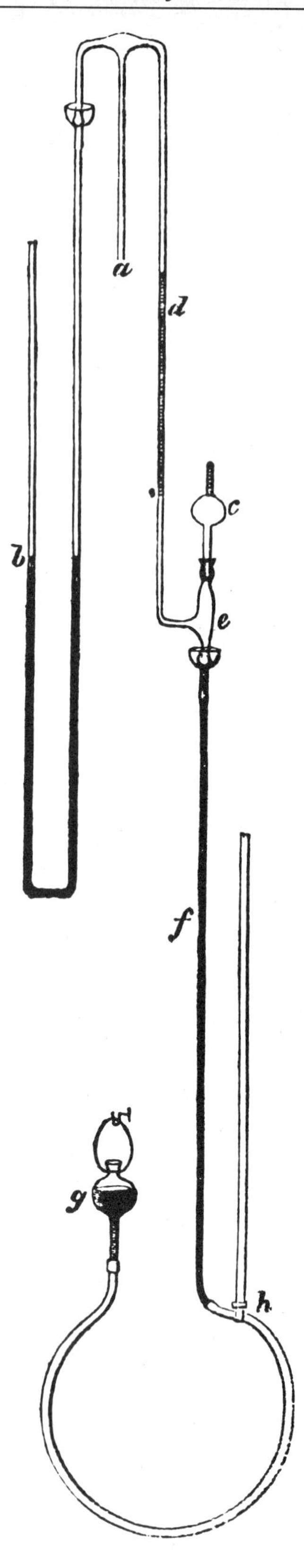

FIG. 3. Original form of the McLeod gauge (Ref. 7): (a) connection to apparatus; (b) siphon barometer (5 mm diameter); (c) globe of 48 ml capacity with volume tube at top; (d) pressure tube; (e) ground glass joint; (f) vertical tube 80 cm long; (g) mercury reservoir; (h) stopcock to control flow of mercury, actuated by a long rod. Tube (d) and the volume tube are of the same diameter.

ity of the gas. This principle of measurement was later extended by Langmuir[10] who developed a gauge using a quartz fiber which was made to oscillate in the gas, the decrement in amplitude of the oscillations gave a measure of gas pressure. Viscosity gauges have not been very widely used and were reviewed a decade ago by Steckelmacher.[11] Interest in viscosity gauges was resurrected in the 1950's by the work of Beams[12] who demonstrated that a magnetically levitated steel ball rotating at about 1 MHz in vacuum could be used as a vacuum gauge, by measuring the decrement in rotational frequency, to pressures as low, in principle, as 10^{-9} Torr. This device was further developed by Fremery[13] and is now widely used as a vacuum transfer standard in the range 10^{-2}–10^{-7} Torr.

The next major advance occurred in 1906 when Pirani developed a vacuum gauge[14] which gave a pressure indication by measuring the thermal conductivity of the gas. The change in resistance of a heated filament in vacuum was used to monitor the heat loss by conduction through the gas and thus to give an indication of pressure. Other forms of thermal conductivity gauges using a thermocouple[15] or thermistor[16] to measure the filament temperature were developed later.

In 1910 Knudsen[17] invented a new type of vacuum gauge based on the radiometer effect. This effect, although not fully understood at the time, was recognized as being pressure dependent. This elegant gauge measures the torque developed by the momentum transfer of hot gas molecules on one side, and cold molecules on the other side of a vane mounted on a sensitive suspension. This gauge is not used now except for very special applications.

The invention of the hot-cathode ionization gauge ushered in the modern era of vacuum measurements. In 1909 von Baeyer[18] showed that a triode vacuum tube could be used as a vacuum gauge. Buckley, in 1916[19] is usually credited with inventing the hot-cathode ionization gauge which, in later developments, extended the lowest measurable pressure to about 10^{-8} Torr. There were no significant improvements in the design of hot-cathode ionization gauges until the Bayard–Alpert gauge in 1950, which will be discussed in the next section.

The cold-cathode discharge gauge was invented by Penning[20] in 1937. In this type of gauge a cold-cathode, Townsend discharge in crossed electric and magnetic fields is established, resulting in a discharge current which is frequently a nonlinear function of pressure. The Penning gauge had a lower limit of about 10^{-6} Torr, later forms of this type of gauge, discussed in the next section, extended the range to 10^{-12} Torr. The Penning gauge and its successors have been widely used because of their ruggedness and simplicity; however, they tend to be nonlinear and occasionally erratic in response so they have not been used much for accurate measurements.

The final step in the historical survey of the period up to 1945 was the development of the mass-spectrometer leak detector. Since this development occurred during the Second World War under the secrecy of the Manhattan project, it is not easy to determine the inventor of the technique. It appears that cold-cathode, mass-spectrometer, leak detectors were developed by Backus[21] and his colleagues in 1943 or 1944. Mass-spectrometer leak detectors became commercially available at the end of the war and the modern devices, though much improved in detail, do not differ fundamentally from the first commercial instruments.

The developments in the period 1660 to 1945 are summarized in Table I.

II. THE MODERN PERIOD: 1945–1983

As we approach, in our historical review, the formation of the American Vacuum Society in 1953 we enter a period of intense research activity concerned with the production and measurement of vacuum. During the 1950's the groundwork was laid for understanding the limitations on the production and measurement of very low pressures which led to the ubiquitous ultrahigh vacuum technology used today.

In the immediate postwar period it was still generally believed that the commonly observed limit to the lowest pressure achievable was set by some unknown process in diffusion pumps which caused their pumping speed to drop precipitously to zero at pressures around 10^{-8} Torr. Manufacturers' literature of the time show pumping speed curves for diffusion pumps going to zero at this pressure. However, there was evidence from papers published in the 1930's (based on the rate of change of work function of metal surfaces after thermal cleaning), suggesting that the pressure in some experiments was considerably lower than the 10^{-8} Torr indicated by an ionization gauge; a clear example was the experiment of Anderson in 1935.[23] It is now quite evident that the diffusion pumps available in th 1930's (when properly trapped and used in baked, glass systems) could achieve pressures of 10^{-10} Torr or better.

The breakthrough occurred as the result of evidence presented by Nottingham at the 1947 Physical Electronics Conference.[24] He predicted that there was a lower limit to the

TABLE I. Developments in vacuum gauges: 1660–1945.

Gauge class	Gauge type	Pressure range (Torr)	First development	Ref.
Pressure				
Liquid column	Mercury manometer	atm–10^{-1}	Boyle ~1660	6
Mechanical	Bourdon	atm–20		
	Diaphragm	atm–10^{-1}		
	Capacitance manometer	10–10^{-4}	Olsen and Hirst 1929	22
Compression	McLeod	10^{2}–10^{-7}	McLeod 1874	7
Viscosity				
Suspended element	Torsion gauges	1–10^{-4}	Sutherland 1897	9
	Oscillating fiber	1–10^{-4}	Langmuir 1913	10
Momentum transfer				
Radiometer	Knudsen	10–10^{-3}	Knudsen 1910	17
Thermal conductivity				
	Pirani	atm–10^{-4}	Pirani 1906	14
	Thermocouple		Voege 1906	15
Ionization				
Hot cathode	Large collector	10^{-3}–10^{-8}	Von Baeyer 1909	18
Cold cathode	Penning	10^{-3}–10^{-6}	Penning 1937	20

pressure measureable with a hot-cathode ionization gauge caused by a pressure-independent photocurrent resulting from soft x rays produced by the ionizing electrons (the x-ray effect). Soon after Nottingham's comments several designs of hot-cathode ionization gauges[25–27] appeared which reduced the flux of x rays at the ion collector either by reducing the collector size or by shielding the collector, thus lowering the measurable pressure limit to about 10^{-10} Torr. The elegant simplicity of the Bayard–Alpert design[25] in 1950 led to its immediate acceptance and it has been the most widely used ionization gauge ever since. The Bayard–Alpert gauge (BAG) and the other technology developed by Alpert's group at the Westinghouse laboratories[28] made possible the rapid expansion of UHV technology which then followed. Figure 4 shows the original design of the Bayard–Alpert gauge with an ion-collector diameter of 200 μm yielding an x-ray limit of about 5×10^{-11} Torr. Alpert proved the existence of a pressure-independent residual current by measuring collector current versus grid potential curves for a range of pressures from 10^{-6} to 5×10^{-11} Torr. It is noteworthy that similar data was published in 1931 by Jaycox and Weinhart,[29] but it took two decades for the observations to result in a new form of ionization gauge.

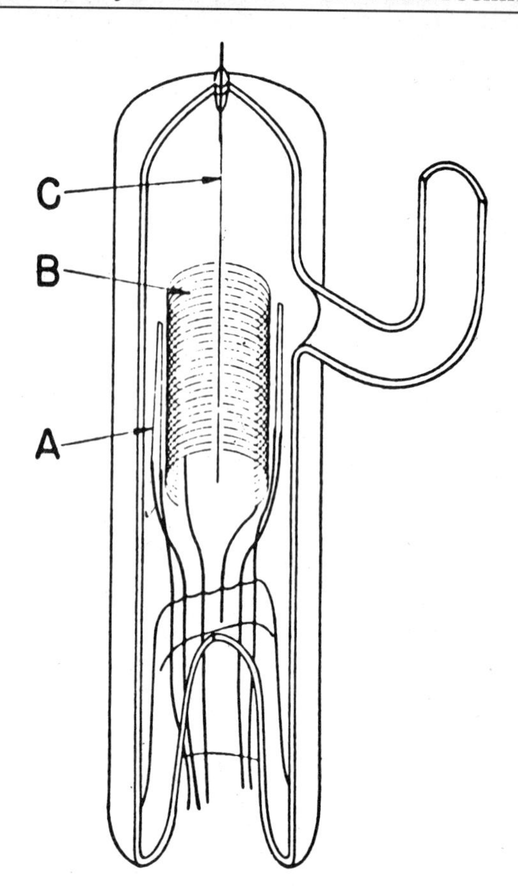

FIG. 4. Original form of Bayard–Alpert ionization gauge (Ref. 25) showing the filament A, grid B, and ion collector C.

In 1963 the reverse x-ray effect was observed[30] where soft x rays striking the walls surrounding a BAG released photoelectrons capable of reaching the ion collector when both the ion collector and the walls were at the same potential. This results in a residual current in the reverse direction to the usual x-ray effect, which is readily prevented by operating the collector at a negative potential with respect to the gauge walls.

In 1960 it was demonstrated that the ion currents in a Bayard–Alpert gauge could be modulated by switching the potential of a modulator electrode with only a minor effect on the residual current[31] (i.e., the pressure-independent current to the ion collector). The modulation method reduced the minimum measurable pressure to about 5×10^{-12} Torr and permitted a simple measurement of the residual current. The modulation method has since been applied to most types of hot-cathode ionization gauges and the various modulation methods have been reviewed.[32,33]

In 1962 it was shown that the residual current in a BAG was greatly increased (by a factor of 400 in some cases) during and after exposure of the gauge to some chemically active gases, in particular, oxygen, carbon monoxide, and water.[34] This effect, capable of causing very large errors in the measurement of pressure, results from electron stimulated desorption (ESD) of positive ions from adsorbed species on the grid.[35] Switching the potential of the modulator electrode does not, to first order, effect the ions released by ESD and thus the modulator method can be used to measure separately both the electron-stimulated ion current and the pressure-dependent ion current. A recent development[36] using the modulation method with phase-sensitive detector on an extractor gauge has achieved complete separation of the gas phase ions from the ESD ions. Figure 5 shows diagrammatically the geometry of various forms of ionization gauge discussed in the next few paragraphs.

Another approach to the problem of reducing the residual current caused by x rays was first suggested by Metson[27] wherein a suppressor electrode, in front of the ion collector is held at a negative potential so as to return all photoelectrons from the collector back to the collector surface. This type of ionization gauge is called a suppressor gauge. Metson's original gauge was limited because the suppressor became a source of photoelectrons able to reach the collector. Later designs[37,38] reduced this problem, but the suppressor gauge has not been widely used.

The BAG has its ion collector contained within the ionization volume causing difficulties in modifying the design to further reduce the lowest measurable pressures by adding additional electrodes or an electron multiplier. A new class of hot-cathode ionization gauge was developed in 1966,[39] called the extractor gauge, to overcome these difficulties. In this type of gauge the ions are extracted from the ionization volume and then focused or deflected onto a collector or electron multiplier. In the original design the ions are extracted from the ionization volume through an aperture and focused on a short, fine collector wire. The x-ray limit of this gauge is below 3×10^{-13} Torr and the residual current, due to ESD from the grid, is greatly reduced (500 times less than in a BAG) because of the low efficiency of collection of energetic ions emitted from the grid surface. Other designs of the extractor gauge have been reported[40,41] and a development of the extractor gauge using a channeltron electron multiplier for ion detection[42] has reduced the low pressure limit to below 10^{-15} Torr. Another form of extractor gauge is the

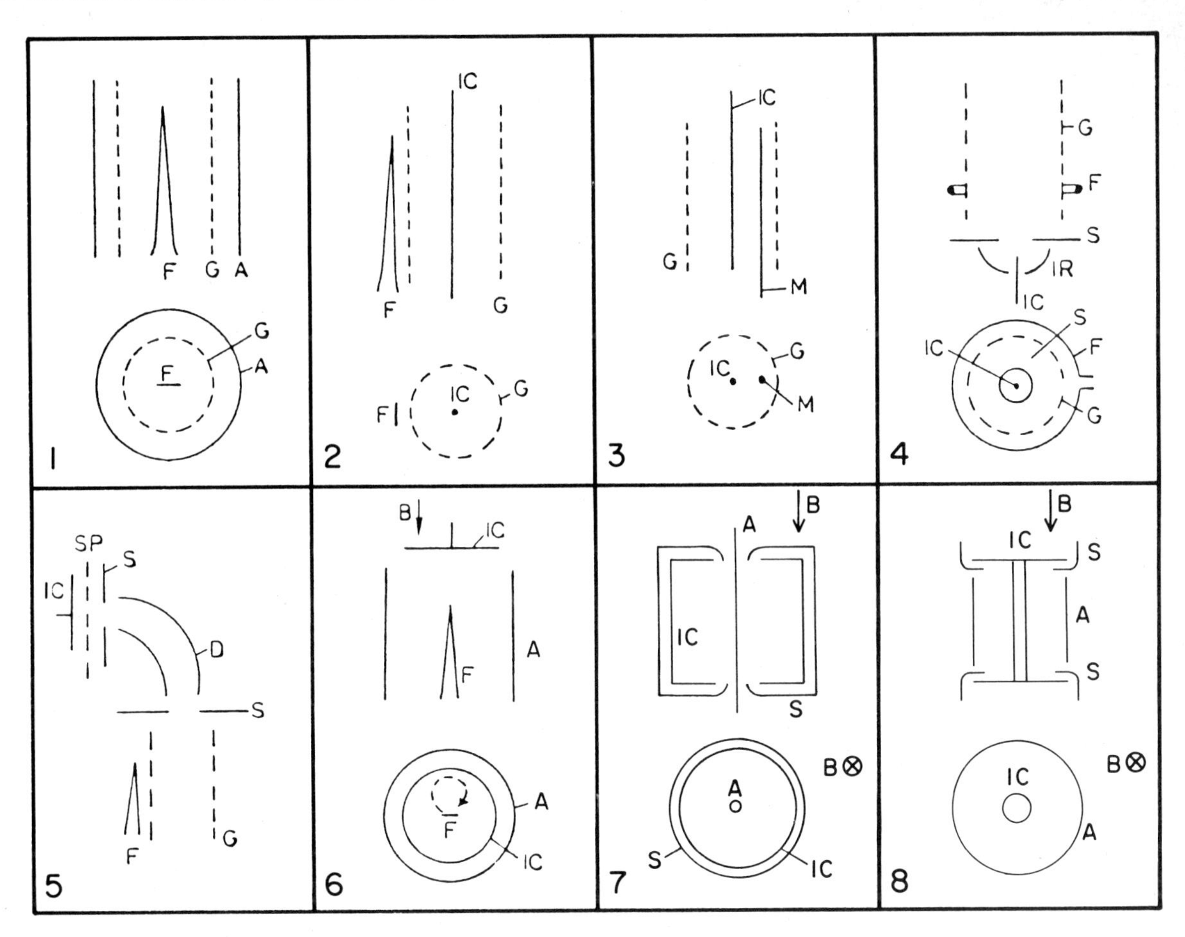

FIG. 5. Schematic diagrams of ionization gauges; (1) conventional gauge; (2) Bayard–Alpert gauge; (3) modulated Bayard–Alpert gauge; (4) extractor gauge; (5) bent-beam gauge; (6) hot-cathode magnetron; (7) inverted-magnetron gauge; (8) magnetron gauge. A–Anode, D–deflector, F–filament, G–grid, IC–ion collector, IR–ion reflector, M–modulator, S–shield, SP–suppressor.

bent-beam gauge[43] developed by Helmer in 1966. In this gauge the ions are extracted and then deflected through 90° onto a collector thereby reducing the x-ray flux at the collector to that reflected off the deflector electrodes. Modifications to the original design have resulted in a gauge[44] capable of measuring pressures to about 10^{-14} Torr without an electron multiplier. Channeltron multipliers have been added to the bent-beam gauge design[45] extending their range to about 10^{-18} Torr.

Another approach to suppressing photoelectrons and increasing the ionizing efficiency by applying a magnetic field to a hot-cathode ionization gauge was first proposed in 1954 by Conn and Daglish.[46] This approach was improved in the design of Lafferty,[47,48] first announced in 1961. Hot-cathode ionization gauges with magnetic fields have not been widely used.

We now turn to the other major direction of gauge development in this period, i.e., the further development of cold-cathode ionization gauges following the initial proposal of Penning.[20] The Penning type of gauge cannot maintain a self-sustained discharge at pressures below about 10^{-6} Torr. It was realized in the 1950's that crossed-field cold-cathode gauges could be designed to reduce the limitations of the Penning gauge by (a) changing the electrode geometry to improve electron trapping, thus making self-sustained discharges at very low pressures possible, and (b) introducing shield electrodes, or appropriate electrode configurations, to prevent any pressure-independent field emission from entering the measuring circuit. With these changes it proved possible to design crossed-field, cold-cathode gauges that would measure to 10^{-14} Torr.

Crossed-field, cold-cathode gauges with an inverted magnetron geometry were first studied by Beck and Brisbane[49] in 1952 and Haefer[50] in 1954. A design suitable for UHV measurements was developed[51] in 1958. This latter gauge has a power-law relation between ion current and pressure ($i^+ = kp^n$), with $n = 1.2$ over a range of 10^{-4}–3×10^{-13} Torr. A later version of the inverted magnetron gauge is claimed to have a linear current versus pressure relationship in the range 10^{-4}–3×10^{-13} Torr.[52]

A cold-cathode magnetron gauge suitable for UHV use[53] was developed in 1959. Most designs of magnetron gauges have a linear current–pressure relation above 5×10^{-10} Torr and are nonlinear below that pressure. One design[54] is reported to be linear from 10^{-5} to 5×10^{-13} Torr, in this design the cathode is a tungsten spiral that can be heated by the passage of current allowing outgassing by electron bombardment. Versions of the magnetron gauge have been used for pressure measurements on the moon's surface.[55]

Cold-cathode ionization gauges (without magnetic field)

using radioactive sources were first developed by Downing and Mellon[56] in 1946 and have proved useful in the pressure range 200–5×10^{-4} Torr.

The only significant advances in nonionization gauges during the period under review have been (a) some important improvements in the design of McLeod and optical/manometer gauges, (b) in the development of the spinning-rotor gauge[57] referred to earlier, and (c) significant improvements in the capacitance manometer.[58] Two processes affecting the accuracy of McLeod gauges have been elucidated and methods to prevent these errors developed. The first process, called the vapor stream effect, was discovered by Ishi[59] and results from the pumping action of the unidirectional flow of mercury vapor into the cold trap which is essential with a McLeod gauge. The error due to this effect can be reduced to $<1\%$ by reducing the conductance between the mercury surface and the trap.[60] The second process liable to cause errors is the thermal transpiration effect across an asymmetrical cold trap, which can cause 5% pressure differentials across a cold trap with arms of different diameters.[61] Improved liquid column manometers using interferometric measurement techniques have been developed[62,63] which are useful for gauge calibration.

Table II compares the properties of modern ionization gauges developed in the period 1945 to the present, only those gauges which are now commercially available or are in fairly common use are listed. The characteristics quoted in each case are for the design with the widest range and best overall performance.

Methods for the calibration of vacuum gauges have improved considerably in the period since 1945. Space does not permit a detailed review of these developments and thus the reader is referred to review articles[64,65] on the subject. One development potentially of considerable significance[66] is a method of generating known pressures of helium in the range 10^{-1}–10^{-20} Torr by physical adsorption on grafoil, the resulting pressure is deduced from the thermodynamic data for the He^4–grafoil system.

Partial pressure measurements (residual gas analysis) are worthy of a historical review of their own and can only be treated briefly here. During the period since 1945 many different types of mass spectrometer have been developed for residual gas analysis (omegatron, time of flight, Farvitron, linear rf., monopole, etc.) but have not received wide acceptance. Only two types of mass spectrometer are now in wide use for residual gas analysis, the sector magnetic type[67] and the quadrupole.[68] Davis[69] has demonstrated that a suitably processed mass spectrometer with an electron multiplier can measure partial pressures down to 10^{-18} Torr. Partial pressure measurements have been recently reviewed by Weston.[70]

III. CONCLUSION

The slow development of vacuum pressure measurement methods from Boyle's first measurement of a vacuum pressure in the late 1650's until the end of the Second World War have been reviewed. An attempt has been made to summa-

TABLE II. Comparison of modern ionization gauges: 1945–1983.

Gauge type	First development	Pressure range[a] (Torr)	ESD effects[b]	I_-(A)[c]	Sensitivity factor[d] ($Torr^{-1}$)	Electron multiplier	Ref.
Hot cathode							
Bayard–Alpert	Bayard and Alpert (Ref. 25) 1950	10^{-4}–4×10^{-12}	Yes	4×10^{-3}	42	No	74
Modulated BAG	Redhead (Ref. 31) 1960	10^{-4}–10^{-13}	No	4×10^{-3}	42	No	74
Extractor	Redhead (Ref. 39) 1966	(a) 2×10^{-4}–1.5×10^{-12}	No	1.85×10^{-3}	6	No	75
		(b) 10^{-5}–10^{-16}	No			Yes	42
Bent beam	Helmer and Hayward (Ref. 43) 1964	(a) 10^{-5}–3×10^{-14}	?	3×10^{-3}	40	No	44
		(b) 10^{-5}–10^{-16}	?			Yes	45
Hot-cathode magnetron	Conn and Daglish (Ref. 46) 1954	10^{-6}–3×10^{-18}	No	3.5×10^{-9}	10^4	Yes	48
					Sensitivity[e] (A $Torr^{-1}$)		
Cold cathode							
Inverted magnetron	Beck and Brisbane (Ref. 49) 1952	10^{-4}–3×10^{-12}	No	...	0.31	No	52
Magnetron	Redhead (Ref. 53) 1959	10^{-5}–5×10^{-13}	No	...	3.0	No	54
Radioactive	Downing and Mellen (Ref. 56) 1946	200–5×10^{-4}	No	...	7×10^{-10}	No	73

[a] Range of linear pressure vs ion-current response.
[b] Indicates whether measurement errors, resulting from ESD effects, are a problem.
[c] Electron emission (A).
[d] Sensitivity factor for nitrogen, $K=(I+/I-)(1/P)(Torr^{-1})$.
[e] Sensitivity for nitrogen (A $Torr^{-1}$).
All pressures are equivalent nitrogen.

rize the significant developments in the period of almost explosive advances since the Second World War. The reader is referred to two general texts[71,72] for more detailed discussions of vacuum gauges and pressure measurements.

On this 30th anniversary of the American Vacuum Society it is encouraging to see the major role played by the society in these modern developments. Many of the original developments and significant advances in the field of vacuum pressure measurements were first reported at the National Symposia of the AVS or in the Journal of Vacuum Science and Technology.

[1] *Saggi di Naturale Esperienze Fatte Nell'* (Accademia del Cimento, Firenze, 1667), p. 35.

[2] W. E. K. Middleton, *The History of the Barometer* (Johns Hopkins, Baltimore, 1964), Chap. 1.

[3] *Oeuvres de Blaise Pascal*, edited by L. Brunshvicg, P. Boutroux, and F. Gazier (Hachette, Paris, 1904–14), Vol. II, p. 165.

[4] Récit de la grande experience de l'equilibre des liqueurs, projectée par le sieur B. P. (Blaise Pascal)..., et fait par le sieur F. P. (Florent Périer) en une des plus hautes montagnes d'Auvergne (Paris 1648).

[5] O. von Guericke, *Experimenta Nova (ut vocantur) Magdeburgica de Vacuo Spatio* (Joannem Janssonium A Waesberge, Amsterdam, 1672), p. 100.

[6] R. Boyle, *New Experiments Physico–Mechanicall, Touching the Spring of the Air and its Effects, Etc.* (London, 1660).

[7] H. G. McLeod, Philos. Mag. **48**, 110 (1874).

[8] E. Thomas and R. Leyniers, *Proc. 6th Int. Vacuum Congress* (Jap. Phys. Soc., Tokyo, 1974), p. 147.

[9] W. Sutherland, Philos. Mag. **43**, 83 (1897).

[10] I. Langmuir, J. Am. Chem. Soc. **35**, 107 (1913).

[11] W. Steckelmacher, Vacuum **23**, 165 (1973).

[12] J. W. Beams, *Trans. 7th National Vacuum Symposium, AVS, 1960* (Pergamon, New York, 1961), p. 1; J. W. Beams, D. M. Spitzer, and J. P. Wade, Rev. Sci. Instrum. **33**, 151 (1962).

[13] J. K. Fremery, J. Vac. Sci. Technol. **9**, 108 (1972).

[14] M. Pirani, Verh. Physik. Ges. **8**, 686 (1906).

[15] W. Voege, Phys. Z **7**, 498 (1906).

[16] J. A. Becker, C. B. Green, and G. L. Pearson, Bell. Syst. Tech. J. **26**, 170 (1947).

[17] M. Knudsen, Ann. Phys. Lpz. **31**, 633 (1910).

[18] O. von Baeyer, Phys. Z **10**, 168 (1909).

[19] O. E. Buckley, Proc. Natl. Acad. Sci. **2**, 683 (1916).

[20] F. M. Penning, Physica **4**, 71 (1937).

[21] R. Loevinger, University of California Radiation Laboratory Report No. RL-20 6 23, 1944.

[22] A. R. Olsen and L. L. Hurst, J. Am. Chem. Soc. **51**, 2378 (1929).

[23] P. A. Anderson, Phys. Rev. **47**, 958 (1935).

[24] W. B. Nottingham, *Proc. 7th Ann. Conf. on Physical Electronics* (M.I.T., Cambridge, Massachusetts, 1947).

[25] R. T. Bayard and D. Alpert, Rev. Sci. Instrum. **21**, 571 (1950).

[26] J. J. Lander, Rev. Sci. Instrum. **21**, 672 (1950).

[27] G. H. Metson, Br. J. Appl. Phys. **2**, 46 (1951).

[28] D. Alpert, J. Appl. Phys. **24**, 860 (1953).

[29] E. K. Jaycox and H. W. Weinhart, Rev. Sci. Instrum. **2**, 401 (1931).

[30] W. H. Hayward, R. L. Jepsen, and P. A. Redhead, *Trans. 10th National Vacuum Symposium, AVS* (MacMillan, London, 1963), p. 228.

[31] P. A. Redhead, Rev. Sci. Instrum. **31**, 343 (1960).

[32] P. A. Redhead, J. Vac. Sci. Technol. **4**, 57 (1967).

[33] G. F. Weston, Vacuum **29**, 277 (1979).

[34] P. A. Redhead, Vacuum **12**, 267 (1962).

[35] P. A. Redhead, Vacuum **13**, 253 (1963).

[36] F. Watanabe, S. Hiramatsu, and H. Ishimaru, Vacuum **33**, 271 (1983).

[37] W. C. Schuemann, Rev. Sci. Instrum. **34**, 700 (1963).

[38] P. A. Redhead and J. P. Hobson, Br. J. Appl. Phys. **16**, 1555 (1964).

[39] P. A. Redhead, J. Vac. Sci. Technol. **3**, 173 (1966).

[40] J. Groszkowski, Le Vide **136**, 240 (1968).

[41] L. G. Pittaway, Philips Res. Rep. **29**, 283 (1974).

[42] D. Blechshmidt, J. Vac. Sci. Technol. **10**, 376 (1973).

[43] J. C. Helmer and W. H. Hayward, Rev. Sci. Instrum. **37**, 1652 (1966).

[44] C. Benvenuti and M. Hauer, Le Vide **2**, 199 (1980).

[45] L. Wang-Kui, Q. Ju-Mei, Z. Qi-Znan, and D. Li-Hua, LeVide **2**, 195 (1980).

[46] G. K. T. Conn and H. N. Daglish, J. Sci. Instrum. **31**, 412 (1954).

[47] J. M. Lafferty, J. Appl. Phys. **32**, 424 (1961).

[48] J. M. Lafferty, *Proc. 4th Int. Vacuum Congress* (Institute of Physics, London, 1968), p. 647.

[49] A. H. Beck and A. D. Brisbane, Vacuum **2**, 137 (1952).

[50] R. Haefer, Acta Phys. Austriaca **9**, 200 (1954).

[51] P. A. Redhead, Can. J. Phys. **36**, 255 (1958).

[52] G. A. Nichiporovich, Instrum. Exp. Tech. No. **6**, 1440 (1967).

[53] P. A. Redhead, Can. J. Phys. **37**, 1260 (1959).

[54] G. A. Nichiporovich and I. F. Khanina, *Proc. 4th Int. Vacuum Congress* (Institute of Physics, London, 1968), p. 666.

[55] S. Johnson, D. E. Evans, and J. M. Carroll, *Lunar Atmosphere Measurements, Proc. 3rd Lunar Science Conf.* (M.I.T., Cambridge, Massachusetts, 1972), Vol. 3.

[56] J. R. Downing and G. Mellen, Rev. Sci. Instrum. **17**, 218 (1946).

[57] G. Reich, J. Vac. Sci. Technol. **20**, 1148 (1982).

[58] J. J. Sullivan, Res. Dev. **27**, 41 (1976).

[59] H. Ishi and K. Nakayama, *Trans. 8th National Vacuum Symposium, AVS* (Pergamon, New York, 1962), p. 519.

[60] C. Meinke and G. Reich, Vakuum Technik **12**, 79 (1963).

[61] T. Edmonds and J. P. Hobson, J. Vac. Sci. Technol. **2**, 182 (1965).

[62] B. Aubrey and S. Choumoff, C. R. Acad. Sci. Paris **261**, 1803 (1965).

[63] A. M. Thomas and J. L. Cross, J. Vac. Sci. Technol. **4**, 1 (1967).

[64] W. Steckelmacher, *Proc.6th Int. Vacuum Congress* (Jap. Phys. Soc., Tokyo, 1974), p. 107.

[65] G. Messer, Vakuum Technik **19**, 343 (1970).

[66] P. Taborek and D. Goodstein, Rev. Sci. Instrum. **50**, 227 (1979).

[67] W. D. Davis and T. A. Vanderslice, *Trans. 7th National Vacuum Symposium, AVS, 1960* (Pergamon, New York, 1961), p. 417.

[68] *Quadrupole Mass Spectrometry and its Applications*, edited by P. H. Dawson (Elsevier, Amsterdam, 1976).

[69] W. D. Davis, *Trans. 9th National Vacuum Symposium, AVS 1962* (MacMillan, New York, 1962), p. 363.

[70] G. F. Weston, Vacuum **30**, 49 (1980).

[71] S. Dushman and J. M. Lafferty, *Scientific Foundations of Vacuum Technique*, 2nd ed. (John Wiley, New York, 1962).

[72] P. A. Redhead, J. P. Hobson, and E. V. Kornelson, *The Physical Basis of Ultrahigh Vacuum* (Chapman and Hall, London, 1968).

[73] L. J. Rigby and C. R. Wright, Vacuum **18**, 274 (1968).

[74] J.-M. Laurent, C. Benvenuti, and F. Scalambrin, *Proc. 7th Int. Vacuum Congress* (Vienna, Dobrozemsky, 1977), Vol. I, p. 113.

[75] Leybold–Heraeus, Extractor Gauge-type IM51.

Atoms and electrons at surfaces: A modern scientific revolution

C. B. Duke

Xerox Webster Research Center, Webster, New York 14580

(Received 8 September 1983; accepted 14 October 1983)

During the past 15 yr surface science has experienced extensive and profound changes. This historical review is devoted to the development of the theme that these changes considered collectively constitute a scientific revolution in the sense described by Thomas Kuhn. The recognition of the consequences of inelastic collisions of fast electrons played a key role in initiating the modern era in surface science. The resulting explosive development of surface characterization spectroscopies is discussed and shown to have led to a fundamental alteration in our perception of a surface or interface. Whereas in the mid-1960's an interface was regarded merely as the boundary between two bulk media, today it is seen as an independent entity: a state of matter determined by its history and exhibiting its own unique composition, structure, and electronic properties. The review concludes with a brief indication of the role of the American Vacuum Society in promoting and facilitating the development of surface science and the diffusion of its results during this formative period.

PACS numbers: 01.65. + g, 68.20. + t, 73.40. − c, 68.90. + g

I. INTRODUCTION

During the past one and one-half decades, surface science has experienced a period of massive development which fits well the description of "scientific revolutions" given by Thomas Kuhn.[1] The basic change in world view accompanying this revolution is the emergence of the interface region between two media as an entity in its own right: i.e., as a state of matter with its own unique composition, structure, chemical bonding, electronic states, and atomistic dynamics, rather than specification as just a group of boundary conditions which define the transition between the two bulk media. This change was stimulated by the rapid evolution of electron spectroscopies during the period 1967–72 and was created via the generation and interpretation of the results of these and other surface characterization spectroscopies during the subsequent decade. The purposes of the present article are to provide a brief outline of the highlights of events in surface science during the period 1967–82 and to indicate why these events led to a scientific revolution in the sense identified by Kuhn.

The surface science revolution exhibits three major dimensions: intellectual, analytical, and practical. The intellectual dimension arises from the emergence of theories which seek to predict the (sometimes nonequilibrium) composition and structure of an interfacial region, and to relate them both to electronic states characteristic of the interface and to the process dynamics which led to its formation. The analytical dimension is a consequence of the remarkable transformation of surface analysis techniques from crude indications of macroscopic interface parameters (e.g., total accumulated charge/area or surface recombination velocity) to quantitative measures of interface composition and structure. Finally, the practical dimension results from the widespread diffusion of surface science techniques and instrumentation from the research laboratory into the metallurgical, chemical, and electronics industries.[2–5] Our attention herein is focused on the second dimension, i.e., the development of quantitative microscopic surface characterization.

II. SURFACE STRUCTURE—A MATURING SPECTROSCOPY

The title of this section is an adaptation of that of a decade-old survey,[6] *Surface Structure—An Emerging Spectroscopy* which appeared in a special issue of *Physics Today* (entitled *Special Report: Vacuum*) which marked the dawn of the modern era in surface science. In this early survey the task of quantitative surface analysis was divided into the determination of four types of information: the composition of the surface region, its atomic geometry, its electronic properties, and the dynamics of atomistic motion in its vicinity. In 1972 the fact that surface-sensitive spectroscopies could provide direct accessibility to these four aspects of surface structure was just becoming evident. Much has happened since then. We proceed by indicating the nature of ensuing developments via the presentation of examples in each of the four areas identified in 1972. Indeed, as we shall see, in 1983 much of the promise recognized in 1972 is being realized: hence the appellation "maturing" in lieu of "emerging."

III. COMPOSITION ANALYSIS: RISE OF THE MODERN ERA

The story of the development of quantitative surface analysis techniques is essentially that of the rise of the modern era in surface science. In textbooks characteristic of the late 1960 era, e.g., the now classic *Semiconductor Surfaces* by Many, Goldstein, and Grover,[7] the concept of a "clean" surface could scarcely be defined because of the lack of suitable analytical techniques. The invention in 1967 of a practical technique[8] for Auger electron spectroscopy (AES) and the subsequent widespread diffusion both of this technique[9] and of x-ray photoelectron spectroscopy[10] (XPS or ESCA, i.e., electron spectroscopy for chemical analysis) dramatically changed the situation in approximately five years.[6] By 1975 the use of both techniques and extensions thereof[11] had become routine to define, to within a few percent, the (average) composition of the uppermost three to five layers of a surface. The further development of AES to yield spatially re-

Originally published in J. Vac. Sci. Tech. A 2(2), 139–143 (1984).

solved images was envisaged at that time[8] and subsequently has become available.[12] The use of incident and/or exit angular dependence and of variations in electron inelastic collision mean free path with energy have yielded extensions of both AES and XPS to estimate the homogeneity of sample surface composition.[13,14]

Two other important classes of techniques for surface composition analysis are based on the use of field-ion microscopy[15] and ion–solid scattering,[5] respectively.The former remain today[16] in approximately the same state as they were 15 yr ago:[15] specialty techniques used in a few research laboratories. The latter, however, have been developed into widely deployed interface characterization tools, especially in the electronics industry.[5,14] The ion-scattering techniques were in their infancy in the early 1970's,[6,17] however, so their subsequent development and diffusion constitute further manifestations of the profound recent changes in surface science.

The intellectual development which triggered the use of AES and XPS for *surface* (as opposed to *bulk*) analysis was the recognition that the short inelastic collision mean free path of "low-energy" electrons ($10\ \mathrm{eV} \lesssim E \lesssim 500\ \mathrm{eV}$) rendered electrons elastically emitted from the solid in this energy range a direct probe of the uppermost few atomic layers of the solid.[18,19] The embodiment of this concept into theories of electron–solid scattering during 1968–72 led to modern models of low-energy electron diffraction[18–20] (LEED) which, in turn, have yielded the vast majority of surface structure determinations.[21–23] Therefore this recognition of the consequences of inelastic electron scattering was the critical change in world view which initiated the scientific revolution in surface analysis and enabled the modern era in surface structure determination.

Today, surface composition analysis is a mature field characterized by widespread diffusion of the basic instrumentation and by current research on detailed topics of interest primarily to small groups of specialists. The routine use of surface analysis instrumentation in the chemical,[2,5] metallurgical, [5,24] and electronics[3,4,25] industries is amply documented in the references cited. The nature of current research is well illustrated by a recent review of one aspect of this topic.[26] Thus, modern activities in surface analysis are characteristic of "normal science" in the sense defined by Kuhn.[1] The paradigm which underlies these activities is the concept of a surface as an identifiable entity, the composition of which is different from that of the corresponding bulk solid and is measurable by surface-sensitive electron and ion spectroscopies. The introduction of such a paradigm is specified by Kuhn[1] as being the essence of a "scientific revolution." The remarkable correspondence between events in surface science during 1967–82 and Kuhn's concept of a scientific revolution is the origin of the title of this essay and the nomenclature used therein.

IV. STRUCTURE ANALYSIS: FROM SYMMETRY TO ATOMIC GEOMETRY

In 1972 the only means of determining surface atomic geometries was direct imaging using the field–ion microscope.[6,15] Even this method was highly qualitative, however, due to uncertainties in the mechanism(s) of the imaging process.[15] At that time LEED was used solely to determine the symmetry of surfaces and, upon occasion, their cleanliness.[6,7,27]

In 1982 more than 100 surface and interface atomic geometries have been determined using analyses of measured LEED intensities as functions of the energy and/or angles of the incident electrons.[21–23] The theoretical advance enabling these structure determinations was the development of computer programs embodying accurate multiple-scattering algorithms describing the diffraction of low-energy electrons.[19,20] Such programs are in routine use today at many laboratories throughout the world. Indeed, thanks to the incredible decrease in the price of computing, the utilization of these complicated multiple-scattering programs in 1982 often costs less than that of comparable single-scattering ("kinematic") programs in 1972. Thus, in 1982 in contrast to 1972, the atomic geometries of single-crystal surfaces are directly accessible to experimental determination provided the surface unit mesh is not too large.

LEED is, of course, not the only technique for determining surface atomic geometries. Methods based on ion–solid scattering, whose potential for this purpose has been recognized for many years,[17] have received considerable attention during the past decade. They remain largely research tools used by the groups which developed them, however, and their quantitative applications are restricted to relatively simple surface geometries like determinations of the expansion or contraction of the uppermost layer spacings at low-index metal surfaces.[28] The development of synchrotron radiation sources[29] has stimulated surface-structure studies based both on surface-sensitive variants of extended x-ray absorption fine structure[30] and on the direct use of x rays.[31] Indeed, with these sources the analysis of diffraction phenomena in photoelectron spectra to determine surface geometries becomes feasible.[32] High-resolution reflection electron energy loss spectroscopy (HREELS), clearly envisioned in the late 1960's[33] but not exploited until the mid 1970's,[34] has proven particularly useful in the study of atomic and molecular adsorption systems.[35] The use of LEED intensity analysis is, however, the major modern source of surface atomic geometries.[21–23,36] Comparison of structures determined by the various different methods reveals, moreover, a good correspondence between the results of LEED intensity analyses and those obtained using other methods once the problems associated with a given structure analysis are well understood.[37]

The results of these surface structure determinations have been enlightening and, in many cases, unexpected. For example, the low-index faces of simple metals usually exhibit contracted top-layer spacings[36] rather than the expansions originally expected:[38] a result first discovered for Al(110) in 1971.[39] Most adsorbates form surface geometries analogous to those found in molecules,[21] although there are some remarkable exceptions to this generalization like Sb on the (110) surfaces of GaAs[40] and possibly O[36,41] and C[36] on the (100) faces of Ni. Similarly, the clean (110) surfaces of zinc blende-structure compound semiconductors reconstruct to form structures intermediate between molecular and crystal geometries for III–V semiconductors.[42] Remarkably, the same surface structures persist for II–VI compounds, correlating with bulk covalent radii rather than bonding ionicities or small-molecule geometries.[43] The theme illustrated by these examples is self-evident: Observed surface structures often differ from those which were expected *a priori*. Thus, the emergence of quantitative surface structure analysis, which itself was enabled by the recognition of the importance of inelastic collision damping, has in turn led to further changes in the world view of surfaces once their actual (as opposed to hypothetical) structures became known. This aspect of current work in surface structure analysis illustrates another feature of Kuhn's concept of a paradigm:[1] i.e., that it delineates a world view that is capable of systematic further elucidation and refinement by future "generations" of practitioners in a given field.

An illuminating example of the nature and consequences of the new world view of surfaces and interfaces is the metal–semiconductor contact. In the mid-1960's such entities were characterized by macroscopic boundary–value parameters:[7] i.e., surface charge, surface barrier height, and surface recombination velocity. Today they are regarded as complex multilayer systems embodying multiple phases, each with its own composition, structure, and electrical properties.[14] Moreover, the measurement of inhomogeneities parallel to such an interface is a testing ground for the most advanced techniques in surface microanalysis.[44] This change in perspective has been wrought by the microscopic characterization of the composition, structure, and electronic properties of the various layers which comprise these contacts, most particularly for Au and Al on GaAs(110)[14,37] and various metals on Si surfaces.[14,44] A practical consequence of such studies is the recognition that as little as a monolayer of metal can produce massive alterations in the electrical properties and/or stability of the contact[14,37]—a result which already is being incorporated into semiconductor device technology. Current concepts of interface stabilization and properties simply cannot be accommodated within the macroscopic-parameter world view characteristic of the period 1930–1970. The concepts themselves are, moreover, in the process of constant refinement by practitioners of semiconductor surface science, process chemistry, and device fabrication. Thus, research in semiconductor contacts and interfaces constitutes almost a textbook example of Kuhn's ideas concerning normal science and the changes wrought therein by the introduction of a new paradigm which signals a scientific revolution.[1]

V. ELECTRONIC STRUCTURE: FROM CONCEPT TO REALISM

Improvements in quantitative surface structure analysis have been paralleled by those in the calculation of the electronic properties of surfaces and interfaces. The schematic models of the pre-1970 era[7,45] have been replaced by realistic and accurate models based on self-consistent solutions to the full valence electron Schrodinger equation[46] and on pseudopotentials.[47–49] Cluster calculations embodying a variety of methods also have been popular.[41,50] These models have been successful at predicting surface geometries,[47,48] surface electronic excitation spectra,[46,51] and surface charge densities.[52] An indication of the magnitude of the change in the realistic and quantitative character of model calculations during the 1967–82 period may be obtained by comparing the self-consistent but largely qualitative jellium models of metal–vacuum interfaces characteristic of the 1967–69 era[45] with today's quantitative models of the surfaces of *d*-band metals.[46] Similarily, the schematic, one dimensional models of surface states in semiconductors discussed by Many, Grover, and Goldstein[7] have been supplanted by quite realistic models of semiconductor surface energies,[47–49] excitation spectra[51] and charge densities.[47,49,52] Thus, a qualitative change in the character of state-of-the-art models of surface electronic structure has occurred during the past one and one-half decades.

A major motivation for the construction of such realistic models has been the development of spectroscopic techniques which can measure directly special features of the electronic structure of surfaces. Historically, these spectroscopies arose from two independent studies of adsorbate-induced resonances in ion-neutralization spectra[53] (1966) and field emission energy distributions[54] (1967), respectively. Accounts of both works may be found in the texts of American Vacuum Society Welch Award addresses.[19,55] Within a few years it was recognized that such resonances can be observed via valence electron photoemission as well, and that any type of surface electronic state, not just adsorbates, can induce them.[56,57] The most recent discoveries pertinent to the experimental determination of the electronic properties of surfaces are that atom solid scattering[52] and vacuum tunneling[58,59] provide sensitive probes of the electronic charge density far outside the surface. Quantitative analyses of such experimental spectra are not yet common, but recent analyses of the atomic diffraction[52] and angular-resolved valence–electron photoemission[51] from GaAs(110) reveal graphically both the current state of the art and the profound influence that reconstructions in the atomic geometries of surfaces (relative to the bulk) can exert on the electronic properties thereof.[37] Thus, both the experimental measurements and their interpretations constitute examples of the ongoing refinement of the surface-as-an-entity paradigm introduced during the 1967–72 formative period in surface science.

VI. ATOMIC DYNAMICS: THE LINK BETWEEN SURFACE SCIENCE AND PROCESS TECHNOLOGY

The status of measurements and models of surface atom dynamics is little changed in 1982 relative to 1967. The refinement of field ion microscopy to measure surface atom diffusion has continued and, indeed, has become the topic of a recent Welch Award address.[16] Considerable technical improvements in HREELS and various surface-sensitive infrared spectroscopies have led to similar refinements in model studies of adsorbate-induced surface vibrations.[60] Nevertheless, these types of experimental measurements were clearly presaged by events which had occurred by 1967,[15,33] and the basic concepts underlying enhanced surface atomic vibrations and diffusion were known at that time.[38] Thus, it seems fair to conclude that the change in paradigm which occurred during 1967–72 and exerted such a profound influence on analyses of surface composition, structure, and electronic properties, has thus far been of peripheral significance in the study of surface atomic dynamics, although the situation may change, as efforts emerge to establish relationships between the kinetics of the formation of an interface and its resulting structure and properties.

This circumstance is unfortunate because an understanding of the dynamics of atoms at surfaces constitutes a key link between surface science and the process technologies which are the backbone of the modern chemical and electronics industries.[61] While progress has been made in characterizing a few catalytic reactions[62] and some of the steps in molecular-beam epitaxy,[63,64] a "yawning chasm" still remains between the surface-science studies on model systems and the process technologies of industrial significance.[61] Spanning this chasm is a major goal for future research in surface science, and constitutes the final step in converting the changed world view of surface science which now prevails in the research community into entities of technological and economic significance of benefit to society at large.

VII. SYNOPSIS: ROLE OF THE AVS

In summary, the ideas and practices which characterize surface science are fundamentally different in 1983 than they were in 1967. Indeed, the changes which have transpired during this decade and a half are of sufficient magnitude as to be characteristic of Kuhn's concept[1] of a scientific revolution: i.e., a fundamental shift in the world view characteristic of an area of scientific inquiry. In this case, the recognition of the consequences of inelastic collision damping plus advances in vacuum instrumentation and electronics spawned the dawn of the modern era in surface science during 1967–72. Moreover, the subsequent decade has been the scene of dramatic refinements in the ideas, instrumentation, and analysis techniques introduced during that period: also a characteristic identified by Kuhn.[1] The formation of the Surface Science Division of the American Vacuum Society (AVS) in 1968 and this division's subsequent founding and sponsorship of the first International Conference on Solid Surfaces in 1971 were harbingers of the future which both presaged and enabled the rapid changes that were to come. One recent major contribution of the AVS in this arena is the founding of the Electronic Materials and Processing Division in 1978: an explicit recognition[4] of the growing demand for the elaboration and diffusion of surface-science ideas, techniques, and instrumentation in the fabrication of modern electronics and electrooptics. Another is the introduction in 1983 of a second section of the Journal of Vacuum Science and Technology entitled *Microelectronics Processing and Phenomena*. Thus, the revolution which started in the late 1960's entered the age of widespread dissemination in the 1980's: a short time by historical standards,[1] but one which is commensurate with today's intensely competitive international economy.[4] Indeed, the AVS has been instrumental in shortening this timescale and has, thereby, been influential not only in developing the knowledge base on which modern society rests but also in the transforming of this knowledge into tangible benefits for mankind: a crucial if not always glamorous ingredient in the preservation of civilized society.

ACKNOWLEDGMENTS

I am indebted to Ms. L. J. Kennedy for assistance with the many versions of this manuscript and to D. M. D. Tabak for his support of my efforts on this somewhat unorthodox enterprise.

[1]Thomas S. Kuhn, *The Structure of Scientific Revolutions* (University of Chicago, Chicago, 1962).
[2]T. E. Fischer, Crit. Rev. Solid State Sci. **6**, 401 (1976).
[3]J. R. Arthur, Jr., Crit. Rev. Solid State Sci. **6**, 413 (1976).
[4]C. B. Duke, J. Vac. Sci. Technol. **17**, 1 (1980).
[5]*Industrial Applications of Surface Analysis*, edited by L. A. Casper and C. J. Powell (ACS Symposium Series, New York, 1982), Vol. 199.
[6]C. B. Duke and R. L. Park, Phys. Today **25(8)**, 23 (1972).
[7]A. Many, Y. Goldstein, and N. B. Grover, *Semiconductor Surfaces* (North-Holland, Amsterdam, 1965).
[8]L. A. Harris, J. Vac. Sci. Technol. **11**, 23 (1974).
[9]J. C. Tracy, in *Electron Emission Spectroscopy*, edited by W. Dekeyser, L. Fiermans, G. Vanderkelen, and J. Vennik (D. Reidel, Dordrecht, 1973), pp. 295–372.
[10]D. T. Clark, in *Electron Emission Spectroscopy*, edited by W. Dekeyser, L. Fiermans, G. Vanderkelen, and J. Vernik (D. Reidel, Dordrecht, 1973), pp. 373–507.
[11]R. L. Park, Phys. Today **28(4)**, 52 (1975).
[12]C. C. Chang, J. Vac. Sci. Technol. **18**, 276 (1981).
[13]D. T. Clark, Crit. Rev. Solid State Mater. Sci. **8**, 1 (1978).
[14]L. J. Brillson, Surf. Sci. Repts. **2**, 123 (1982).
[15]E. W. Mueller, in *Proceedings of the International School of Physics "Enrico Fermi," Course 53* (Editrice Compositori, Bologna, 1974), pp. 23–51.
[16]G. Ehrlich, J. Vac. Sci. Technol. **17**, 9 (1980).
[17]J. A. Davies, J. Vac. Sci. Technol. **8**, 487 (1971).
[18]C. B. Duke and C. W. Tucker, Jr., Surf. Sci. **15**, 231 (1969).
[19]C. B. Duke, J. Vac. Sci. Technol. **15**, 157 (1978).
[20]T. N. Rhodin and S. S. Y. Tong, Phys. Today **28(10)**, 23 (1975).
[21]G. A. Somorjai and M. A. Van Hove, *Adsorbed Monolayers on Solid Surfaces* (Springer, Berlin, 1979).
[22]C. B. Duke, Adv. Ceram. **4**, 1 (1983).
[23]A. Kahn, Surf. Sci. Repts. **3**, 106 (1983).
[24]E. Bauer, Appl. Surf. Sci. **11/12**, 479 (1982).
[25]J. N. Ramsey, J. Vac. Sci. Technol. **A1**, 721 (1983).
[26]G. G. Kleiman, Appl. Surf. Sci. **11/12**, 730 (1982).
[27]W. D. Robertson, J. Vac. Sci. Technol. **8**, 403 (1971).
[28]L. C. Feldman, Crit. Rev. Solid State Mater. Sci. **10**, 143 (1981).
[29]A. Biedenstock and H. Winick, Phys. Today **36(6)**, 48 (1983).
[30]T. L. Einstein, Appl. Surf. Sci. **11/12**, 42 (1982).
[31]C. J. Sparks, Jr., Phys. Today **34(5)**, 40 (1981).
[32]D. A. Shirley, Crit. Rev. Solid State Mater. Sci. **10**, 373 (1982).
[33]F. M. Propst and T. C. Piper, J. Vac. Sci. Technol. **4**, 53 (1967).
[34]H. Froitzheim, in *Electron Spectroscopy for Surface Analysis*, edited by H. Ibach (Springer, Berlin, 1977), pp. 205–250.
[35]P. A. Thiry, J. Electron Spectrosc. Relat. Phenom. **30**, 261 (1983).
[36]P. M. Marcus and F. Jona, Appl. Surf. Sci. **11/12**, 20 (1982).
[37]C. B. Duke, Appl. Surf. Sci. **11/12**, 1 (1982).
[38]C. B. Duke, Ann. Rev. Mater. Sci. **1**, 165 (1971).
[39]G.E. Laramore and C. B. Duke, Phys. Rev. **B5**, 267 (1972).
[40]C. B. Duke, A. Paton, W. K. Ford, A. Kahn, and J. Carelli, Phys. Rev. **B26**, 803 (1982).
[41]T. H. Upton and W. A. Goddard, III, Crit. Rev. Solid State Mater. Sci. **10**, 261 (1981).
[42]C. B. Duke, A. R. Lubinsky, B. W. Lee, and P. Mark, J. Vac. Sci. Technol. **13**, 761 (1976).
[43]C. B. Duke, J. Vac. Sci. Technol. **B1**, 732 (1983).

[44]J. M. Poate and K. N. Tu, Phys. Today **33(5)**, 34 (1980).
[45]C. B. Duke, J. Vac. Sci. Technol. **6**, 142 (1969).
[46]J. R. Smith, F. J. Arlinghaus, and J. G. Gay, J. Vac. Sci. Technol. **18**, 411 (1981).
[47]J. E. Northrup and M. L. Cohen, J. Vac. Sci. Technol. **21**, 333 (1982).
[48]J. Ihm and J. D. Joannopoulos, J. Vac. Sci. Technol. **21**, 340 (1982).
[49]I. P. Batra, J. Vac. Sci. Technol. **B1**, 558 (1983).
[50]F. Herman, J. Vac. Sci. Technol. **16**, 1101 (1979).
[51]A. Zunger, Phys. Rev. **B22**, 959 (1980).
[52]D. R. Hamann, Phys. Rev. Lett. **46**, 1227 (1981).
[53]H. D. Hagstrom, Phys. Rev. **150**, 495 (1966).
[54]C. B. Duke and M. E. Alferieff, J. Chem. Phys. **46**, 493 (1967).
[55]H. D. Hagstrum, J. Vac. Sci. Technol. **12**, 7 (1975).
[56]D. E. Eastman and M. I. Nathan, Phys. Today **28(4)**, 44 (1975).
[57]E. W. Plummer, J. W. Gadzuk, and D. R.Penn, Phys. Today **28(4)**, 63 (1975).
[58]J. Tersoff and D. R. Hamann, Phys. Rev. Lett. **50**, 1998 (1983).
[59]N. Garcia, C. Ocal and F. Flores, Phys. Rev. Lett. **50**, 2003 (1983).
[60]C. R. Brundle and H. Morawitz, J. Electron Spectrosc. Relat. Phenom. **29/30** (1983).
[61]C. B. Duke, J. Electron. Spectrosc. Relat. Phenom. **29**, 1 (1983).
[62]G. Ertl, Crit. Rev. Solid State Mater. Sci. **10**, 349 (1982).
[63]J. R. Arthur, J. Vac. Sci. Technol. **16**, 273 (1979).
[64]C. T. Foxen, Crit. Rev. Solid State Mater. Sci. **10**, 235 (1981).

The future of vacuum technology

J. P. Hobson

Electrical Engineering Division, National Research Council, Ottawa, Ontario, K1A OR6, Canada

(Received 11 August 1983; accepted 18 October 1983)

For 30 years vacuum technology has responded to spontaneous efforts to extend the limits of vacuum production and measurement and to the demands of diverse fields of application. In the future it is predicted that: vacuum pumps will see incremental improvements, particularly cryogenic pumps; more fundamental and less empirical research will take place on seals and outgassing leading to major improvements; vacuum gauges will improve incrementally; calibration will improve greatly and will be extended to 10^{-13} Pa (10^{-15} Torr); calibration methods will utilize cryogenics; leak detectors with sensitivities approaching 10^{-13} cm^3 STP s^{-1} will appear commercially; a gauge to assess directly the interaction of ambient gases on surfaces will be developed; more special purpose vacuum facilities, such as storage rings and fusion reactors, will be built; portable systems at ultrahigh vacuum will multiply; microelectronics under vacuum will master the control of fabrication from 10 to 10 000 Å and new products will appear, in particular three-dimensional structures; there will be a synthesis of vacuum technology with lasers, superconductors, catalysis, and solar energy; pervading all these developments will be an ever-increasing application of microprocessors and computers to vacuum systems of all kinds.

PACS numbers: 07.30. – t

I. INTRODUCTION

On the occasion of this 30th National Symposium of the American Vacuum Society, and in particular in this session on the History of Vacuum Science and Technology, it is appropriate that one paper be devoted to the future of vacuum technology. However, it should be noted that this paper is unlike a scientific paper in the normal sense, wherein the evidence for the conclusions is clearly and objectively presented, and predictions of the future, if any, are limited. In this paper evidence for the conclusions certainly exists but is massive in scope, the detailed relationships between the evidence and the conclusions are, to a large degree, subjective; and finally, the entire purpose of the paper is to predict the future. The paper is thus a sort of essay rather than a true scientific paper, although some basis for the conclusions is to be found in a paper by the author[1] entitled "The Limits of Vacuum Production and Measurement." It may be noted that the title of this paper does not include "vacuum science," only "vacuum technology." While the main reason for this is that the session organizer asked for this title, nevertheless, after some consideration the author has decided to retain it. Interest in vacuum has always been, and will remain, primarily in what can be done with it (technology) rather that in its detailed physical mechanisms (science). The number of users of vacuum systems will always far exceed the number of vacuum scientists and engineers. What is done in vacuum systems will continue to upstage the design and operation of the systems themselves. This is not to say that the stature of the vacuum scientist and engineer has not risen in the last 30 years. It has gone from that of second class citizen, whose responsibility terminated where the pumping pipe entered the experiment, to full partner in a complex team of experts. What can be done with vacuum technology today is based inextricably upon the physical understanding and imagination of scientists and engineers. It is the understanding of this role of the vacuum scientist and engineer that has made the American Vacuum Society the fastest growing member of the American Institute of Physics, and has made the courses given by the American Vacuum Society the most advanced and focused anywhere, ahead of educational institutions. In dramatic sequence the space program, all manner of surface investigations (including fundamental studies, catalysis, and solar energy), high energy accelerators and storage rings, the microelectronics industry, the search for economical thermonuclear fusion, have all called on vacuum technology for solutions to their immediate problems. A great cross fertilization has already taken place, although surprising gaps have been left. However, already in this brief introduction we know as we examine below the various subfields of vacuum technology, that we must be prepared in any prediction of the future to synthesize the conclusions from many fields, not only at a technical level, but perhaps also at the level of society itself. Because it is a convenient date not too far away we define the year 2000 as the future.

II. VACUUM PUMPS

The speed of vacuum pumps is inherently limited by the expression

$$S = 3.64\,(T/M)^{1/2}\,\mathrm{l\,s^{-1}\,cm^{-2}}, \qquad (1)$$

with T absolute temperature and M molecular weight of gas being pumped. With the exception of rotary mechanical pumps all the ultrahigh vacuum and high vacuum pumps (see Fig. 1) can reach close to the limits of Eq. (1) in ideal circumstances, but practical constraints often reduce pumping speeds to perhaps 1/3 or 1/4 these values. Thus, with existing apendage pumps only incremental improvements in speed are foreseen (see Table I). Extensions in size, both larger and smaller, in orientation, along with reductions in pow-

Originally published in J. Vac. Sci. Tech. A **2**(2), 144–149 (1984).

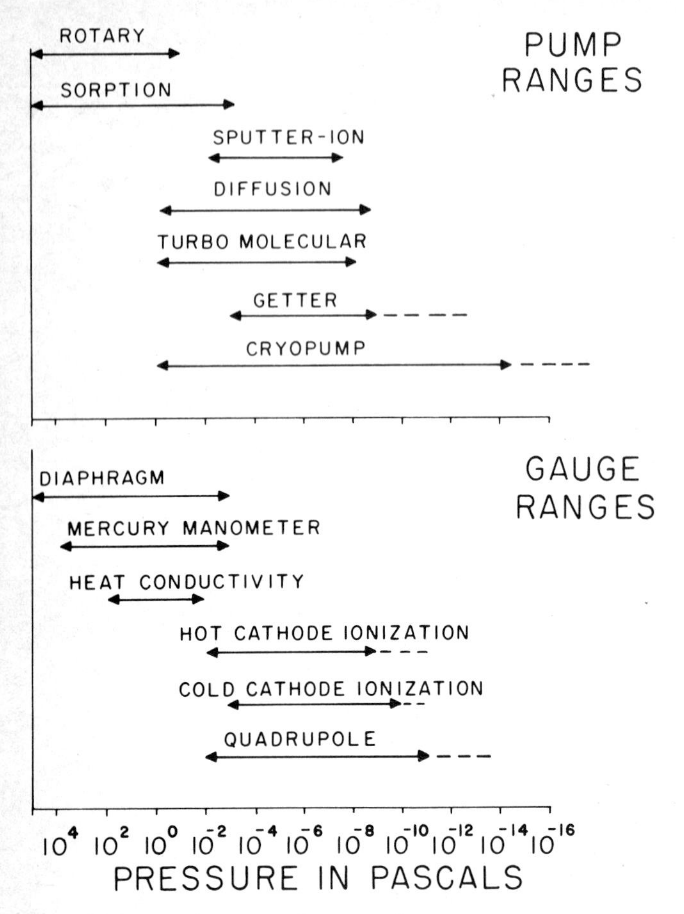

FIG. 1. Main operating ranges of pumps and gauges.

er, and a diversification in applications may be foreseen. New pump materials: fluids in diffusion pumps, rotor materials and lubricants in turbomolecular pumps, cathode materials in ion pumps, pumping surfaces in all types of getter pumps, improvements of efficiency in cryopumps with increases in sorption capacity—all these seem certain in the future. The past 30 years has seen the turbomolecular pump, the ion pump, the nonevaporable getter pump, and the cryopump all move from relative obscurity to important roles today. What are the chances that new types of pumps will develop similarly in the future? In the author's view they are not great, primarily because new principles are required, and while there have been new pumping principles suggested,[9,10] they seem unlikely to develop beyond specialized applications. In terms of the lowest pressures achievable (see Fig. 1) the cryopump would seem to have the greatest promise since it can universally pump all gases if the temperature is low enough, and it has been shown that even helium, the most difficult gas to pump cryogenically, can be pumped to extremely low pressures ($P \leqslant 10^{-20}$ Pa) in significant amounts on a porous surface[11] at 4.2 K, which is a temperature readily achieved (as long as the world's supply of helium lasts). Predictions for appendage vacuum pumps are given in Table I. However, in the area of pumps which are an integral part of the experiment (i.e., nonappendage pumps), advances are currently taking place and considerable developments may be expected. Examples are given later under the appropriate application (Sec. V).

TABLE I. The future of appendage vacuum pumps.

Pumps	Prediction
New types of pumps	No new major types will appear.
Rotary and sorption	Incremental improvements.
Diffusion and turbomolecular (Refs. 2 and 3)	Minor improvements and reduction in base pressure to 10^{-10} Pa.
Ion (Refs. 4 and 5) evaporable getter, nonevaporable getter (Ref. 6)	Improved materials providing higher capacities and somewhat higher speeds and reduced base pressures. Broader range of operating temperatures.
Cryopumps (Refs. 7 and 8)	Greater use of liquid helium temperatures will yield large reductions in base pressure. Capacities of cryosurfaces will increase.

III. VACUUM MATERIALS

The material dominating the construction of vacuum systems today is stainless steel, which appears to have gained prominence in the early 1960's in response to the demands of the space program. The choice was made at a time when exhaustive measurements had not been made on the outgassing properties of materials. Stainless steel was known to be structurally strong, machinable, and capable of being outgassed to adequate levels with acceptable temperature treatments. Bellows and diaphragms of stainless steel were developed and welding techniques found for junctions to kovar and thus to transparent glass windows, and to ceramic for electrical feedthroughs. Stainless steel was matched with gaskets of copper in the widely used Conflat flange. Glass itself, particularly Pyrex, is a "complete" vacuum material in that entire systems can be constructed from it.[12] It was used extensively in the 1950's and early 1960's for experimental purposes and remains a flexible research material, but has lost ground to stainless steel, essentially because of its fragility and minimal machinability. Aluminum and aluminum alloys are perhaps the third group of complete vacuum material, developed in particular by the Japanese in their High Energy Physics Institute.[13] Aluminum has a great advantage in being cheap and readily extruded to yield piping of arbitary cross-sectional configuration. Since long lengths of such tubing are to be found in large accelerators and storage rings, it is here that aluminum - based systems have seen the most development. It has been found that aluminum outgassing is reduced by the synchrotron radiation itself in electron storage rings. Achievable outgassing rates with stainless steel and aluminum are in the range of 10^{-14}–10^{-12} Pa l cm^{-2} s^{-1} and consist mostly of H_2.[14,15] True outgassing rates of glass are comparable and permeation of atmospheric helium through Pyrex is about 5×10^{-13} Pa l cm^{-2} s^{-1}. The search for vacuum materials with specialized properties is currently intense in the research and development leading to thermonuclear fusion. New material properties for mechanical stability under intense neutron and ion fluxes are being sought, as well as means for reducing both spontaneous outgassing under plasma fluxes. Wall discharge cleaning in both argon and hydrogen are under active study. Outgassing rates will be reduced to 10^{-14} Pa l cm^{-2} s^{-1} more routinely in the future.

It is inevitable that in this intense search for new materials for particular purposes that vacuum technology will be changed, but perhaps once again in an incremental rather than a qualitative way. At this stage in the development of vacuum technology with the first thrust of the search for specialized vacuum materials past, the search will be less empirical and more fundamental.

Permanent seals (welds, glass-to-metal seals, explosive bonds, etc.), semipermanent seals (demountable flanges), and temporary seals (valve seals) are an area where significant developments may be expected, both in new materials and in design. The Conflat seal, reliable in general, can still suffer from leaks following a bake, particularly in the smaller sizes. A general theory of seals is beginning to emerge,[16] which may well place the science of seals on a more theoretical and predictable rather than empirical basis. New combinations of materials which can be hermetically sealed together are being found (e.g., aluminum can be explosively bonded to stainless steel). The whole science of valves which seal repeatedly to conductances below 10^{-13} l s^{-1} is only beginning to emerge. Here the leak paths are of atomic dimensions. An immense data base of the adherence of one material to another is emerging from the microelectronic industry, which will very likely have a spin-off into vacuum technology.

IV. VACUUM MEASUREMENT

Vacuum is here defined as all pressures below atmospheric (i.e., $P \leqslant 10^5$ Pa or 760 Torr). Figure 1 shows the main existing gauges available for measurement. The subatmospheric pressure range will be divided into three regions: 10^{+5}–10^{-1} Pa; 10^{-1}–10^{-7} Pa; $\leqslant 10^{-7}$ Pa. Predictions are given in Table II.

TABLE II. The future of vacuum gauges.

Gages	Prediction
Diaphragm	A simple, reliable, widely used total pressure gauge for the range 10^5 to 10 Pa will be developed.
High pressure mass analyzer	A simple widely used mass analyzer for the pressure range 10 to 10^{-1} Pa will be developed.
Heat conductivity	Incremental improvements.
Hot cathode ionization (Refs. 17 and 18)	Incremental improvement.
Cold cathode ionization (Refs. 19 and 20)	Improvements in stability and extension of linearity by an order of magnitude to 10^{-10} Pa.
Low pressure mass analyzer (Refs. 17 and 21)	Major improvements in outgassing of ion sources. Improvements in partial pressure sensitivity to 10^{-15} Pa using counting detection.
Gauge calibration (Refs. 23 and 24)	Will be extended to lower pressures ($\sim 10^{-13}$ Pa) using cryopumps. Major improvements in reproducibility and absolute accuracy. The spinning rotor gauge will be widely used as a secondary standard.

A. Low vacuum (10^{+5}–10^{-1} Pa)

This range is characterized by applications such as freeze drying, sintering, sputtering, reactive ion etching, etc. While gauges measuring true force (e.g., the diaphragm manometer or the McLeod gauge) exist down to 10^{-4} Pa, there is a surprising lack of a simple, widely used, direct reading vacuum gauge in the range from atmosphere to 1.3×10^2 Pa (1 Torr). This is not to say that gauges do not exist in this range, only that they are not routine.

On the other hand, the pressure range 10^2 to 10^{-1} Pa is universally and simply serviced by the thermal conductivity gauge — a rugged, cheap, and adequately accurate instrument, in which no major improvements are foreseen in the future. The need for mass analysis in this range has not been urgent to date; however, many semiconductor processes require carefully controlled gas purity in the range 10 to 10^{-1} Pa, and the development of an appropriate mass analyzer may be anticipated.

B. High vacuum ($10^{-7} \leqslant P \leqslant 10^{-1}$ Pa)

In this region and below, pressures (actually densities) are measured by the ionization gauge. At 5×10^{-2} Pa the mean free path of molecules in the gas phase at room temperature is about 10 cm. Thus, as the pressure falls below 10^{-2} Pa, molecule–wall collisions become more frequent than collisions in the gas phase. Thus, with falling pressure events become more and more determined by the interaction of molecules, electrons, ions, and photons with surfaces. Simply stated, this is the central problem of vacuum measurement at low pressure. In the high vacuum range it is successfully managed by controlling particle trajectories to emphasize the desired gas phase ionization and to suppress the surface reactions. Mass spectrometers of various designs make this problem simpler, and electron multipliers as detectors are not essential. In general, apart from the question of quantitative gauge calibration, vacuum measurement is satisfactory in the high vacuum range, and major changes are not anticipated in the future.

C. Ultrahigh vacuum ($P \leqslant 10^{-7}$ Pa)

Here surface preparation becomes important to reduce unwanted surface contributions. However, finally, the design problem of a hot cathode total pressure gauge is one of identifying unambigously the small gas phase signal amidst several possible surface contributions. The present limit of these gauges is about 10^{-10} Pa without multipliers and is not expected to be reduced much in the future. The cold cathode gauge, inherently very attractive at low pressures because of its low power consumption, suffers much less from surface effects (if we exclude electrical leakage across insulators), but has difficulty maintaining a stable discharge and linearity as the gas density diminishes. It can be used with some caution to 10^{-10} Pa and might well be developed to become more stable and reliable in the future.

As noted above the mass analyzer inherently solves some

of the difficulties of the hot cathode total pressure gauge, as well as providing much more information about the gas phase. For UHV, multiplier detectors are important and are almost universally used. With them the partial pressure sensitivities of modern commercial mass spectrometers are about 10^{-12} Pa; although frequently this is a misleading figure since the ion sources of these instruments cannot be outgassed to reach these levels. A major improvement in the latter area is foreseen in the future with a more modest improvement in the partial pressure sensitivity. The quadrupole mass analyzer dominates this field today and the simplicity of its construction with no magnetic field required is expected to maintain this dominaton into the future. However, to the best of the author's knowledge, the lowest partial pressure sensitivity reported was with a magnetic sector instrument many years ago[21] (10^{-15} Pa).

Frequently at UHV the experimenter is less interested in the pressure in the gas phase than in how the gases affect the conditions upon a surface of interest. Simple methods for answering this question are at present not well developed and might well be the subject of future activity. The field emission pattern of a surface changes with the adsorption of gases and the time taken between flashing clean and the appearance of a certain pattern could be used as a measure of the contaminating power of the gas phase.[22]

D. Gauge calibration

Truly quantitative vacuum gauge calibration is still in its infancy. A number of countries have set up gauge calibration laboratories, and the intercomparison of secondary standards is underway.[23,24] In Europe agreement has been found to $\pm$ 2.5% over the pressure range 10^{-4} to 10^{-1}Pa. However, an intercomparison in the U.S. has led to less satisfactory results. The uncertainties increase as the pressure diminishes. The lowest pressure today at which a quantitative gauge sensitivity has been claimed[25] is $\pm$ 30% in the 10^{-10} Pa range.

A method has been proposed[26] for achieving quantitative pressures of helium of 10^{-18} Pa (10^{-20} Torr) using the physcal adsorption of helium at temperatures $\leqslant 4.2$ K. This cryogenic method could be developed for gauge calibration, although care would be needed with corrections for thermal transpiration. At 10^{-14} Pa there is only 1 molecule cm^{-3} at room temperature. Hence, counting techniques rather than the measurement of continuous currents will be necessary. It would seem possible to do direct experiments to verify the statistical laws for the desorption of molecules from surfaces, etc. A major expansion of gauge calibration into the UHV range is predicted for the future.

E. Leak detection

The present limiting sensitivity of leak detection[27] is in the range of 10^{-14} cm^3 STP s^{-1} which is well below the normal commercial leak detector limit of 10^{-10} cm^3 STP s^{-1}. It is predicted that the latter will be reduced to 10^{-13} cm^3 STP s^{-1} and that large leaks of 10^{-3} cm^3 STP s^{-1} and above will be detected with a sensor other than the main mass spectrometer.

V. VACUUM APPLICATIONS

A. Space science

Pressures encountered naturally in the space program extend through the entire subatmospheric range under discussion (10^{-5} Pa at shuttle altitude, 10^{-8}–10^{-10} Pa on the moon's surface, 10^{-13} Pa in geostationary orbit), and it was the demands of the space program that provided the first major postwar stimulus to vacuum technology. Space simulation chambers for test, new pumps, gauges, and materials were all developed rapidly and successfully. At shuttle altitudes the pressure is about 10^{-5} Pa and the full difficulties of friction at UHV are not encountered. The Canadarm works successfully.[28] While it seems unlikely that the space vacuum at shuttle altitudes will be used as a vacuum system, there is a reasonable probability that manufacturing under vacuum in the microgravity of shuttle orbit will develop.[29] While the space program will continue to stimulate vacuum technology in specialized ways, its main impact is over and likely will not return.

B. Surface science

As noted in the Introduction there is a synergistic relationship between vacuum technology and surface science.[30] Vacuum technology provides the means for performing modern experiments in surface science, while surface science provides vacuum technology with the fundamental information necessary to improve its products, as well as a demand for new products (e.g., UHV rotary drives, long stroke bellows, sample carousels, etc.). A great number of different types of surface analytic instruments operating at UHV are today available commercially and a number of others are to be found in the laboratory. The most commonly used are LEED, AES, RHEED, SIMS, ESCA, UPS, XPS, TEAM, etc. Most commonly, three or four of these instruments are mounted in the same vacuum system and their beams can be directed at a single target, which is often a single crystal. This forms the well-known multiport surface analytic system. In recent years the growing of epitaxial films on samples (MBE) has become important and is approaching the production facility stage. In the future the various steps of production will be verified on line with surface analytic instruments. It is generally important that surfaces to which some change has been made not be exposed to the atmosphere. Thus, analysis has to be done either within the same chamber with some sort of mechanical motion,[31] or the sample has to be transferred for surface analysis with a vacuum transfer device.[32] To avoid the long delays in pumping vacuum systems from atmosphere to very low pressures, load lock sample entry systems have been developed. A general development of all these aspects of surface science is to be expected in the future and many ingenious designs may be anticipated. Surface science has been driven in part by important industrial fields such as catalysis,[33] solar energy,[34] corrosion, etc.

C. Accelerators and storage rings

The impact of large accelerators and storage rings upon vacuum technology was difficult to foresee 20 years ago. In hindsight the impact has been of the greatest importance, essentially because the ring engineers and scientists, particularly at CERN, soon realized that vacuum and surface conditions played a first-order role in the operation of proton storage rings, and every advance they made in these areas led to an increase in stored beam current and thus to an increased sensitivity in the detection of new particles. It was discovered that if the residual pressure was reduced below about 10^{-9} Pa, a circulating proton beam became its own ion pump. At Brookhaven[35] and in Japan[36] the concept of the distributed ion pump around the circumference of the beam tube, utilizing the magnetic fields necessary for beam bending, has been used in the National Synchrotron facilities. The photodesorption of gases from the walls caused by synchrotron radiation became of major concern and study. In the Tevatron (10^{12} eV) at Fermilab[37] the proton storage ring 6 km in circumference successfully operates with 92% of its length at 4.6 K (i.e., a vacuum system with 92% of its wall surface, a powerful cryopump with no baffles), a temperature derived from the bending and focusing superconducting magnets. The next accelerating storage ring is considered likely to be of energy 2×10^{13} eV, 160 km in circumference, to be located in a desert and to be called the "Desertron."

TABLE III. Vacuum application in the future.

Application	Prediction
Space Science	Will be a customer of state-of-the-art vacuum technology but not a driver. May demand space manufacturing facilities under vacuum, at zero gravity.
Surface Science	Will experience major advances in complexity of instrumentation and in automatic control; a limited number of new instruments will come into general use. There will be many applications to industrial processes.
Accelerators and storage rings	The "Desertron," 160 km in circumference, will be built using cold bore. Superconductivity and cryopumping will combine in many other large installations.
Fusion technology	Will be the area of the most intense search for new (vacuum) materials and ways of using them. This field will be the main driver of new developments in vacuum technology.
Semiconductor technology and thin films	An intense synthesis of existing technology with automation will occur. Production lines will have many surface science instruments on line for quality control. Molecular beam epitaxy will develop into a means of building new three-dimensional devices.

D. Fusion technology

The most formidable challenge facing vacuum technology today is undoubtedly the fusion program.[38] There are several very formidable and currently unanswered problems. Even the most efficient configuration: inertial confinement, magnetic mirror, and tokamak, is not yet clear. Many of these major problems belong in the domain of vacuum technology, along with a whole series of others outside this domain, all of which must be solved simultaneously before thermonuclear fusion becomes a reality. All are under intense study around the world. The full panoply of vacuum technology and surface science is being directed toward their solution. Immense liquid helium pumps for the hydrogen isotopes are being developed for neutral beam injection into the plasma[39]; samples of material are being inserted into discharges and withdrawn under vacuum for surface analysis,[31] methods of leak detection using directional detectors within the vacuum chamber are being evaluated because the exterior of the vacuum vessel is too cluttered to make conventional leak testing practical; and so forth. What will the outcome be? That is an important question not only for vacuum technology but for all of mankind. Feasibility or lack of it should be established before the year 2000.

E. Semiconductor technology and thin films

While neither the general vacuum range used in the semiconductor and thin film industries[40,41] nor its specific vacuum problems appear as severe as those of other areas, nevertheless, it is in the modern fabrication of integrated circuits and other related thin films devices that the greatest synthesis of vacuum processes and control takes place.

A modern VLSI (very large scale integration) circuit is fabricated in some 200 separate steps, the majority of which are done under vacuum. Current production aims at lithography dimensions down to 1 μ (10 000 Å). The wavelength of visible yellow light is about 5000 Å and hence, further major advances in lithography will use x-ray, electron, or ion beams of high resolution. These are the basic tools of surface science which have already been carried to resolutions of 10 Å in the case of electron beams. For lithography they will require modification and above all positional computerized control, but beyond this the basic technology already developed should be applicable. Integrated circuit production does involve substantial mechanical motion of the wafer under vacuum. This too requires computer control, which is under extensive current development. Of course all the other applications of vacuum technology that we have mentioned are experiencing computer control and frequently involve mechanical motions under vacuum, but it is thought that the semiconductor industry will carry these developments to their highest level. It is predicted that by the year 2000 the most impressive developments in semiconductor technology will have been in three-dimensional structures built up a layer at a time utilizing an extension of current MBE instruments.

VI. SUMMARY

Tables I–III provide the author's main predictions in the area of vacuum technology and its applications. There are other fields undergoing intense development: lasers, superconductors, computers, microprocessors, robotics, information science, medical engineering, biotechnology, etc. Vacuum technology will interact dynamically with all of them.

[1] J. P. Hobson, Proc. 9th Int. Vac. Cong. Invited Speakers' Vol, p. 35, (1983).
[2] N. T. M. Dennis, B. H. Colwell, L. Laurenson, and J. R. H. Newton, Vacuum **28**, 551 (1978).
[3] L. Maurice, P. Duval, and G. Gorinas, J. Vac. Sci. Technol. **16**, 941 (1979).
[4] S. Komiya, H. Sato, and C. Hayashi, J. Vac. Sci. Technol. **3**, 300 (1966).
[5] D. G. Bills, J. Vac. Sci. Technol. **10**, 65 (1973).
[6] C. Boffito, B. Ferrario, P. della Porta, and L. Rossi. J. Vac. Sci. Technol. **18**, 1117 (1981).

[7]R. A. Haefer, J. Phys. E **14**, 399 (1981).
[8]C. Benvenuti and M. Firth, Vacuum **29**, 427 (1979).
[9]A. I. Livshitz, M. E. Notkin, Yu. M. Pustovoit, and A. A. Samartsev, Vacuum **29**, 113 (1979).
[10]J. P. Hobson, J. Vac. Sci. Technol. **7**, 351 (1970).
[11]J. . Hobson, J. Vac. Sci. Technol. **3**, 281 (1966).
[12]P. A. Redhead, J. P. Hobson, and E. V. Kornelsen, Can. J. Phys. **40**, 1814 (1962).
[13]H. Ishimaru, K. Narushima, H. Nakanishi, and G. Horiskoshi, Proc. 8th Int. Vac. Cong. **2**, 176 (1980) (and more recent papers).
[14]R. Calder and G. Lewin, Br. J. Appl. Phys. **18**, 1459 (1967).
[15]G. Grosse and G. Messer, Proc. 8th Int. Vac. Cong. **2**, 399 (1980).
[16]A. Roth, J. Vac. Sci. Technol. A **1**, 211 (1983).
[17]J. H. Leck, Contemp. Phys. **20**, 401 (1979).
[18]J. M. Lafferty, J. Vac. Sci. Technol. **9**, 101 (1972).
[19]P. J. Bryant, W. W. Longley, and G. M. Gosselin, J. Vac. Sci. Technol. **3**, 62 (1966).
[20]J. H. Singleton, J. Vac. Sci. Technol. **6**, 316 (1969).
[21]W. D. Davis, Trans. Am. Vac. Soc. Vac. Symp. **9**, 438 (1962).
[22]L. de Chernatony, Vacuum **29**, 389 (1979).
[23]K. F. Poulter, A. Calcatelli, P. S. Choumoff, B. Iapteff, G. Messer, and G. Grosse, J. Vac. Sci. Technol. **17**, 679 (1980).
[24]I. Warshawsky, J. Vac. Sci. Technol. **20**, 75 (1982).
[25]G. Grosse and G. Messer, Vak. Tech. **30**, 226 (1981).
[26]P. Taborek and D. Goodstein, Rev. Sci. Instrum. **50**, 227 (1979).
[27]C. R. Winkelman and H. G. Davidson, Vacuum **29**, 361 (1979).
[28]B. A. Aikenhead, R. G. Daniell, and E. M. Davis, J. Vac. Sci. Technol. A **1**, 126 (1983).
[29]C. Joyce, New Scientist **98**, 607 (1983).
[30]Y. Margoninski, Vacuum **28**, 515 (1978).
[31]R. E. Clausing, L. Heatherly, and L. C. Emerson, J. Vac. Sci. Technol. **16**, 708 (1979).
[32]J. P. Hobson and E. V. Kornelsen, J. Vac. Sci. Technol. **16**, 701 (1979).
[33]A. L. Cabrera, N. D. Spencer, E. Kozak, P. W. Davies, G. A. Somorjai, Rev. Sci. Instrum. **53**, 1888 (1982).
[34]D. E. Carlson, J. Vac. Sci. Technol. **20**, 290 (1982).
[35]J. C. Schuchman, J. B. Godel, W. Jordan, and T. Oversluizen, J. Vac. Sci. Technol. **16**, 720 (1979).
[36]H. Ishimaru, N. Nakanishi, and G. Horikoshi, Proc. 8th Int. Vac. Cong., **2**, 331 (1980).
[37]C. L. Bartelson, H. Jöstlein, G. M. Lee, P. J. Limon, and L. D. Sauer, J. Vac. Sci. Technol. A **1**, 187 (1983).
[38]Phys. Today **36**, 17 (1983).
[39]J. T. Coupland and D. P. Hammond, Vacuum **32**, 613 (1982).
[40]J. L. Vossen, J. Vac. Sci. Technol. **18**, 135 (1981).
[41]J. F. O'Hanlon, J. Vac. Sci. Technol. A **1**, 228 (1983).

2

A Special History Exhibit

A unique exhibit of historic vacuum apparatus was organized to commemorate the 30th Anniversary of the American Vacuum Society. Prior to its display at the 30th National Symposium of the AVS in Boston (Nov. 1–3, 1983), the entire exhibit was on display at Boston's Museum of Science for a six week period. In conjunction with the exhibit, a special reenactment of the "Magdeburg hemispheres" experiment, cosponsored by Boston's Museum of Science and the AVS, was held in Boston in October 1983. The reenactment is described in a paper by M. Hablanian and C. Hemeon, immediately following the detailed listing of the items in the exhibit.

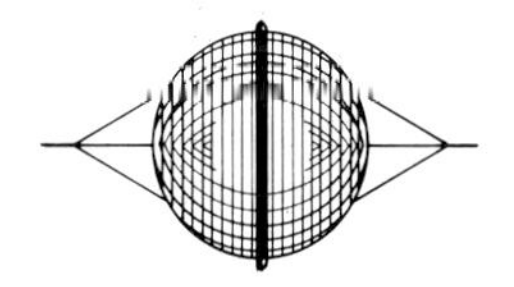

Magdeburg Hemispheres. Fig. 1
Replicas of the famous Magdeburg hemispheres are shown. The hemispheres were constructed in about 1654 by Otto von Guericke, the inventor of the air pump (vacuum pump). Also shown in Fig. 1 is a reproduction of an air pump built by von Guericke and used for his experiments. In a dramatic display of the force of atmospheric pressure, von Guericke attached teams of horses to the evacuated hemispheres; in repeat experiments the horses were generally unable to pull them apart. (See pp. 67–69 for a translated description of his experiments.)
(Hemispheres on loan from the Deutsches Museum, Munich, Germany)

Otto von Guericke's Book: *EXPERIMENTA NOVA (UT VOCANTUR) MAGDEBURGICA DE VACUO SPATIO.* Amsterdam, J. Jansson, 1672.
This classic book was the product of von Guericke's old age, following his busy career as political leader and engineer in the free city of Magdeburg. It describes the results of a quarter century of spare time experimenting—not only the famous demonstration that a vacuum can exist, but other remarkable facts such as electrical repulsion and the elasticity of air (von Guericke originally thought air would settle in his vessel and must be pumped out the bottom). Since his youth von Guericke had been fascinated by the endless starry heavens, and he longed to understand the true nature of the primordial empty space. Because of such philosophizing, his book was not fully appreciated by the younger generation of scientists. (A translation of key chapters of this volume is given on pp. 67–69.)
(On loan from the Niels Bohr Library of the Center for History of Physics, American Institute of Physics; presented to the Library by Bell Telephone Laboratories in honor of Lloyd Espenschied)

Small Brass Magdeburg Hemispheres, mid 19th century, maker unknown. Fig. 2
The famous Magdeburg experiment was miniaturized in the 19th century for convenient classroom demonstration in the form of hemispheres such as these. After being carefully fitted together with an airtight gasket (now lost), they would be screwed to the air pump and evacuated. The instructor would then close the cock, remove the apparatus from the pump and show how strongly atmospheric pressure kept the hemispheres together, and how easily they came apart once the air was readmitted.
(The Smithsonian Institution Photo No. 59527-A)

Vacuum Pump, late nineteenth century. Fig. 3
Piston-type air pumps of this design, which was introduced before the middle of the nineteenth century, were used for school demonstration experiments in physics classes. Some experiments were carried out inside a bell jar, whereas for others the jar was removed and the apparatus, such as the Magdeburg Hemispheres, the Guinea and Feather Tube, or the Electric Egg, displayed nearby, was screwed into the aperture in the plate. More elaborate pumps were fitted with accessories such as a mercury gauge and arrangements for compressing air.
(The Smithsonian Institution Photo No. 80-18004)

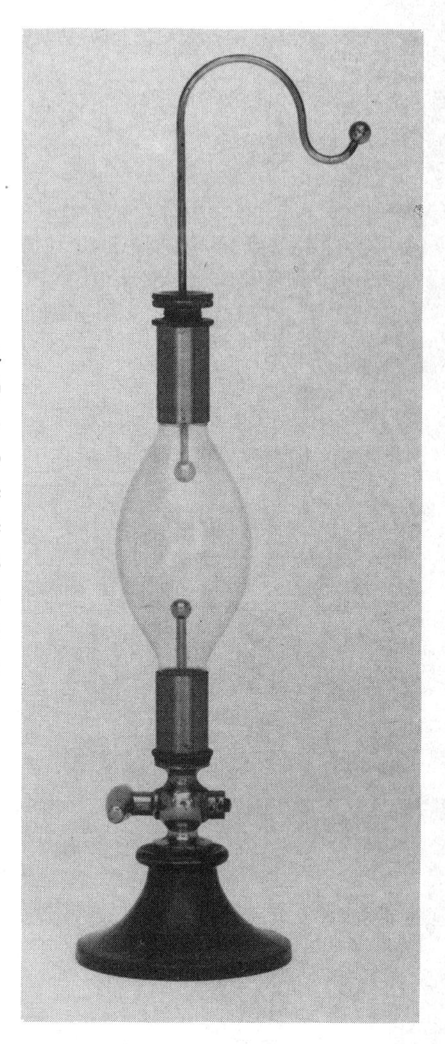

Electric Egg, circa 1900, by Max Kohl, Chemnitz, Germany. Fig. 4
In the middle of the 18th century, when interest in the remarkable phenomena of electricity was rapidly growing, a French experimenter (the Abbé Nollet) devised apparatus of this type for exhibiting the behavior of an electric discharge at low pressures. The base and the upper knob would be connected to an electrostatic generator, making sparks pass between the terminals inside the egg. When the air was pumped out the form and behavior of the discharge underwent curious changes. The resistance of the gap at various pressures could be investigated by sliding the upper knob up or down, and in later years physicists experimented with various gases. These researches led directly to the important investigations carried out with vacuum tubes of more elaborate construction in the late 19th century.
(The Smithsonian Institution Photo No. 64594)

Guinea and Feather Tube, mid 19th century, maker unknown.
To show that the resistance of the air to falling bodies disappears in a vacuum, a lecturer would insert a guinea (a small gold coin) and a feather into a tube such as this, and quickly turn it vertical to show that in air the guinea falls faster than the feather. Then he would exhaust the tube and this time feather and coin would drop together. This experiment goes back at least as far as 1662, when Christian Huygens performed it to the delight of his contemporaries.

(The above four items were on loan from the Smithsonian Institution, National Museum of American History)

Nineteenth Century Vacuum Chamber.
Vacuum Chamber. Used in the Physics Laboratory as an educational demonstration apparatus at the Massachusetts Institute of Technology, circa 1885. J. H. Temple, Boston, maker.
(On loan from The MIT Museum)

Case Western Reserve University Vacuum Pumping Station. Fig. 5
The 19th century brass-metal-glass bell jar system with hand cranked brass piston pump has a patent date of 19 November 1867. The manufacturer is unknown and the glass bell jar is not original with the system. Two photographs were obtained from the CWRU Archives which show the pumping station in use during the period 1885–1895 in the third floor of the Adelbert College building.
The pumping station may have existed at Western Reserve College in Hudson, Ohio prior to 1885 but we have no documentation to that effect. Little is known of the use of the vacuum pumping station. Dr. Frank P. Whitman was the first Perkins Professor of Physics and Astronomy and was "universally loved and respected by the faculty and students" but Professor Whitman was not primarily a research scholar. The photographs show a laboratory and teaching arrangement using the pumping station for demonstration experiments.
(*On loan from Case Western Reserve University; photo by R. Hoffman*)

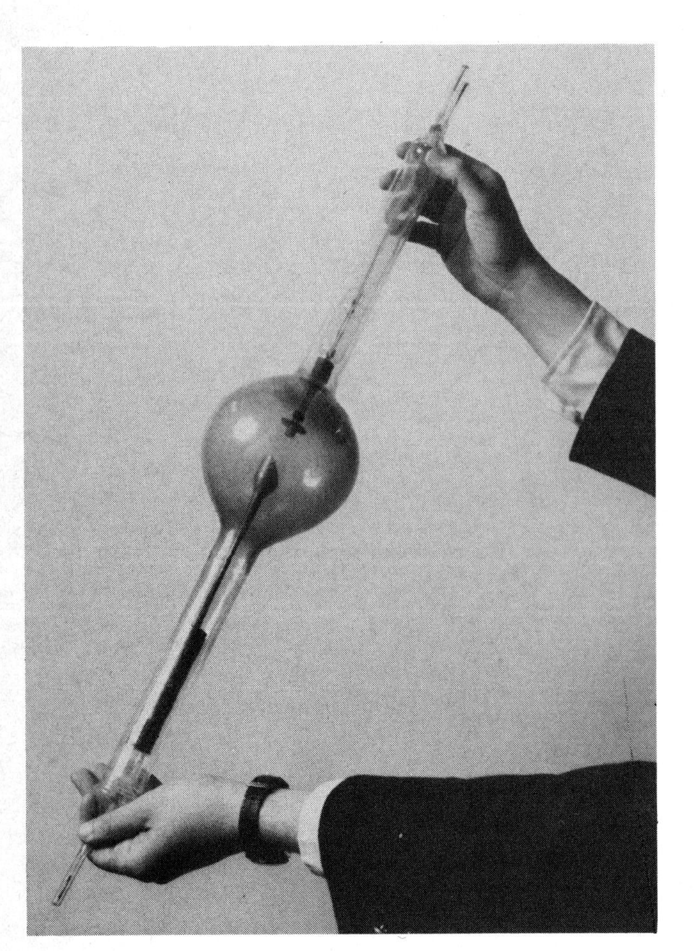

Early Coolidge X-Ray Tube. Fig. 6
Coolidge evacuated x-ray tubes to high vacuum, and was able to operate them by the "hot cathode" principle. This led to the invention, in 1913, of a fundamentally new type of x-ray tube which has become the world's standard design.
(*On loan from General Electric Company*)

Geryk-Oil-Air Pump (piston type), 1904. Fig. 7
This pump was manufactured by Pfeiffer in approximately 1904, under license from the British company Pulsometer Engineering Corporation, in accordance with a patent by Fleuss. Used to evacuate electric bulbs, it was the first vacuum pump used for industrial applications. Specifications: Two-cylinder piston pump, displacement 200 cm^3 per stroke, ultimate pressure 0.0002 mm Hg (2×10^{-4} Torr), hand or motor driven, price in 1904: 925 marks.
(On loan from Balzers; photo compliments of J. Freeman)

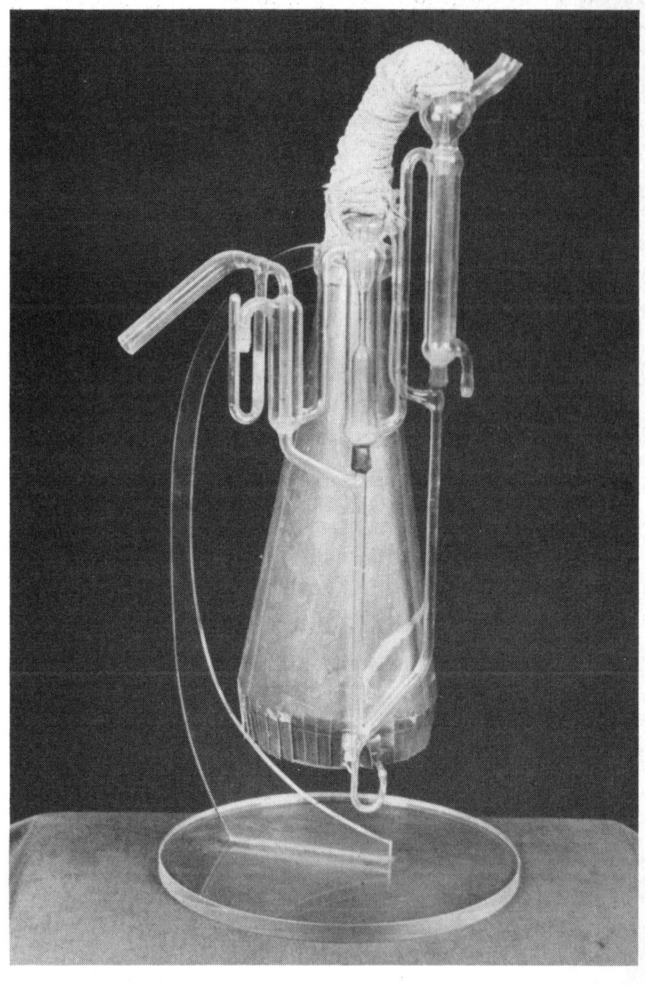

Stimson Two-Stage Diffusion Vacuum Pump. Fig. 8
A major improvement to the single-stage mercury vapor diffusion vacuum pump developed by Langmuir in 1916 was introduced in 1917 by H. F. Stimson of the National Bureau of Standards. The addition of a second stage allowed the pump to function against a much higher discharge pressure than was previously possible. This is the second two-stage pump built in the NBS glass-making shop under Stimson's direction.
(*On loan from National Bureau of Standards Museum*)

Röntgen Two-Stage Rotary Vane Pump. 1925.
Manufactured by Pfeiffer in 1925, this type of oil-sealed pump was widely used in the pharmaceutical industry, in physics research, and in the manufacture of Röntgen tubes. Specifications: Two-stage rotary vane pump, 1.77 *cfm displacement, ultimate partial pressure* 10^{-5} *Torr.*
(On loan from Balzers)

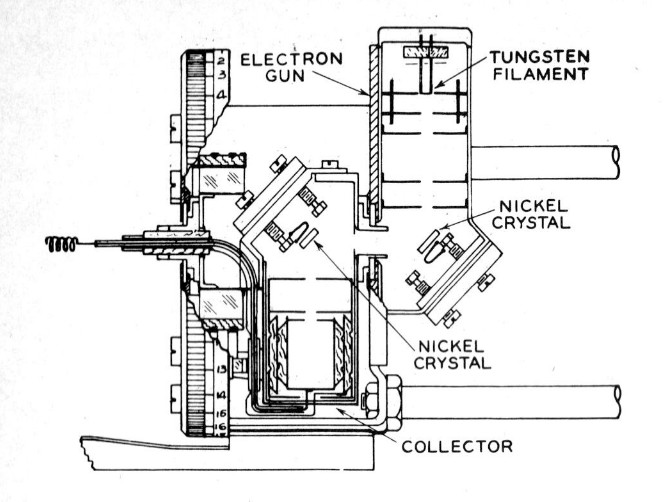

Davisson and Germer Electron Polarization Tube. Figs. 9 & 10
This experimental tube was designed by Davisson and Germer, the famous Bell Laboratories team whose work, published in 1927, earned Davisson a share in the 1937 Nobel Prize in physics for the demonstration of the wave nature of the electron. In this tube an electron beam reflected from one crystal of nickel is reflected a second time from another such crystal. The electron intensity after the second reflection is measured as the second crystal and its associated collector are rotated about the axis of the electron beam passing between the crystal surfaces. Arguing by analogy to light waves, Davisson and Germer expected to observe two maxima and two minima per 360° rotation of the second crystal. No such variation was found, indicating no polarization analogous to that of light waves. Others have pointed out later, however, that the single maximum and minimum per 360° rotation to be seen in the data of Davisson and Germer might indicate the detection of electron spin polarization but could also result from a very small (less than 1°) misalignment of the crystal axes relative to one another.
(*On loan from AT&T*)

Turbo Molecular Pump, Type TVP 500.
In 1957, *Messrs. Pfeiffer demonstrated the turbo molecular pump developed by Willi Becker. This was the first pump of this type which was reliable in operation and developed according to the experiences with molecular pumps by Gaede, Holweck, and Siegbahn. It was called a turbomolecular pump because its design is similar to a turbine. In* 1958, *the sale of this pump was started. This new type of pump has decisively influenced the development of high and ultra-high vacuum techniques.*

Technical Data of the TVP 500

Pumping speed:	140 l/s
Ultimate pressure:	$<1\times10^{-9}$ mbar
No. revolutions of rotor:	16 000 l/min
Mechanical drive weight:	96 kg

(On loan from Balzers)

History of TV Camera Tube Development.
When television graduated from the research laboratory and was ready to enter the world of commercial broadcasting, RCA provided key developments. In the studio RCA provided new camera tubes, at the transmitter new antennas, and at the home receiver a choice of picture tubes that rapidly grew in size from 7 to 25 in. This display traces the history of camera tube development. It starts with the 1943 model of the Image Orthicon, which was widely used in TV broadcasting because of its greatly improved sensitivity over the earlier Orthicon. Three RCA scientists were responsible for this development: Al Rose, Paul Weimer, and Harold Law. In 1941–45 RCA developed several special versions of the Image Orthicon for military application. While photoconducting materials were known even before the photoemitters (which were used in the Image Orthicon), they did not come into very prominent use for camera tubes until the invention of the Vidicon. The display features several examples of Vidicons, ranging from the 1 in. tube (in 1950) to the ½ in. tube (in 1955) to the special Silicon Intensifier Vidicon (1970) which was used by NASA for its mission of the Apollo Moon Walks.
(On loan from RCA Laboratories)

The History of Television. Fig. 11
The history of television can be traced by its milestone-inventions during the past 60 years. RCA, through its David Sarnoff Research Center in Princeton, New Jersey and its various Product Divisions, has been a pioneer and major contributor in every important development. For instance, the first all-electronic camera tube known as the Image Iconoscope was invented by Vladimir Zworykin in 1923 and further improved by him and his co-workers at RCA in 1934 and 1939. Three samples of these tubes are shown in our display. Other discoveries and inventions by RCA scientists (like George Morton and Jan Rajchman) in the field of secondary emission and photoconduction led to the development of the Photo Multiplier Tube, the Image Amplifier Tube, and the Image Converter Tube, which are also represented in our display. At the same time, during the 1930's, research and development was carried on for Kinescope Display Tubes. The first Kinescope using magnetic deflection was developed in 1930, followed by the first Tricolor Kinescope in 1931, and a Projection Kinescope for home projection TV receivers.
(On loan from RCA Laboratories)

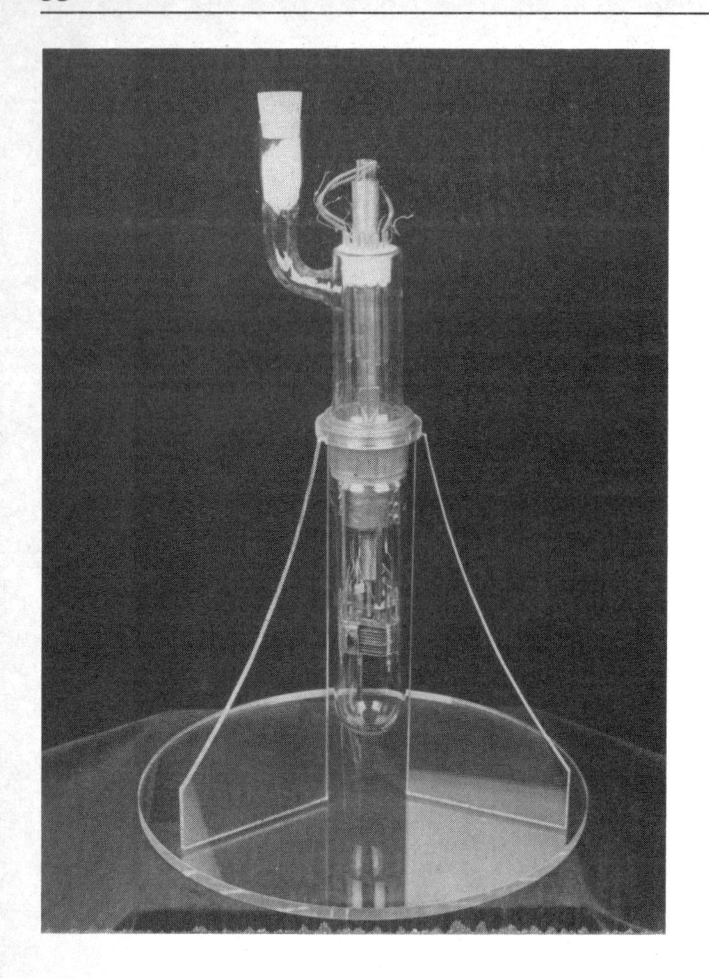

Omegatron. Fig. 12
Basically a miniature cyclotron, the omegatron made possible the determination of the values of several important atomic constants with higher precision than ever before. When developed by Thomas and Hipple at NBS in 1950, it permitted for the first time the direct determination of the faraday by physical methods.
(*On loan from National Bureau of Standards Museum*)

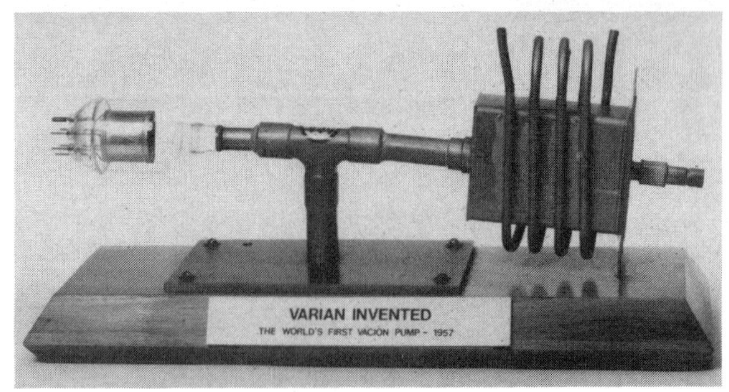

The First VacIon Pump. Fig. 13
In 1957, Varian Associates invented the VacIon pump to improve the life and reliability of their main product, microwave tubes. The VacIon pump maintains a vacuum in the tube through sputter-ion pumping, which is electronic and fluid free. The technique eliminates the potential of oil-contamination of the electron-emitting cathode, the heart of a microwave tube.
In the following decades, the VacIon pump has proved to be even more significant to science and industry with the development of new applications, such as in high-energy particle accelerators, electron microscopes, and surface science experiments. The pump has also contributed to the U.S. space program by making possible the construction of chambers that simulate the conditions of outer space.
Displayed here is the first model in which high-speed pumping was obtained. The copper box contains a multicellular anode sandwiched between two titanium plates and the poles of a permanent magnet (not shown). In operation the tenuous vacuum supports a gas discharge within the anode. The sputtering action of the discharge continuously deposits a titanium thin film which removes gas by chemical reaction.
(*On loan from Varian*)

An Early Traveling-Wave Tube Amplifier.
In the traveling-wave tube the electromagnetic wave is slowed by its travel along a helix until it is nearly synchronous with an electron beam directed along the axis of the helix. This structure avoids the frequency and bandwidth limitations inherent in the resonant circuits used in other vacuum tube amplifiers. Conceived in England by Rudolph Kompfner during World War II, it was subjected to intensive theoretical study and experimental development at Bell Laboratories by John R. Pierce and Kompfner. The traveling-wave amplifier is unmatched among electron tubes in its capability of high gain at high frequency and large bandwidth. It is today an essential component in orbiting communications satellites.
(On loan from AT&T)

Early Mass Spectrometer Tube.
60° mass spectrometer tube. Built in about 1947 in the physics shops at the University of Minnesota for research on isotopes and analyses of gases.
(Loaned by Professor Alfred O. Nier, Regents Professor of Physics Emeritus, University of Minnesota)

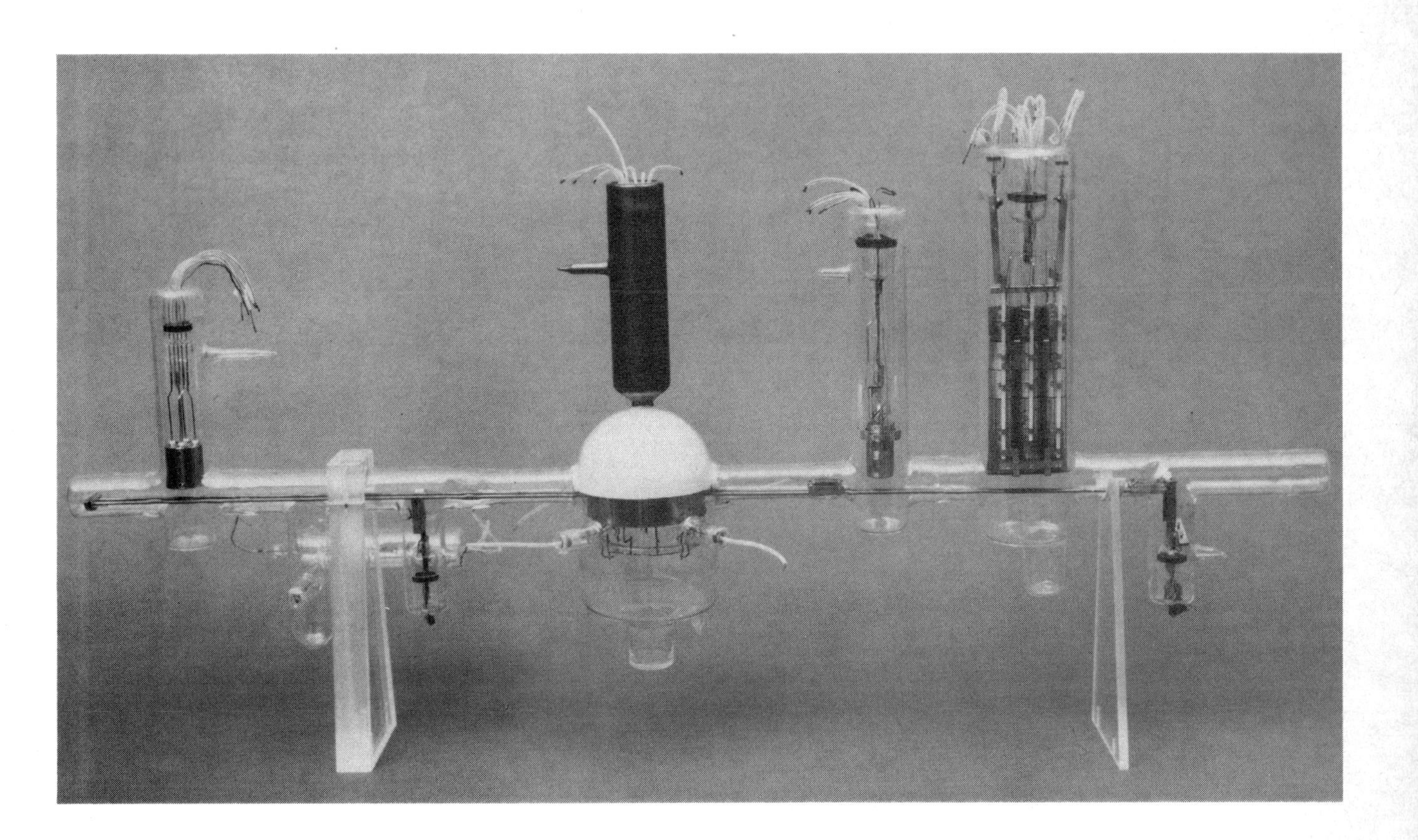

First LEED/Auger Spectrometer. Fig. 14
This Ultra-High Vacuum system became the first LEED/Auger Spectrometer on March 30, 1967 when Roland Weber[1] observed the differentiated Auger peaks of silicon and cesium. L. A. Harris[2] made the first practical Auger Electron Spectroscopy measurements in 1966 using an electrostatic velocity analyzer, and Weber's demonstration of the use of the LEED apparatus for this purpose opened up a new era in surface physics.
This sytem was originally built by Paul Palmberg[3] and used for his graduate research in W. T. Peria's Physical Electronics Laboratory. Weber modified the system and also used it for his graduate research of alkali metal layers on semiconductor surfaces.

[1]R. E. Weber and W. T. Peria, J. Appl. Phys. **38**, 4355 (1967).
[2]L. A. Harris, J. Appl. Phys. **39**, 1419 (1968).
[3]P. W. Palmberg and W. T. Peria, Surf. Sci. **6**, 57 (1967).
(On loan from Perkin-Elmer, Physical Electronics Division)

Bakeable Valves for Ultrahigh Vacuum Systems. Fig. 15
In order to achieve very low pressures on a routine basis, a vacuum system must be baked to as high as 400 °C before use. The first widely used all-metal valve which could meet such a requirement was developed by Daniel Alpert at the Westinghouse Research Laboratories in 1951. The display includes one of the original valves and an example of a slightly later version which was easier to assemble and use. The second generation of valves developed by Alpert's group could be made with a much higher conductance (2 in. diam opening) and were completely demountable to permit rapid servicing. An example of a valve with a $\frac{1}{2}$ in. diam. opening is shown.
(*On loan from Westinghouse Research Laboratories*)

Comments about Magdeburg Hemispheres Reenactment

M. H. Hablanian and C. H. Hemeon
Varian Associates, Lexington, MA

On October 22, 1983, Boston's Museum of Science and the New England chapter of the AVS recreated von Guericke's famous Magdeburg Hemispheres experiment. This was done in association with the historical exhibit commemorating the 30th Anniversary of the American Vacuum Society. The event was shown on a television monitor at the exhibit.

Briefly, the first three attempts to separate the hemispheres were not successful, but on the fourth try, the horses were able to separate the two halves of the chamber.

Most comments overheard at the exhibit from people watching the proceedings on the screen were that the separation was due to a lateral shift of one side relative to the other. Specifically, this explanation is incorrect because lateral restraints were incorporated internally into one of the chambers to inhibit shifting due to a possible uneven transient load during an initial jerk. The explanation for separation lies in the fact that an impulse load developed by the horses was actually great enough to overcome the force of the atmospheric pressure.

Even though the demonstration in Boston was done in the spirit of fun, a number of technical problems had to be considered during the design of the modern version of the hemispheres. They may be of some interest to the readers.

The secret and the drama of the demonstration lies in the question of coordination of the effort of the two teams of horses as well as the simultaneous application of the pulling force by each horse. If von Guericke's experiment were to be scaled down to only one horse (with a smaller chamber) and if one side of the evacuated chamber were attached to a tree, the horse would have separated the chambers each time. von Guericke may have been a better showman than a scientist, but, to be fair, it must be noted that in his time Newton's laws were unknown; force and momentum were usually confused and energy considerations in impulse load calculations were not appreciated.

As designers of the modern hemispheres, we were so acutely aware of the possibility of embarrassment despite the use of only three horses on each side that safety chains were provided to prevent injury to the horses or to their handlers in case the half-chambers flew through the air after separation.

The following considerations may be illustrative. von Guericke usually used eight horses on each side and hemispheres of 20 in. in diameter. The shape of the chambers is, of course, immaterial. Only the projected force acting on the area enclosed by the sealing circle should be considered. This gives an atmospheric pressure force of 4600 lb and the challenge of only 575 lb per horse. It would seem that even Don Quixote's Rosinante should have been able to produce that much force. As a rule of thumb, a horse can produce a pulling force (standing on soft, grassy ground) roughly equivalent to its own weight.[a] von Guericke undoubtedly was clever enough to use light-weight carriage horses. At the Boston demonstration, two teams of three horses each were used, each horse weighing over 1600 lb.

[a] *The Draft Horse Primer*, Maurice Telleen (Rodale Press, 1977).

Reenactment of the Magdeburg Hemispheres experiment, conducted at Boston on 22 October 1983.

There was another important distinction between von Guericke's and our demonstrations. The more horses are used, the more difficult it is to have them pull in unison and, more importantly, to have the two opposing teams apply a peak force simultaneously. The horses used in Boston (Eastern Draft Horse Association) are trained for and accustomed to participation in pulling contests. They usually produce a great exertion for about 10 s pulling continuously a loaded sled approximately 30 ft. They seem to be trained to move initially with sudden acceleration, presumably to overcome the static friction force and dislodge the sled. When given a signal, they practically leap forward, producing a substantial impact force when the chains connecting their harnesses and the hemispheres become taut. In the photograph taken at the site, one of the horses appears to have only one hoof on the ground. If we make what would appear to be conservative assumptions for acceleration, velocity, and time of interaction, we can estimate that the impact force can easily exceed the force of atmospheric pressure.

The chambers used in Boston were 24 in. in diameter. We did not want to make them larger for two reasons. First, to be reasonably close to the original (20 in.), especially because of smaller number of horses used; second, because we wanted the hemispheres to be light enough for handling during the demonstrations. We used two 0.25-in.-thick, deep-dished steel pieces without flanges. A commercially available bell-jar gasket was used as the seal.

The nominal force of atmospheric pressure for a 24-in.-diameter sealing circle is about 6550 lb. Separation is likely to occur with a force somewhat lower than that because the gasket between hemispheres may start leaking when most of the atmospheric force is counterbalanced by the pulling force. We used the 0.25-in.-thick chamber edges for sealing, without flanges, giving approximately 350 psi initial sealing pressure on the rubber gasket. von Guericke must have had difficulties with the seal because he used a greased leather gasket and rather thick flange with a wide sealing surface, giving an initial sealing pressure roughly one quarter of ours. In the model displayed in the museum, the flange surface was ~ 1 in. wide.

One final comment: Why was the gasket propelled into the air after separation of the chambers? (see figure). One possibility is that the adhesion between the gasket and the edge of the mating chamber was stronger than the adhesion to the edge to which the gasket was attached. But, more likely, the air pressure distributions around the L-shaped gasket at the time of separation may have produced a net outward force.

In conclusion, the lessons from the recreation of the Magdeburg experiment are as follows:

1. Use small, tired, uncoordinated horses. 2. Feed the horses as little as possible two days before the test. 3. Build in some cushion into the harness. For example, use nylon straps or ordinary ropes instead of chains. 4. Tell the drivers not to synchronize the forward signals to the two teams. 5. If money is no object, use a cylinder with a well-sealed piston instead of hemispheres to eliminate the effect of impact forces. 6. Finally, always use a safety chain or strap to keep things together in case of separation.

Acknowledgments: John Sullivan (MKS) conceived the idea of the demonstration and participated in all phase of the project. Ted Madey (NBS) set the project into motion and obtained the necessary funds from the AVS and its New England chapter (with the help of Chairman Tom Shaughnessy). Larry Bell of Boston's Museum of Science organized the demonstration, obtained the teams of horses, provided the television tape, and served as the Master of Ceremonies. Cal Hemeon designed the details and supervised the construction of the hemispheres.

3

Introduction to Reproductions of Historical Papers

The following pages contain translations or direct reproductions of papers which describe key experiments and concepts in the history of vacuum science and technology. Some of the papers (e.g., von Guericke's account of the Magdeburg hemispheres experiment, Boyle's description of his pump, and McLeod's discussion of his vacuum gauge) are true classics. More modern papers include Buckley's description of an ionization gauge (the first such paper in English), and a sample of Langmuir's extensive works which illustrates the clarity of his thoughts in an early surface-science experiment. Two short sections from the 1911 Encyclopedia Brittanica summarize the "state of the art" in Röntgen rays and vacuum technology at that time.

We have also included several review articles and sections of books which neatly summarize important experimental or conceptual developments. Andrade's paper traces the history of the vacuum pump, and deKosky's account of the work of William Crookes illustrates progress in vacuum science in the late 1800's. A few pages from Dushman's 1922 volume illustrate the remarkable advances in vacuum technology in the early twentieth century. Gehrenbeck's paper describes the scientific and technological background of the Davisson–Germer experiment in electron diffraction, and the Redhead–Steinhertz paper is a lucid, readable description of ultrahigh vacuum technology of the 1960's. The landmark 1950 paper by Bayard and Alpert describes the gauge which revolutionized vacuum measurements. We close with Edward's imaginative, futuristic account of a large-scale challenge in vacuum technology as applied to high-speed transportation. Enjoy!

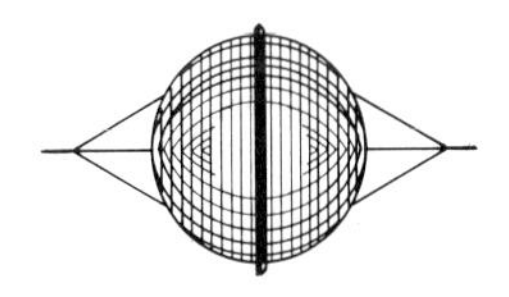

OTTO DE GUERICKE
Serenifs: ac Potentifs: Elector: Brandeb
Confiliarius et Civitat: Magdeb. Conful:

Otto von Guericke's Account of the Magdeburg Hemispheres Experiment

(translated by Roger Sherman)

The following translation of von Guericke's account of the Magdeburg hemispheres experiment was made by Roger Sherman, of the National Museum of American History, Washington, DC. It was taken from *Otto von Guerickes Neue (sogenannte) Magdeburger Versuche über den Leeren Raum* übersetzt und herausgegeben von Hans Schimank, published in Düsseldorf, 1968. The original work, in Latin, was published in Amsterdam in 1672 as *Experimenta Nova (ut Vocantur) Magdeburgica de Vacuo Spatio.* (Opposite is an engraved portrait of von Guericke taken from this volume.)

Third Book
My Own Experiments

Chapter 23

Experiment that shows how two hemispheres stick so firmly to each other because of the pressure of the air that 16 horses cannot tear them apart [see figure on following page].

I had two copper hemispheres or shells A and B made, of about 3/4 of an ell in diameter, and received them 67/100 of an ell in diameter (for the workmen are not accustomed to make things as accurately as they are asked to). They fitted very closely to each other, and to one of them was soldered a cock or rather that other kind of valve H, which allowed the air inside to be pumped out, and prevented the exterior air from entering, as we have already described in Chapter 8 above, and as the present Figure II shows. In addition 4 iron rings N,N,N,N are soldered to it, to which, as the picture shows, horses can be harnessed. I also had a leather ring D sewn, which was thoroughly saturated with a mixture of wax and turpentine so that no air would be let through.

These hemispheres were fitted together with the leather ring in between, and then the air was quickly pumped out (with the help of the connecting tube described in Chapter 8 and illustrated in Plate 7, Fig. IV, f,g,h,i) [Ed. note: cf. Fig. 3, on p. 18 of this volume, for an engraving of von Guericke's pump]. Then I saw how great was the force with which the shells pressed against the ring! And thus they stuck so firmly to each other by the effect of the pressure of the air, that 16 horses could not at all, or only with great difficulty, tear them apart. But if, occasionally, by the utmost exertion they succeed in pulling them apart, then there is a report like a gunshot.

But as soon as the air is let in by opening the cock H, anyone can merely separate and pull apart the shells with his bare hands. But if one desires to know exactly how great is the weight which presses the hemispheres so strongly together, one should calculate as in the preceding chapter the weight of a column of air 67/100 of an ell in diamter, a figure which we chose for the example there in order to make the matter here more comprehensible.

Thus one finds there that the weight of this column is 2686 or 2687 pounds, and consequently the air pressure pushes one of the hemispheres with a weight of 2686 pounds against the other. The latter in turn presses with the same weight in the opposite direction, so that 8 horses, when they tear them apart, sustain or exert a pull of 2686 pounds. The remaining 8 on the other side exert an equally great opposing pull.

Originally published in *Otto von Guericke's Neue (sogenannte) Magdeburger Versuche über den Leeren Raum*, edited by Hans Schimank (VDI-Verlag, Düsseldorf, 1968). Translated into English with the kind permission of VDI-Verlag GmbH.

Drawing of the Magdeburg Hemispheres experiment that originally appeared in von Guericke's *Experimenta Nova (ut Vocantur) Magdeburgica de Vacuo Spatio* (Amsterdam, 1672).

Now it is true that 8 horses can without special difficulty pull a wagon loaded with 2686 pounds. But in our case the pull is far stronger because it acts against the entire column of air and, as it were, more contrary to nature, than does the pull for drawing a load by means of a vehicle.

[Otto von Guericke goes on to suggest that the experiment could be done differently by hanging a weight from the hemispheres. The weight necessary would vary slightly acording to the changes in atmospheric pressure. He explains how to calculate the weight of the entire atmosphere, and deduces that there is nothing else outside surrounding the atmosphere.]

Chapter 24

A further experiment of this kind, in which 24 horses cannot separate the hemispheres, but a puff of air can. When the aforementioned hemispheres are torn asunder, something almost always gets damaged, and especially when they drop down and strike the ground, that is to say, if the pull of the horses tears them apart, they easily lose their perfect roundness. Therefore I had two more, larger, hemispheres made, a full ell in diameter. (But because coppersmiths can rarely make such things according to the specified dimensions, I found that the diameter of my new hemispheres amounted to only 95/100 of an ell.) My intention was that 24 horses should be unable to separate them, but that anyone could puff them apart.
The calculation goes according to the method described in Chapter 22.

[There follow two different computations, each giving a result of 5399 pounds. Therefore 34 horses will be necessary to separate the larger hemispheres, and 24 will certainly be unable to do it. But if the air is let back in, and then a syringe is attached, three or four strokes will cause the hemispheres to fall apart. Thus a single man can puff apart what 24 horses could not pull apart.]

The SECOND CONTINUATION of

PHYSICO-MECHANICAL EXPERIMENTS.

ICONISME I.

The deſcription of the engine, with a double tube, for the exhauſting of the air.

AA ARE two pumps made of braſs.
BB are two plugs hollow within, and open below.

CC are two holes in the upper part of the plugs, with valves opening outwardly, that they may afford paſſage to the air to go out, and hinder it from coming in.

DDDD are iron rods ſerving to move the plugs, and annexed to them, by means of the gnomons FF.

EE are two flat iron ſtirrups at the top of the rods DD, on which the operator muſt ſtand to ſet a work the engine.

GGG is a cord joined to the two ſtirrups, and compaſſing the pully H.

LL are two valves at the bottom of the pumps, opening inwardly, for the admiſſion of the air out of the tube MM.

MM is a tube reaching from both pumps to the plate OO, by means of the curvature PP QQ; which curvature ought to be of ſo great length, that the tube PQQ may not hinder the exerciſer of the pumps, but that he may conveniently ſtand on the ſtirrups EE.

OO is a plate bored in the middle, on which the receivers, to be evacuated, are to be put; as R for example.

BEFORE this engine can be fit for uſe, it is to be put into a frame of wood to ſupport it, as is ſhewed in the ſecond ſcheme, and as much water is to be poured through the hole Q in the plate OO into the pumps, as is ſufficient to fill the cavities of the plugs, and a little more; and then ſomebody muſt ſtand on the two iron ſtirrups EE, and muſt alternately depreſs and elevate them. For by this means it will come to paſs, that the plugs, following the motion of the ſtirrups in their aſcent, will leave the ſpace in the bottom of the pumps empty, and ſeeing all other paſſage is intercluded from the air, that air alone which is contained in the receiver R, is conveyed into the aforeſaid pumps by the tube Q Q PP M, and opens the valve L, which, being preſently ſhut, hinders the ſame air from making a regreſs; wherefore the plug, afterwards deſcending, compreſſeth that air, whence of neceſſity, the valve C muſt be opened, and all the air muſt paſs out at it, *viz.* becauſe the water, in the bottom of the pumps, doth exactly fill all the ſpaces, and doth alſo regurgitate through the valve C.

HERE we may obſerve, that this double engine is, upon many occaſions, to be preferred before a ſingle one, (that is moved with the foot) for it doth not only produce a double effect, but performs it alſo much more eaſily; for in thoſe engines, which are furniſhed but with one tube, whilſt the plug is drawn up to evacuate the pump,

Originally published in *The Works of Robert Boyle, Vol. IV* (London, 1772), pp. 510–513 from an article written in 1660.

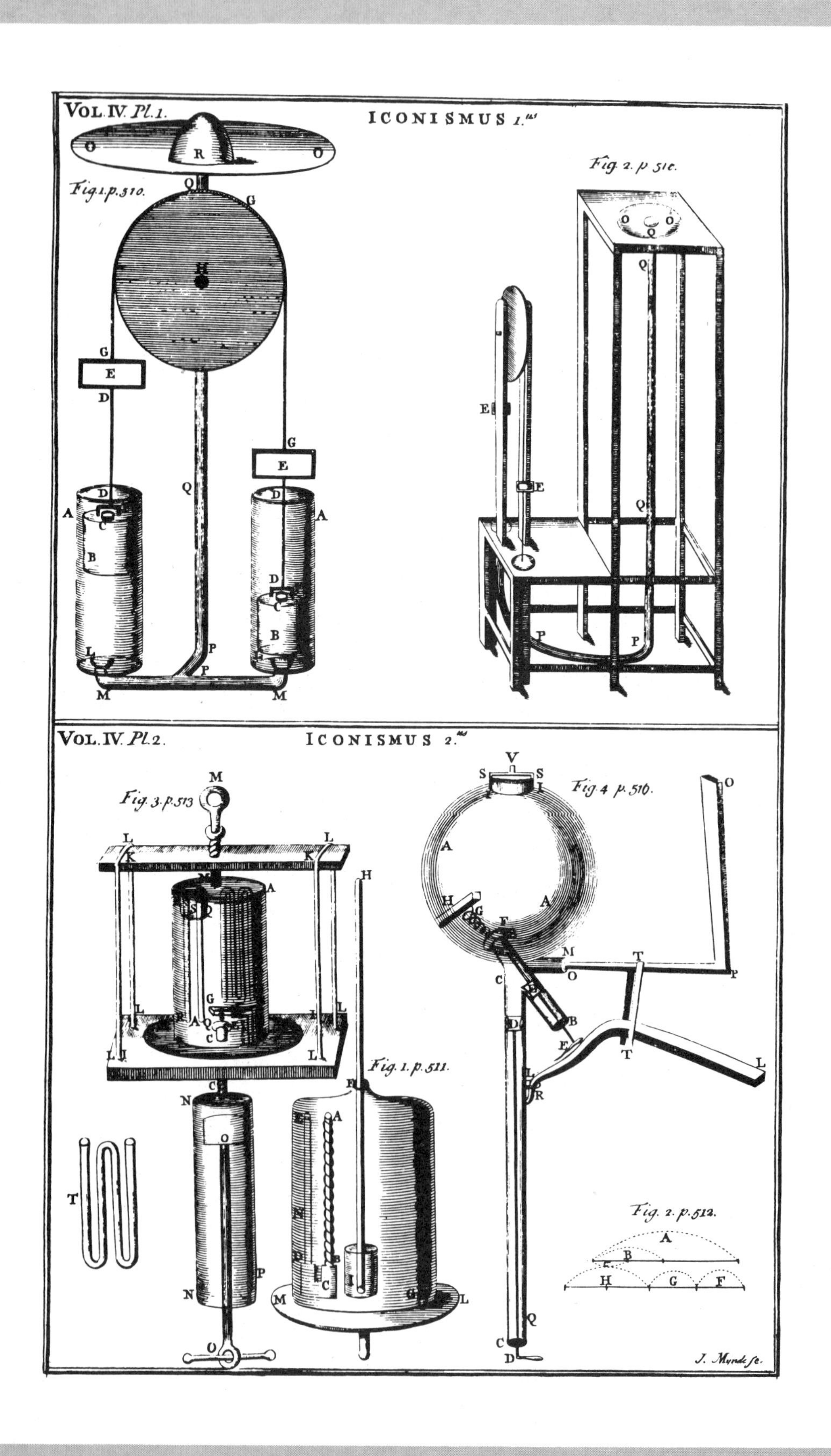

VOL. IV. Pl. 1.
ICONISMUS 1.
Fig. 1. p. 510.
Fig. 2. p. 510.
VOL. IV. Pl. 2.
ICONISMUS 2.
Fig. 3. p. 513
Fig. 4 p. 510.
Fig. 1. p. 511.
Fig. 2. p. 512.
J. Mynde sc.

the whole pillar of the air, incumbent on the plug, is to be elevated by force; and again, when the plug returns back, it is also by force to be restrained, lest it should be too swiftly impelled by the air, and so break the bottom of the engine; but in these double engines, the plyer of them is in a manner wholly free from that toil. For in the first suctions, the plugs are easily lifted up, because the air, immediately derived from the receiver R into the pumps, presseth the plugs downwards, almost as strongly as the external air incumbent on the opposite part; and when the quantity of the internal air is diminished, it comes to pass, that the plug to be depressed tends downwards with so much the greater force; and so, by means of the cord GGG compassing the pully, draws the other plug upwards, and, at the same time, hinders it from too much velocity of descent. And by this means both plugs, at one and the same time, will be helpful to him, that exerciseth the pumps.

SEEING the plugs make but a very small resistance, a man may easily judge, that the two pumps of this engine may be plyed with greater ease, and also with more speed, than one pump in single engines can; so that this engine is of great use in order to those experiments, which cannot be well made but with velocity and speed.

ICONISME II.

The description of the mercurial gage.

THE first description of a mercurial gage, to discover the degrees both of the rarefied and condensed air, may be seen about the begining of the Continuation of our Physico-mechanical Experiments; but those gages, which I used in the following experiments, are declared in the subsequent scheme.

Fig. I. THE whole gage ABCDE consists of three glass tubes, all very well fastned and cemented together, yet so, that a passage is open from one to the other; the first of these tubes AB being open at the extreme A, is of less capacity than the tube BCD, but of greater than the tube ED. The tube BCD is crooked in the middle, and the tube ED ought to be hermetically sealed, at the extreme E, but the part BCD must first be filled with mercury.

THIS instrument, thus prepared, if it be put into a receiver, out of which the air is afterwards to be extracted, it will come to pass, that the air remaining in the part ED, will, by its spring, compress the mercury DCB, and force it to ascend into the part BA, and itself will be dilated in the cavity DC. If then the proportions be duly observed between the bigness and length of the tubes, as shall be declared hereafter, when the air is extracted, the mercury will almost reach to the top A, and the air in the other leg, being so dilated, that in cannot sustain a greater body of mercury, will be kept included in that place.

BUT that this instrument may exactly tell the quantity of the air produced in its receiver, the tubes AB ED are to be distinguished by marks into several parts; and when the Torricellian experiment is tried, above the plain plate LM of the pneumatick engine, as you may see in the figure, a receiver FGE is to be taken, being perforated in the top F, and the tube HI is to be transmitted through the hole, that so the receiver may be applied to the plate; and then the hole F being stopped, and the gage ABCDE being put into the receiver, the air is to be exhausted; the air then being dilated in the receiver, the mercury cannot be sustained so high in the tube HI, but must descend by degrees, and at the same time the air of the tube ED, drives the mercury, by little and little, into the tube AB. When then the mercury in the tube HI descends to the height of 29 digits (I take digits for inches throughout all this tract) and stays at that height, if we mark to what height the mercury hath ascended into the tube AB, we may know, that as often as the mercury in our gage shall rest at that height, the air in the same receiver will be able to ſuſtain only 29 digits of mercury; ſo that the place in the gage, or in the paper ſemblably divided, muſt be marked with the figure 29. And ſo further, every digit of the mercury in the tube HI may be marked in our mercurial gage, and the part AB will be fit to ſhew all the degrees of the rarefied air.

BUT now if the air be condenſed in the receiver above its wonted preſſure, and all ways of its eſcape be ſtopped, you may immediately know it by the tube ED; for the mercury will be impelled into it by the incumbent air, through the open hole, ſo much the higher, as the compreſſion of the air in the receiver ſhall be the greater; and how great that is, and what an altitude of the mercury it can ſuſtain, may eaſily enough be found out, if the computation be made after the manner following:

IT is evident, from the experiments long ſince publiſhed by Mr. *Boyle*, in his anſwer to *Linus*, that the ſpace poſſeſſed by the air, is diminiſhed in the ſame proportion, as the compreſſing force is increaſed, and *vice verſâ*.

Fig. 2. LET then (for example) the ſpace A be poſſeſſed by a certain quantity of air, when (for inſtance) the compreſſing force is F: if now we encreaſe that force by the addition of G, which is equal to it, it will happen, that our ſelf-ſame quantity of air will be reduced to half its ſpace; ſo that B, the remaining ſpace, will be the half of the total ſpace A, even as the former preſſure F is the half of total preſſure F+G. So further, if we encreaſe the preſſure more, by the addition of H, ſo that the firſt preſſure F is only $\frac{1}{3}$ of the total preſſure F+G+H, it will come to paſs, that the air can poſſeſs only the ſpace C, which is $\frac{1}{3}$ of the total ſpace A. And ſo afterwards, the remaining ſpace will be in the ſame proportion to the total ſpace, as the firſt preſſure is to the total preſſure.

THE remaining ſpace : the total ſpace :: the firſt preſſure : the total preſſure.

So that three of thoſe terms or quantities being known, it will be eaſy to find out a fourth, by the rule of proportion. For inſtance, in our gage let the tube ED be the total ſpace, in which the air is compreſſed by the wonted preſſure of the air, which in *England* is wont to be equivalent to 30 digits of mercury, or thereabouts; and therefore the firſt preſſure will be 30 digits of mercury. Now, if that preſſure be encreaſed, and the air be reduced into a narrower ſpace, ſuppoſe into the ſpace NE; if I would find out the quantity of this preſſure, I meaſure the remaining ſpace NE exactly, and I conſtitute that, ſuppoſe ſix digits or inches, for the firſt term of proportion; the ſecond term will be the total ſpace DE, ſuppoſe 12 digits; the third term will be the height of 30 digits of the mercury, which was the firſt preſſure; and ſo the fourth term, or total preſſure will be found to be 60 digits of mercury; whence I may conclude, that the preſſure of the air in the receiver can ſuſtain the mercury to the height of 60 digits: and ſo of the reſt.

FROM the ſame principle before laid down, it will be eaſy to collect, what ought to be the proportion between the largeneſs of the tubes AB and ED. For that depends on the length of the legs, which the higher they are, ſo much the better they can reſtrain and keep in the air, being but a little dilated in the ſealed part. For inſtance, let the length AB be of 10 inches, which height of the mercury is $\frac{1}{3}$ of the accuſtomed preſſure, it will be ſufficient, that the tube HB be twice as big as the tube ED; for after the mercury hath aſcended to the top of the tube AB, the air included in the other leg, expanding itſelf into the ſpace, forſaken by the mercury, will poſſeſs three times more than its former ſpace, and ſo $\frac{1}{3}$ of the firſt preſſure, which is 10 digits, will be ſufficient to curb its ſpring. But if the legs were of leſs length, then the mercury would be expelled by the included air, at leaſt in part. And therefore the bigneſs of the tube AB ought to have a greater proportion to the bigness of the tube ED, that the ascending mercury may afford greater place to the air to be dilated, and so, the spring of the air being weakened, the weight of the mercury cannot be overcome. And that would happen so, if the height of the gage be to the height of 30 digits, in the same proportion which the first space of the air is in, to the total space, which the air would possess *in vacuo*, according to the principle before laid down.

It is better, that the height of the tube be longer than shorter; because if it be shorter, the mercury will be expelled in part, and so will not be able to shew all the degrees of rarefaction; but if it be longer, this only will happen, that the mercury will not reach to the top, and so the gage will nevertheless indicate all the variations, though they be less sensible ones.

But the tube DC ought to contain that quantity of mercury, at the least, which may be sufficient to fill the tube AB, before any way of eruption be opened for the air included in the tube ED. If the capacity of it be much greater, the matter is not much; nor need we be very solicitous concerning the figure of this tube.

ICONISME II.

A description of the engine to compress the air.

Fig. 3. AA Is a glass vessel, whose orifice is exquisitely fitted to the plain plate BB.

BB Is a plain plate of brass, made to cover the vessel AA exactly.

CC Is a small tube of brass, passing through the middle of the said plate, and fastened thereunto.

E Is a little valve, opening inwardly, to shut the small tube C aforesaid.

F Is the spring depressing the valve E.

GGG Is the gnomon fastened to the plate BB, made for restraining the spring F.

II Is a square lath, sustaining the plate BB, and bored through in the middle to transmit the little tube C.

LLL LLL Are two iron wires, which passing through the holes in the lath II, and compassing the upper part of the iron plate KK, do hinder the said plate, that it cannot be much moved from the lath.

KK Is an iron plate, with a hole in the middle, formed into a female-screw, to receive the male-screw MM.

MM Is an iron screw, whose use is, straitly to conjoin the receiver AA with the plate BB. And lest the brass vessel should be broken, it is convenient to put some wood with leather between the screw and the upper part of the receiver: also leather is to be put upon the plate BB, both to prevent the breaking of the glass, and also for the more exact shutting of the receiver.

NN Is a pump fastened to the tube C, below the plate BB.

OO Is the sucker or plug of the pump NN.

P Is a little hole in the lower part of the pump, by which the air enters into it, when the plug is brought to the lowest part thereof.

Now if we would compress the air by the help of this engine, we put the bodies, about which the experiment is to be made, into the receiver AA; and laying it on the plate BB, we firmly bind it thereto by the help of the screw MM. This being done, the sucker or plug OO is to be drawn, till the external air, by the hole P, can fill all the upper part of the pump: then if the plug be drawn upwards, it will come to pass, that the air, finding no other way of egress, will open the valve E, and enter into the receiver AA, from whence there is no regress, because the valve E is presently depressed by the spring F, and doth shut the hole C. And so we may iterate the compression of the air into the vessel AA, as often as we please, and the quantity thereof is easily known by the mercurial gages.

But I am wont so to fashion the pump, that it may be fitted by a screw to the tube C; for so when one receiver is full, we may take away the pump, and use it to fill other receivers.

The History of the Vacuum Pump

by E. N. DA C. ANDRADE

Imperial College of Science and Technology, London, United Kingdom

The first man to make a vacuum pump was Guericke, somewhere about 1654. It was a crude pump, with joints kept airtight by water immersion. Soon after Boyle produced a much improved piston pump, made by Robert Hooke. About the same time the Accademia del Cimento used the vacuous space in the enlarged head of a mercury barometer to carry out experiments: this may be considered as a one-stroke mercury pump. Pumps on the Boyle pattern were made by Huygens, Papin and Senguerd.

In 1682 Boyle described a two-cylinder pump, in which the pressure of the atmosphere on one piston did most of the work required to raise the other. A much improved pump of this type was made by Haukesbee about 1705 and this remained essentially the pattern of cylinder vacuum pumps until the end of the nineteenth century, when Fleuss introduced the cylinder oil pump. For producing high vacua for the classic research on discharge in gases, pumps of the Geissler and Toepler type, based on continuous repetition of Torricelli's experiment, were used, which were extremely slow in action.

A new era was initiated in 1905, when Kaufmann introduced a continuously rotating mercury pump, which was almost at once superseded by Gaede's much speedier rotary mercury pump. Fore pumps of box type, the prototypes of which were water-pumps of the sixteenth and seventeenth century, were brought in by Gaede and others. In 1912 Gaede invented a pump based on a fundamentally new principle, that a surface moving rapidly parallel to itself drags with it gas molecules. This so-called molecular pump produced extremely low pressures. The last inventive step was also made by Gaede, when in 1915 he utilized the carrying along of gas molecules by a jet of mercury vapour to effect a high vacuum. These vapour-stream pumps, oil and mercury, have been developed on an enormous scale of recent years.

THE first vacuum pump, in the sense of a machine by which air could be progressively removed from a closed vessel, was undoubtedly that invented by Otto von Guericke somewhere about 1650, but before that, in 1644, Evangelista Torricelli had carried out his famous experiment with the barometer tubes, showing that the mercury stood at the same height whether the upper end of the tube had the ordinary form or was blown out to a sphere. From this, he correctly concluded that the space above the mercury was empty. This method was some 20 years later used to produce a vacuous space for experiment, as shown in Fig. 1, which is taken from the *Saggi di Naturali Esperienzi*, published by the Accademia del Cimento in 1667. In (a) a small ball of bitumen is being heated by sunlight concentrated by a concave mirror, to show that smoke falls *in vacuo*; in (b) where the evacuated space is closed by a glass lid, another innovation, it is shown that a tied lamb's bladder, containing a trace of air, swells out to a sphere *in vacuo*. This method which, comparing it with the mercury pumps of the nineteenth century, may be called a one-stroke mercury pump, was used in 1786 by Rumford to create a vacuum for investigating "the propagation of heat in various substances." He specified that the mercury had been previously freed of air and moisture by boiling. Properly handled, the method is clearly capable of giving a much better vacuum than the early cylinder pumps, with water-saturated washers and such like, but it is not easy to evacuate large spaces by this means, and the space must not contain constructions of metals or other substances either fragile or attacked by mercury.

Guericke's pump was first described in Kaspar Schott's *Mechanica Hydraulica-Pneumatica*, 1657, as figured in Fig. 2. It consisted simply of a cylindrical tube with two valves one at *H*, halfway along it, which opened outwards to the atmosphere under internal pressure and one at *I*, the lower end of the cylinder, opening when the piston was withdrawn and closing when it was returned. This is clearly not good design, even if the valves were efficient—and we are not told how they were made. A tap *E* could be used to shut off the evacuated vessel, which could then be removed, as shown in the background. Water was freely used to make the joints air-tight.

Guericke himself first published an account of his work much later, in 1672, in his famous book *Experimenta Nova (ut vocantur) Magdeburgica de Vacuo Spatio* (The New, commonly called Magdeburg, experiments on Empty Space) generally referred to as *De Vacuo Spatio*. The pump figured

Originally published in Adv. Vac. Sci. Tech. 1, 14–20 (1960).

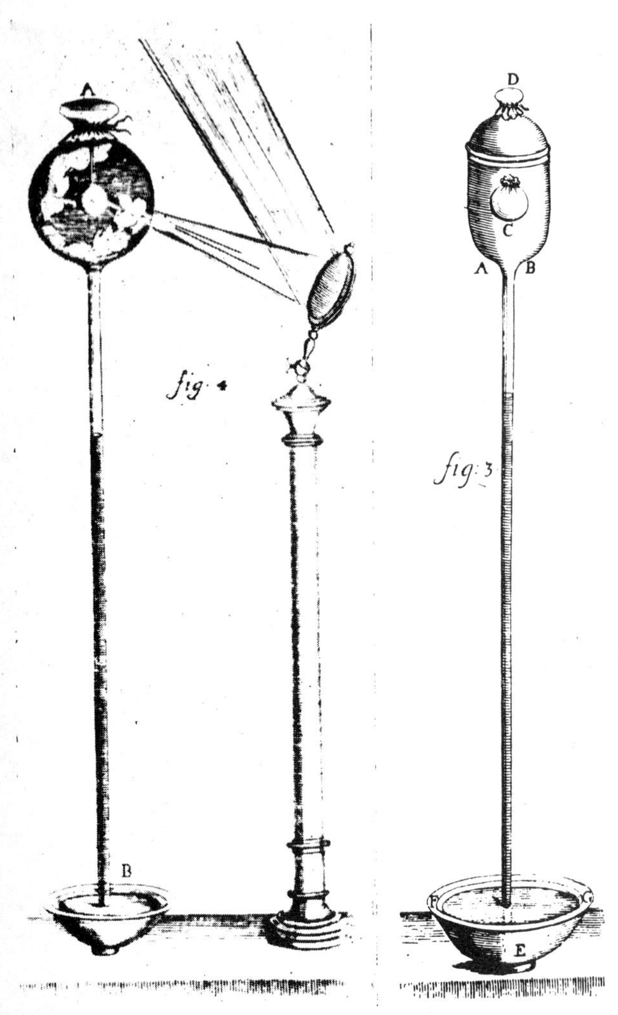

FIG. 1(a) & (b). The vacuum experiments of the Accademia del Cimento.

there (Fig. 3) is of much better design and probably much later construction than that figured by Schott. The essential cylinder, *gh*, was supported by a tripod : it is shown separately in "Fig. III". The piston running in this cylinder was of wood, made air-tight with wet string. For use in the early stages of pumping, a leather valve, *z*, closed by a spring, was provided. When the pressure of the air to be expelled became too low to force it open, use was made of the peg, *m*, taken out by hand during the upstroke and replaced during the downstroke. Essential was the tap, *gr*, used as in Boyle's pump, to be next described. It must be remembered that the account of Boyle's pump appeared in a Latin edition in 1661, 11 years before *De Vacuo Spatio* and that Guericke may well have seen it. The conical vessel, *xx*, at the upper end of the cylinder and that shown in "Fig. VI", designed to be hung so as to enclose the lower end of the cylinder, were to be filled with water, which was relied on to make the joints air-tight. The spherical glass receiver *L* was cemented to the brass attachment carrying the tap. The famous Magdeburg experiment with the two hemispheres enclosing an exhausted space, which two teams of horses could not pull apart, was carried out in 1654, probably with the type of pump figured by Schott.

Robert Boyle's pump, of which he published an account, with the illustration reproduced in Fig. 4, in 1660, was actually made by Robert Hooke, which accounts for the excellent design and workmanship, evidenced by the fact that there was no need of water to keep the joints tight. In the vertical brass cylinder ran a "sucker", "4 4" in the diagram, which

FIG. 2. Schott's representation of Guericke's early pump.

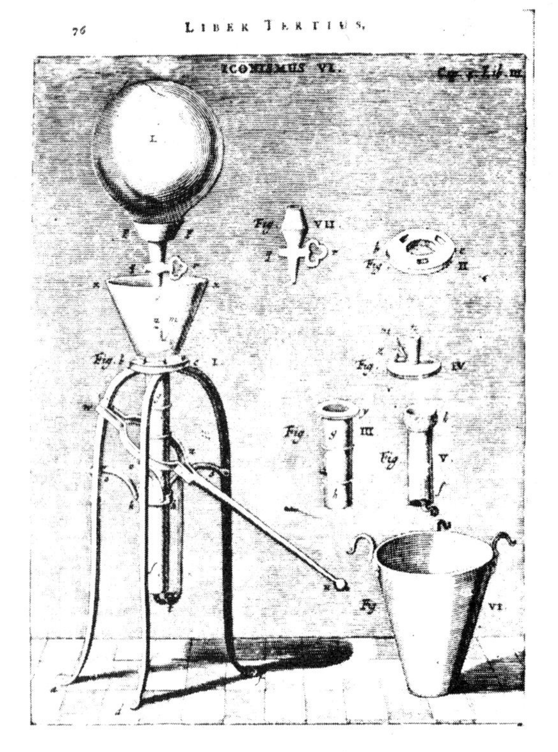

FIG. 3. Guericke's pump, from *De Vacuo Spatio*.

was presumably of wood, since a disc of oiled leather was nailed to it, for air-tightness. This was driven up and down by a rack and pinion. Oil, or an emulsion of oil and water, was put on the piston for easy running. The receiver was a large glass sphere with an opening at the top, cemented into which was a brass ring made to receive a brass cover, shown in detail at the upper right-hand corner. All this bespeaks excellent handicraft.

Essential for the working was the hole at the upper shoulder of the piston, shown closed by a peg. To operate the pump, the tap below the receiver was closed, the peg removed and the piston run up, expelling the air. The peg was then replaced, the tap opened and the piston withdrawn, sucking air from the receiver. The tap was then closed, the peg removed and the process repeated. This is how Guericke's last pump operated. It is clear that this pump of Boyle's was an excellent instrument, well adapted for experiment.

Boyle's second pump, described in a book published in 1669, is shown in Fig. 5. The single cylinder was of the type of the first pump, the peg being on the end of the long stick shown on the right of the rack : the whole was immersed in water for air-tightness, no doubt because, Hooke being no longer in Boyle's service, the workmanship was not up to standard. Today it is hardly necessary to stress the importance of perfection of finish in a mechanical pump. The advantageous new feature was the flat plate to which the bell-jar for experiment could be cemented, a feature anticipated in a pump made by Huygens, of which, however, he published no account, so that possibly Boyle's innovation was an independent one.

Just as Hooke had worked for Boyle, so Denis Papin, also a genius, worked for Huygens and made, to Huygens' design, a pump, similar in many respects to Boyle's first pump, with a plate for the receiver, of which he published an account in 1674. At the same time he described a pump of his own design, shown in Fig. 6, remarkable for having a two-way tap, *G*, drawn separately to the right. One channel was an ordinary hole at right-angles to the axis of the turning part, the other a groove along *aa* for communication with the air. The stirrup was another novel feature : with these early pumps with wide cylinders it required considerable force to overcome the atmospheric pressure.

This trouble was avoided by providing two cylinders, the pressure of the atmosphere on one piston helping to withdraw the other. The first such pump, illustrated in Fig. 7, was described by Boyle in 1682, but he clearly stated that it was the invention of Papin, who was working for him at the time.

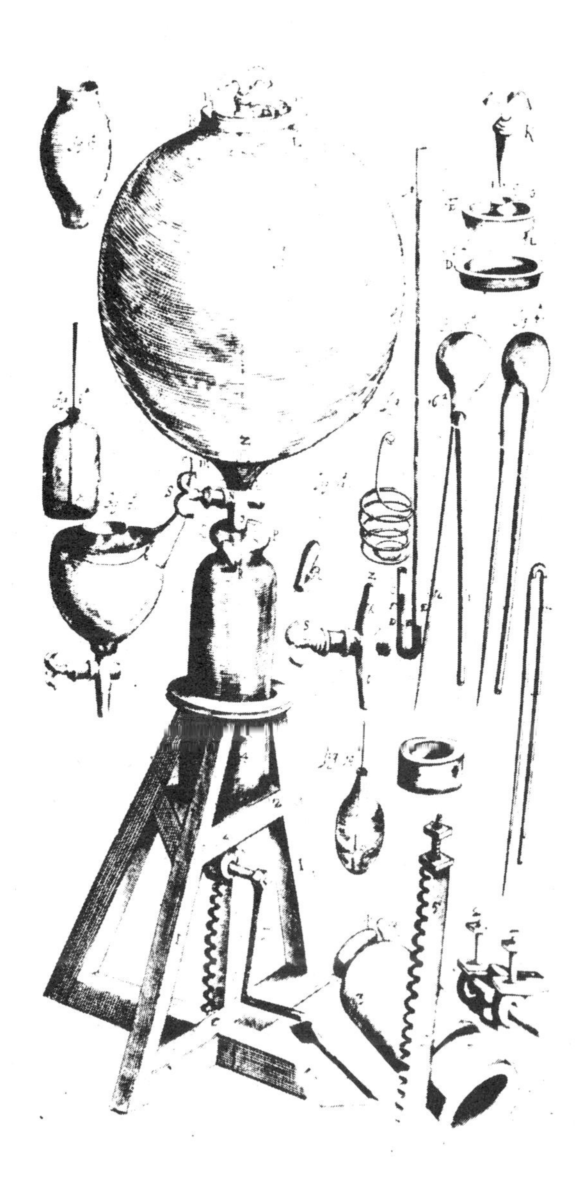

FIG. 4. Robert Boyle's first pump.

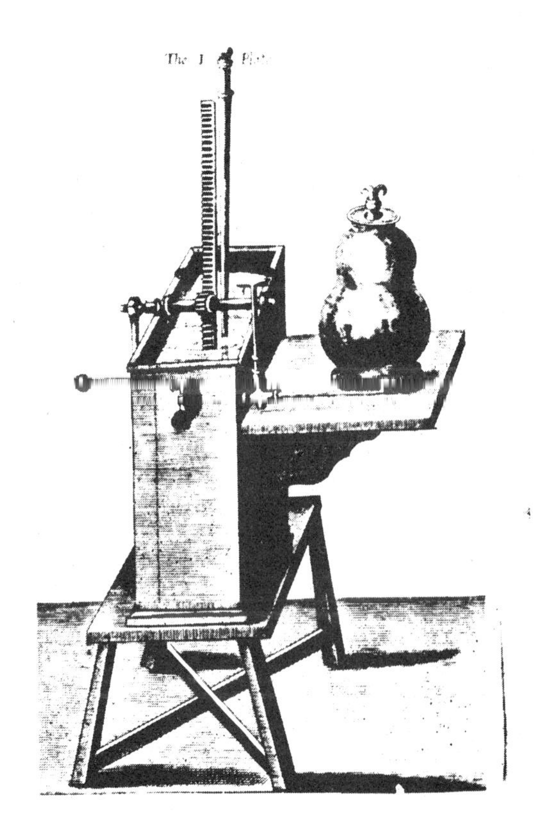

FIG. 5. Boyle's second pump.

The history of the vacuum pump 17

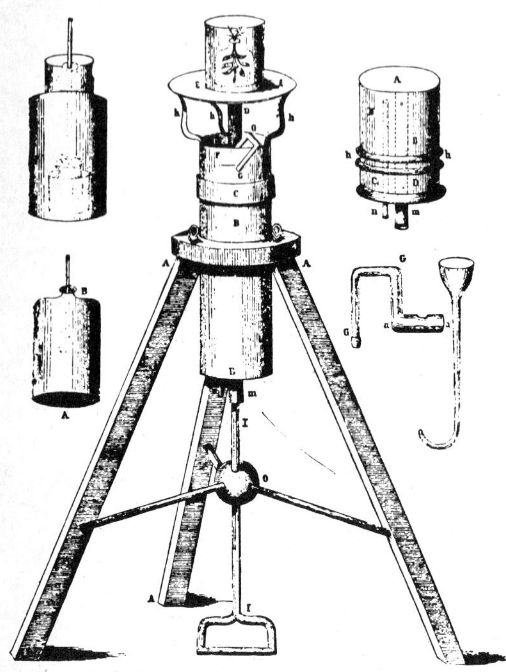

FIG. 6. Papin's pump with two-way tap.

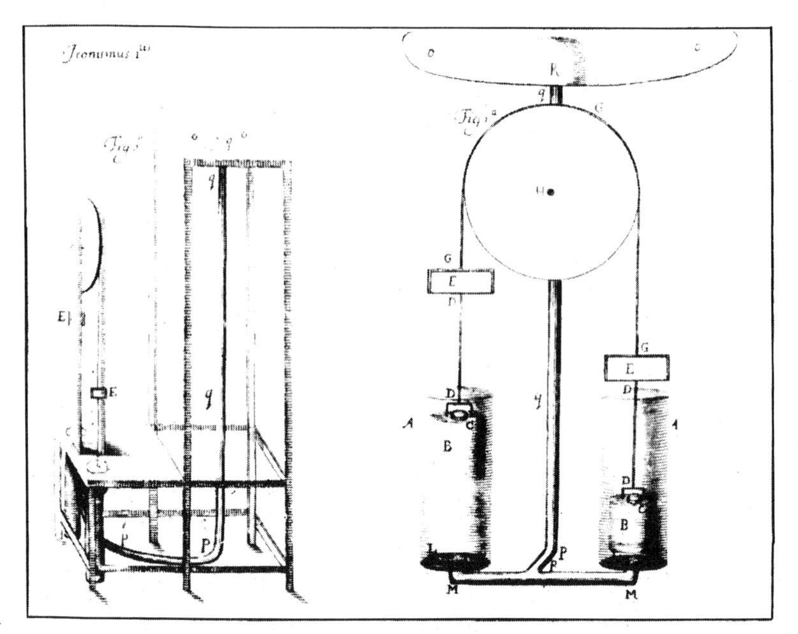

FIG. 7. Boyle's two-cylinder pump.

FIG. 8. Francis Hauksbee's pump.

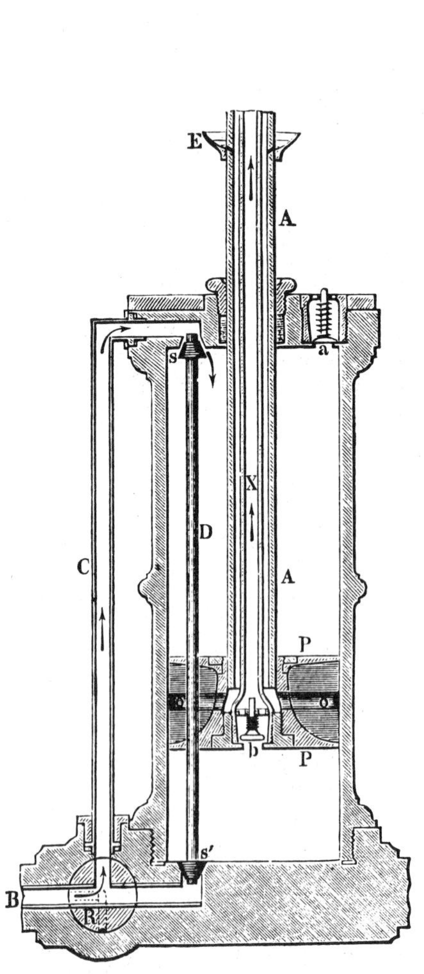

FIG. 9. Bianchi pump.

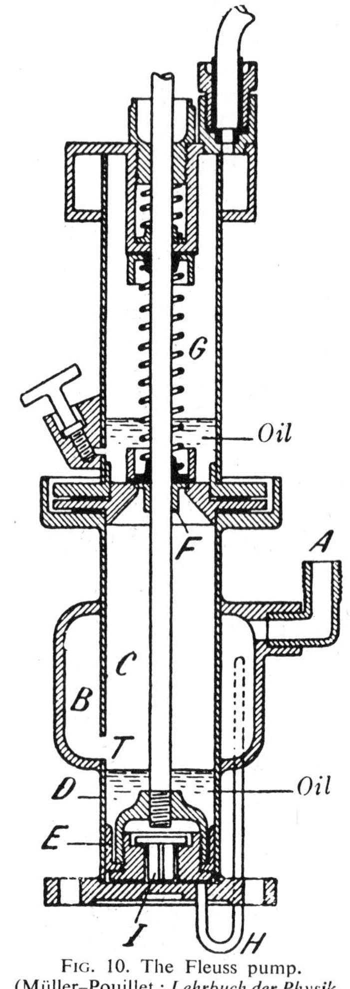

FIG. 10. The Fleuss pump. (Müller–Pouillet : *Lehrbuch der Physik*, vol. 1, 1906, Fig. 514, Vieweg.)

The pistons were coupled together by a cord passing over a pulley wheel. A two-cylinder pump of much better design was produced by Francis Hauksbee and described in 1709. His illustration, reproduced in Fig. 8, corresponds exactly to a pump still in the possession of the Royal Society. The two pistons are worked by rack and pinion, so arranged that as one descends the other ascends. The valves were made of bladder. The arrangement of the bell-jar, with a mercury column manometer directly beneath it, was a very convenient one. This pump, when put in order some years ago according to Hauksbee's description, gave a vacuum within about 1 in. of mercury of perfect, better than which, as wet leather was used for certain washers, could not be expected.

Hauksbee's design remained, in principle, unchanged until well on in the nineteenth century : in fact, pumps of this type, with small modifications and improvements, were still being made at the end of that century. In 1856, Bianchi introduced a pump with a single cylinder, Fig. 9, so designed that the air from the receiver was drawn in alternately above and below the piston as it played in the cylinder, with the effect that, as far as the mechanical work was concerned, the pressure difference in both up and down stroke was negligible. This pump does not seem to have come into general use, perhaps on account of difficulties of construction.

A substantial improvement in the vacuum produced by cylinder pumps was made by Fleuss in 1892 with what he called the Geryk pump, in honour of the Magdeburg pioneer. The new feature was that the dead space which is still left when the piston has advanced to the end of the cylinder was filled with oil, whence these pumps were sometimes called oil pumps. As shown in Fig. 10, halfway up the cylinder there is a partition, supporting a spring-controlled valve through which the piston-rod passes. Oil is contained above this and above the piston. At the end of its upward stroke, the piston, with its superincumbent oil and air, lifts the valve and forces out the air. The valve does not close until the piston has descended a short distance, which ensures a correct distribution of the oil. A special oil, of low vapour pressure, although not as good as the modern apiezon oils, was used and it is claimed that pressures as low as 0.0002 mm mercury were attained. In any case, these pumps were at one time widely used in the electric lamp industry.

In the middle of the nineteenth century, when high vacua—which in those times might mean a few times 10^{-3} mm of mercury—were required for electrical discharge tubes, the mercury pump began to be developed. That used by Geissler in 1855, said to be a modification of one devised in principle by Swedenborg, was the first of a large class. It consisted of a bulb connected by a flexible tube to an open reservoir of mercury, this bulb being provided with a two-way tap by means of which it could be connected either to the outside air or to the vessel to be exhausted. The tap being turned so as to give passage to the air and to cut off the vessel, the reservoir was raised until all the air had been expelled from the bulb : the tap was then turned so as to cut off passage

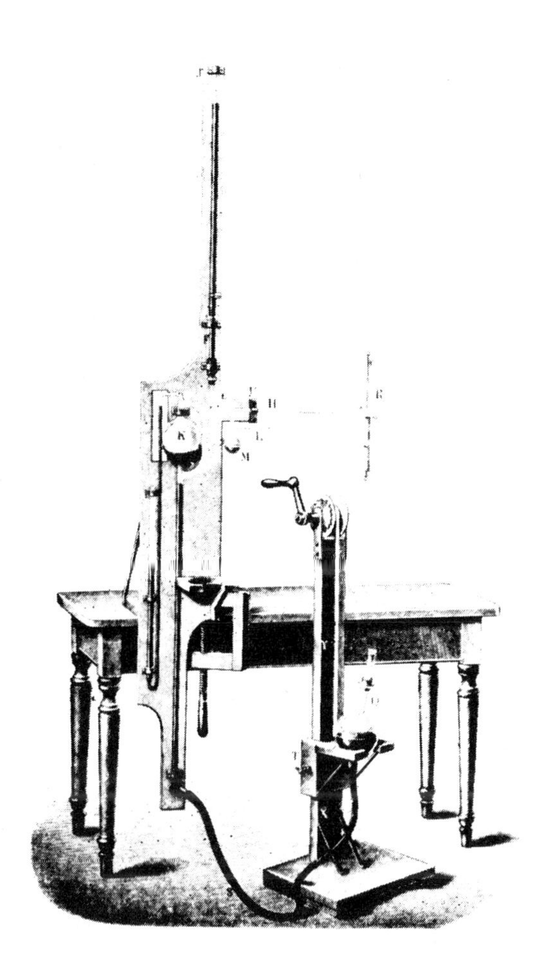

FIG. 11. Töpler pump, as used towards the end of the nineteenth century. (Müller-Pouillet ; *Lehrbuch der Physik*, vol. 1, 1906, Fig. 527.)

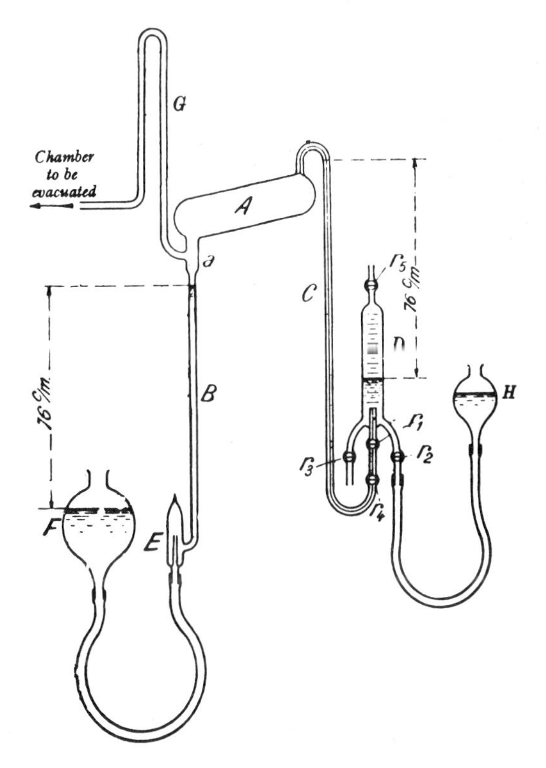

FIG. 12. Improved form of Töpler pump. (Dunoyer ; *Vacuum Practice*. G. Bell, 1926, Fig. 1.)

to the air and to connect to the vessel and the reservoir lowered. The process was repeated as often as desired. The vessel was thus effectively connected to a Torricellian vacuum at each stroke, the fraction of air removed depending upon the relative volume of bulb and vessel to be exhausted.

A great improvement was effected in 1862 by Töpler, who retained the mercury reservoir and flexible tube but substituted for the tap an arrangement of tubes by which passage was made for the air to be expelled, and the vessel to be exhausted was connected to the Torricellian vacuum, automatically at the appropriate position of the mercury level. This went through improvements at the hands of Neesen and Bessel–Hagen which brought it to the form shown in Fig. 11. *R* is the vessel to be exhausted (a Geissler tube), *K* is the bulb in which the Torricellian vacuum is produced, *P* is the joint at which the rising mercury cuts off connection of *K* and *R*, *OBC* the tube through which the air is expelled. *M* is a bulb containing phosphorus pentoxide. It was with pumps of this general type that exhaustion was effected for many of the early classical experiments on cathode rays, Röentgen rays and electronics in general. The latest type, which was in use well into the present century, is illustrated in Fig. 12. One advantage of this pattern is the cylindrical inclined bulb, *A*, in place of the old spherical bulbs. This gives a big volume for a small vertical displacement and also lessens the mechanical shock of the rising, uncushioned mercury when the pressure is low. Anyone who has used one of the old Töpler pumps systematically knows how easy it was to knock the head off, if the mercury reservoir was raised without due caution.

Mention must be made of the Sprengel mercury pump, invented in 1873, which was also extensively used by the classic experimenters on the electric discharge. In this, the air was carried away by a succession of falling mercury pellets, between which it was trapped. It was with an improved form of this pump, devised by Gimingham and having a number of tubes for the mercury fall instead of one, that William Crookes, one of the most skilled workers on vacuum physics in the second half of the nineteenth century, carried out, for instance, his classic work on the radiometer. He used phosphorus pentoxide to absorb water vapour, precipitated sulphur to stop mercury vapour and reduced copper to stop the sulphur vapour entering his evacuated space. He claims to have reduced the pressure to 0.4×10^{-6} atm, say 3×10^{-3} mm of mercury, and says, " Formerly, an air-pump, which would diminish the volume of air in the receiver 1000 times, was said to produce a vacuum", adding that the Sprengel pump had already rarified air a few hundred thousand times, which in those days was very good performance.

In 1905, a new period was ushered in by the invention of the rotary vacuum pump of W. Kaufmann, in which a continuous rotation, effected by an electric motor or by hand, carried away the air. It is illustrated in Fig. 13. It needed a backing pump, to reduce the pressure to 2 cm of mercury or so, a feature of all modern high-vacuum pumps. The principle is clear : the air is trapped by the mercury in the inclined Archimedean spiral and delivered to the roughly evacuated space. Only a relatively few of such pumps were, however, made, for at the end of the same year, Gaede produced his rotary mercury vacuum pump, which, on account of its simplicity and speed of working, was an immediate success.

The general scheme is shown in Fig. 14, *G* is a cast-iron cylindrical container, with a thick glass front *B*, within which a porcelain drum, shown in section, rotates. This drum consists of three similar compartments. With rotation in a counter-clockwise direction, the air in W_2 is displaced by the mercury along the passage between the walls Z_1 and Z_2, until it is delivered into the space above the mercury in the outer container. Here it is removed, through s_2, by the backing pump, which produces the low pressure, 1 cm of mercury or so, needed if the mercury level inside and outside the drum is not to be too large. As the drum rotates, the

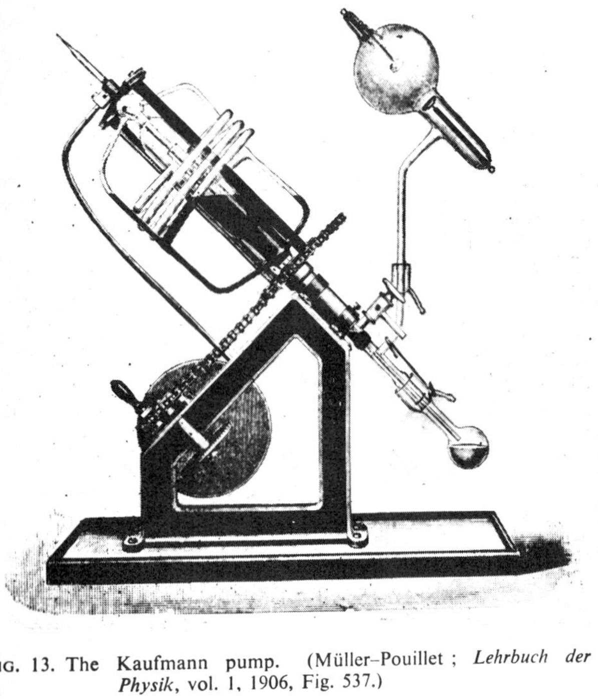

FIG. 13. The Kaufmann pump. (Müller–Pouillet ; *Lehrbuch der Physik*, vol. 1, 1906, Fig. 537.)

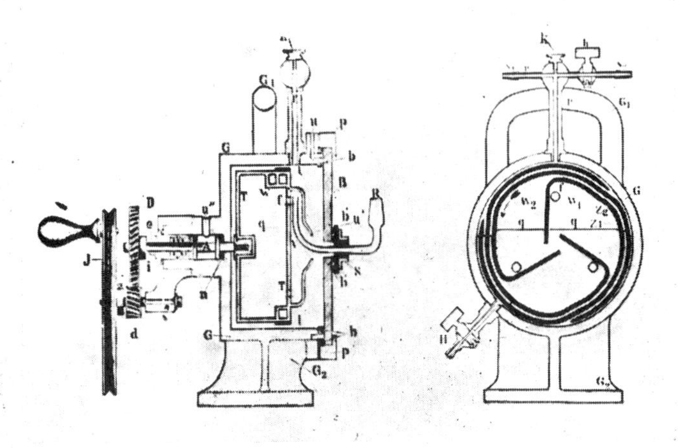

FIG. 14. Gaede's rotary vacuum pump.

volume of air in W_1, entering through the passage *f*, from the vessel to be exhausted, increases, to be expelled in its turn by further rotation, and so on. By 1910 such pumps were to be found in constant use in every continental laboratory devoted to vacuum physics. They could produce a vacuum of 10^{-6} mm of mercury.

For backing pump a mechanical rotary pump was generally used. Gaede's so-called *Kapselpumpe*, or box pump, is shown in cross-section in Fig. 15. There is an eccentrically-

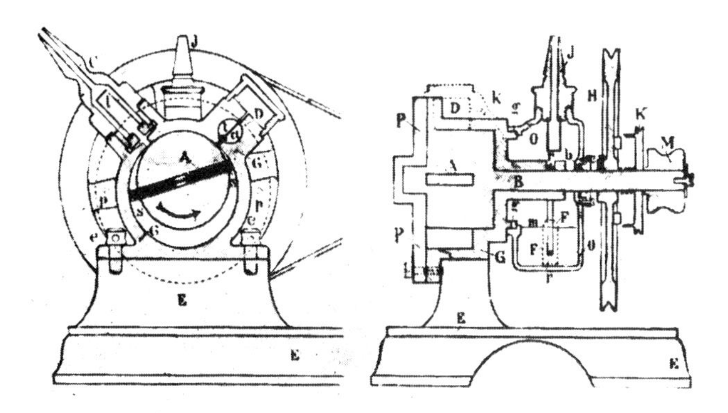

Fig. 15. Gaede's box-pump.

mounted rotating cylinder, with vanes maintained in contact with the wall of the container by springs : the mode of operation is clear. There was nothing new about the general

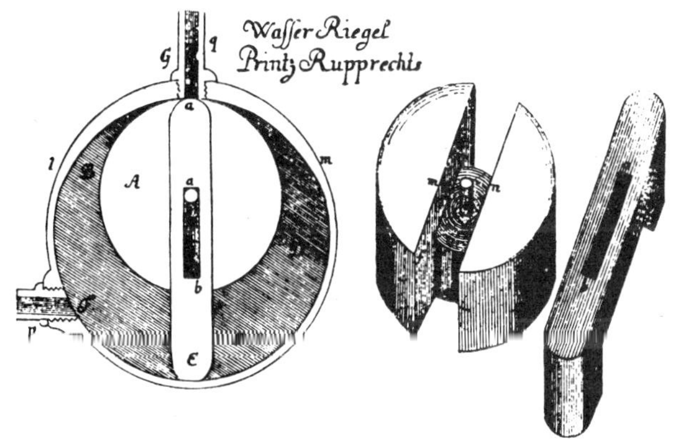

Fig. 16. Rupert's water-bolt.

design, as can be seen by comparing Fig. 15 with Fig. 16, which shows Prince Rupert's " Water-Bolt," a water pump invented some 250 years earlier, and itself not unlike a device figured by Ramelli in 1588. Gaede's early box pump was the forerunner of a large range of recent pumps of this general pattern. It produced a vacuum of about 0.01 mm of mercury : the new pumps, with improved design and special oils, can do somewhat better than that.

In 1912, Gaede made another fundamental advance by his invention of the molecular pump. This depends for its action on the peculiar properties of gases at pressures so low that the mean free path is larger than the width of the passage that is under consideration : in this connection it may be recalled that the mean free path of the nitrogen molecule at a pressure of 0.001 mm of mercury is about 7 cm. At the low pressures in question, the molecules striking a wall behave as if the gas condensed on the wall and evaporated again. If, therefore, the wall is moving in its own plane, the velocity of the wall must be added to that of the molecules. In Gaede's molecular pump, which demanded a comparatively high fore-vacuum, a cylinder rotating at high speed gave the rapid surface movement required. A number of grooves in the cylinder, each divided by a tongue projecting from the housing, were connected in series, to increase the effect. The pump had the advantage that it dealt with condensable vapours as well as with ordinary gases.

The construction demanded a very high degree of mechanical skill and it was, in consequence, expensive. It never came into general use because in 1915 Gaede invented the vapour stream pump, which is simple to make and very efficient. A form of the molecular pump devised by Holweck, which was somewhat simpler to make, is, however, in its modern form used for certain purposes.

The history of the vapour stream pump of Gaede and its development by Langmuir and many others to the position which it now occupies is too recent a story, and one too familiar to all present, to be included in this brief sketch. I should like to conclude with a tribute to Gaede, who originated so much, if not all, of the modern equipment for winning a vacuum. He combined deep theoretical knowledge with clear physical insight and a genius for design. Lenard in 1926 wrote to him telling him that he had proposed him for the Nobel prize : " Denn, mein lieber Gaede, was wären wir alle ohne Sie?" It would have been an award which would have given pleasure to many physicists. With his inventive genius Gaede combined a character of singular courage and probity. He was of old German descent, but he was denounced by the Nazis as politically unreliable and friendly to Jews, was removed from his university post and, in general, shamefully treated for his open desire for decency. He died in 1945 of diphtheria and undernourishment. All honour to his memory.

William Crookes and the Quest for Absolute Vacuum in the 1870s

ROBERT K. DEKOSKY
Department of History, University of Kansas,
Lawrence, Kansas 66044, USA

Received 29 July 1982

Summary

This essay examines the technical evolution and scientific context of William Crookes's effort to achieve an absolute vacuum in the 1870s. Prior to late 1876, along with interrogation of the radiometer effect, the quest for perfect vacuum was a major motive of his research programme. At this time, no absolutely dependable method existed to determine exactly the pressures at extreme rarefactions. Crookes therefore employed changes in radiometric, viscous and electrical effects with changing pressure in order to monitor the progress of exhaustion. After late 1876, his research priorities shifted because he had reached a plateau of technical accomplishment in the effort to attain extreme vacua, and because observed effects in vacua—particularly electrical—assumed an importance in their own right, and as bases for elucidation and defence of his concept of a 'fourth state of matter' at very low pressures.

Contents

1. The work to 1876

Previous analyses of William Crookes's work in vacuo have emphasized rightly, that his concept of a fourth state of matter provided him with a theoretical unification of investigations into what, in retrospect, are unrelated phenomena.[1] This essay does not challenge that view. But it draws from manuscript evidence the conviction that Crookes's desire to achieve a vacuum 'approaching perfection'—and the technical difficulty of determining accurately the success of this undertaking—also stimulated his investigation of radiometric, viscous and electrical properties of rarified gases in the 1870s. It will deal with the technical evolution and scientific context of Crookes's effort to attain an absolute vacuum.

The background is well known and does not require detailed restatement. In 1861, Crookes had discovered the element thallium by spectroscopic means.[2] He studied the properties of thallium throughout the 1860s, and his efforts to determine very accurately its atomic weight by weighing in vacuo exposed what seemed to be a thermal

[1] Robert K. DeKosky, 'William Crookes and the Fourth State of Matter', *Isis*. (1976), 36–60. A. E. Woodruff, 'William Crookes and the Radiometer', *Isis*, 57 (1966), 188–98.
[2] W. Crookes, 'On the Existence of a New Element, Probably of the Sulphur Group', *Chemical News*, 3 (1861), 193–4.

Originally published in Ann. Sci. **40**, 1–18 (1983).

influence on weight; this encouraged his further investigation of physics at extreme exhaustions.[3] By the early autumn of 1871, Crookes and his young assistant Charles Gimingham were studying energetically the effect in vacuo.

Two interrelated technical problems confronted Crookes and Gimingham. First, they had to achieve as extreme an exhaustion as possible. Second, they required a means to indicate the extent of the vacuum accomplished. From the beginning, they utilized a Sprengel pump to achieve high levels of exhaustion. In 1865, Hermann Sprengel had devised a means to produce vacua by allowing mercury to fall down a barometric tube.[4] His pump differed from the equally popular Geissler pump in which the operator first drove mercury up a barometric tube—forcing air out through an upper opening in a three-way tap—and then, turning the tap to connect the tube with a receiver, permitted the mercury to fall down the tube and thus draw air from the receiver.[5] A third pump in wide use, the Töpler, combined both these principles.[6]

Figure 1 depicts the Sprengel arrangement contrived by Crookes and Gimingham.[7] Mercury fell from reservoir *A* down the flexible tube *g* and was led up through *f* and around *h*. As it then fell through *N*, it broke into drops. The receiver—shown partially—extended laterally from the upper right portion of tube *b*. As *A* emptied, liberation of the detente *T* enabled the reservoir to be lowered toward *L* at a rate controlled by the chain and wheel mechanism at *mk*. At the bottom, *A* could receive mercury through tap *I* from reservoir *H*. When raised mechanically to its original position, reservoir *A* could serve once again as the source of mercury for the fall-tube. Thus *gfhN* served as a barometric siphon preventing air within *A* from entering *N* after mercury had emptied from *A*. Barometer *ee* dipped into the same vessel as the gauge barometer *P*. Scale *dd* permitted a quantitative comparison between the two and thus a reading of the pressure in the apparatus.

[3] W. Crookes, 'Researches on the Atomic Weight of Thallium', *Philosophical Transactions of the Royal Society*, 163 (1873), 277–330. (Hereafter designated as *Phil. Trans.*) 'On Attraction and Repulsion Resulting from Radiation', *Phil. Trans.*, 164 (1874), 501–27.

[4] Hermann Sprengel, 'Researches on the Vacuum', *Journal of the Chemical Society*, 3rd Series, 18 (1865), 9–21. K. R. Webb, 'Sprengel and the Vacuum Pump (1865)', *Chemistry in Britain*, 1 (1965), 569–71. Sprengel, a Heidelberg Ph.D., had come to England in 1859 and assisted Benjamin Brodie, Professor of Chemistry at Oxford, until 1862. At that time he moved to London, where, for the next several years, he undertook research at the Royal College of Chemistry and at Guys and St Bartholomew's Hospitals. Very probably he constructed his first pump in the laboratory at St Bartholomew's, where the lecturer in chemistry, William Odling, arranged for him to have facilities. Almost immediately the Sprengel pump attracted experimenters: Graham used it in his researches on gaseous diffusion. Thomas Graham, 'On the Absorption and Dialytic Separation of Gases by Colloid Septa', *Phil. Trans.*, 156 (1866), 399–439. Bunsen applied its principle to accelerate filtration. R. Bunsen, 'On the Washing of Precipitates', *Philosophical Magazine*, 4th Series (1869), 1–18.

[5] As early as the 1720s, an alternative to the mechanical air pump appeared in which the operator drove mercury up a barometric tube. Emmanuel Swedenborg, *Miscellanea observata circa res naturales, el praesertim circa mineralia, ignem, et montium strata* (Leipzig, 1722), p. 101. For a discussion of subsequent precedents to the Geissler pump, see Sylvanus Thompson, 'The Development of the Mercurial Air-Pump', *Journal of the Society of Arts*, 36 (1887–8), 21–5. According to Thompson (pp. 25–6) the first public mention of Geissler's pump came in a pamphlet entitled 'Ueber das geschichtete elektrische Licht' published in Berlin in 1858 by W. H. T. Mayer. Geissler was a mechanic at the University of Bonn who established a workshop for constructing chemical and physical instruments. Hans Kangro, 'Johann Heinrich W. Geissler', *Dictionary of Scientific Biography* (New York, 1972), v, 340–1.

[6] The idea underlying this device originated as early as 1828 in the invention by Mile of Warsaw, and lay in undeserved obscurity until Töpler reincarnated it in 1862. J. Mile, 'Neue hydrostatische Luftpumpe ohne Kolben, Hähne, Klappen, und Stöpsel', *Dingler's Polytechnisches Journal*, 30 (1828), 1–6. A. Töpler, 'Ueber eine einfache Barometer-Luftpumpe ohne Hähne, Ventile, und Schädlichen Raum', *Dingler's Polytechnisches Journal*, 163 (1862), 426–32.

[7] Crookes (footnote 3), 295. This was a more mechanized version of the arrangement employed earlier in the autumn of 1871. It came into use in the spring of 1872.

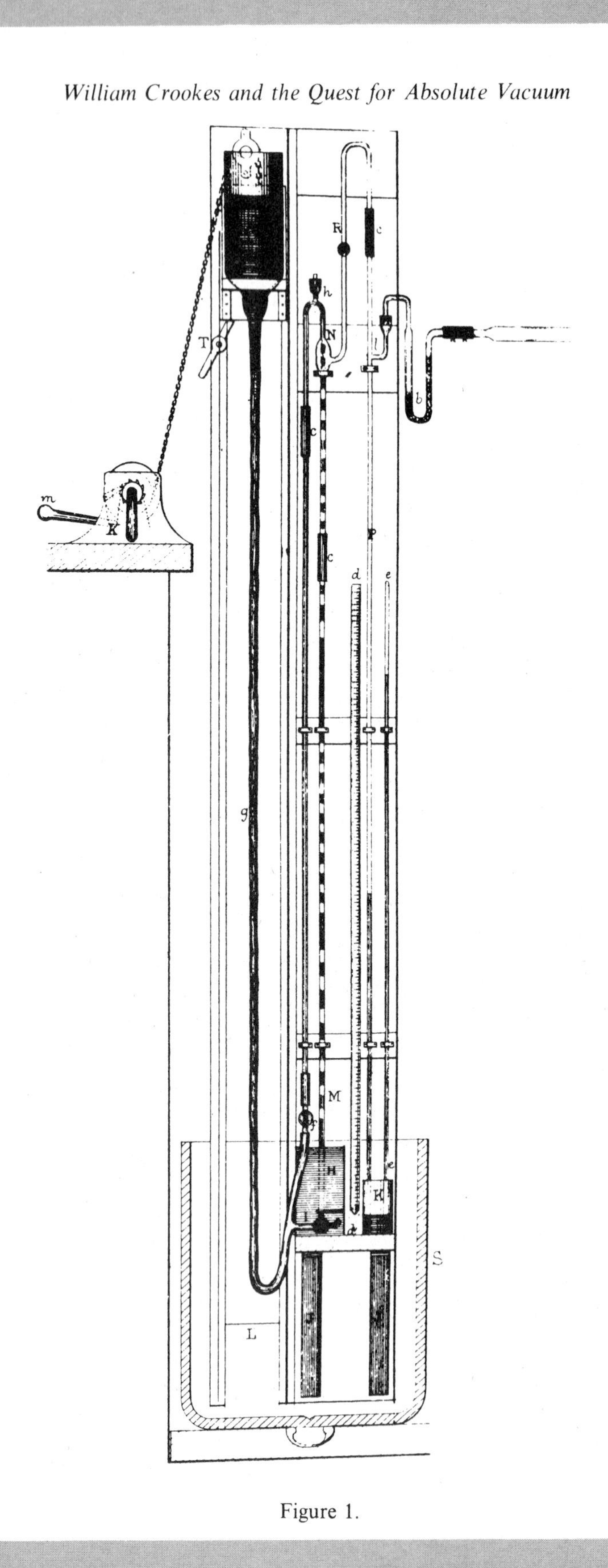

Figure 1.

Two other features of this early arrangement merit attention. Tube *b*, filled with small glass beads and sulphuric acid, interposed between the pump and receiver in order to absorb moisture. This procedure of removing residual vapour, chemically together with the pump action, traces to an innovation of Victor Regnault in 1845.[8] In addition, leakage at joints *h*, *l*, and *c*, *c*, *c*, was reduced by the use of 'mercury joints'. Sprengel himself had realized that joints composed of rubber tubing were inadequate for high-vacua work if these connected the glass tubes and were bound about the glass with tightened coils of wire. Though he employed these traditional joints in his original pump, by the late 1860s, he introduced the new mercury joint that became popular in the 1870s.[9] Sprengel ground the conical end of a glass tube to fit into a conical cup at the end of the other tube. He placed mercury in the cup to thwart any contact between the junction and the external air. In the Crookes–Gimingham apparatus only the joint at the entrance to the receiver remained of the traditional construction.

Crookes's estimation of the vacuum that his apparatus could produce was over-optimistic in the autumn of 1871. He congratulated Gimingham on 13 October for 'the very good vacuum' with 'no air in it' that the Sprengel pump yielded,[10] and referred six days later to 'the perfect vacuum' in which 'air currents cannot exist'.[11] As he was soon to discover, this early apparatus still left significant amounts of air in the receiver. Dependable means to determine pressures in highly evacuated receivers were unavailable at this time. Of course, one could follow accurately the decrease in pressure in the receiver of a Sprengel pump by comparing the height of mercury rising at the lower portion of the fall-tube with the height of a Torricellian barometer. But when the levels of these two columns neared equality, direct visual measurement proved wholly inadequate: the eye cannot detect accurately differences in levels of mercury smaller than 0·1 mm.[12] Sprengel had suggested an alternative method by which to estimate pressures at extreme exhaustions. This procedure derived from Dumas' method for determining vapour densities. Sprengel exhausted a receiver bulb extended on one side into a capillary tube. He then contrived to fill the apparatus with mercury, compressing the residual gas into a small section of the capillary. Determining the weight of a volume of mercury equal to the volume of the air compressed into the capillary, he recognized that the weight of this small volume of mercury bore the same proportion to the weight of mercury filling the entire volume of bulb and capillary as the weight of air remaining in the receiver after exhaustion (compressed by mercury to its small volume at atmospheric pressure) bore to the weight of air prior to exhaustion (at atmospheric pressure). The method was exceedingly delicate, but Sprengel claimed that it indicated his instrumentation could exhaust a receiver to about 10^{-6} atm.[13]

In order to follow the progress of exhaustion, Crookes, from 1871, had used the discharge from an induction coil across the evacuated space.[14] Faraday's discovery, in 1838, of the negative dark space in discharge tubes had rekindled interest in luminous

[8] V. Regnault, 'Études sur L'Hygrometrie', *Annales de Chimie et de Physique*, 3rd Series, 15 (1845), 190.
[9] W. N. Hartley, 'Experiments Concerning the Evolution of Life from Lifeless Matter', *Proceedings of the Royal Society*, 20 (1871–2), 141.
[10] Crookes to Gimingham, 13 October 1871 (Science Museum Library archive, hereafter designated by SML).
[11] Crookes to Gimingham, 19 October 1871 (SML archive).
[12] J. H. Leck, *Pressure Measurement in Vacuum Systems*, 2nd edition (London, 1964), p. 1.
[13] Sprengel (footnote 4), 18–19.
[14] Crookes to Gimingham (footnote 10).

effects due to conduction of electricity through gases at low pressure.[15] In 1852, W. R. Grove found that at low pressures the discharge was 'striated by transverse, non-luminous bands'.[16] J. P. Gassiot detected changes in the striae with changes in pressure over the range of pressure in which the striae appeared; and he even produced a Torricellian vacuum at a temperature below the freezing point of mercury that inhibited the passage of an electrical discharge.[17] This seemed to some people[18] a vindication of Faraday's thesis that electrical discharge necessarily involved particles of matter,[19] and understandably it seemed a proper test for vacua approaching 'perfection'.

After experimenting with several ways to study the movements of bodies exposed to heat in the evacuated receiver, Crookes obtained consistent results by balancing on a needle point a beam of straw with pith-ball ends.[20] By the autumn of 1872, he became convinced that at extreme exhaustions any residual movement of the beam was caused by heat—in his view an interfering cause—and not by currents of air in the receiver.[21] All the same, his correspondence with Gimingham reveals that Crookes conceived two explanations for the behaviour of the straw balance in the first half of 1873: the vertical movements of the pith terminals could derive from a modification of the gravitational force by heat, or perhaps the heat itself could exert direct mechanical action on the masses.[22] By August 1873, two experimental findings committed him to the latter explanation. First, he had observed in a Sprengel vacuum that heat consistently repelled a pith terminal when the source was below and above the pith.[23] Second, instead of supporting beams on needle-points—which restricted movement to the vertical direction—he began to suspend the beams at their centres by a long cocoon-silk fibre permitting movements in a horizontal plane;[24] here again, heat sources manifestly repelled the masses in a Sprengel vacuum.

To eliminate any possibility that air currents caused these movements, Crookes had resolved to study the effects in a vacuum 'so nearly perfect that it would not carry a current from a Ruhmkorff's coil'.[25] To accomplish this, with the earlier Sprengel arrangement he incorporated a chemical method for exhaustion used previously by Thomas Andrews in 1852[26] and several years later by Gassiot in his experiments on gaseous discharge.[27] Figure 2 illustrates the apparatus he attached to the Sprengel pump through the mercury joint *e*. Carbon dioxide could flow from below through *h* into tube *d* through a stopper fitting into a funnel joint. The tube *d* contained concentrated sulphuric acid. Tube *c* enclosed a copper boat filled with sticks of caustic

[15] Michael Faraday, *Experimental Researches in Electricity* (London, 1839, reprinted in New York, 1965), I, paragraphs 1544–60.

[16] W. R. Grove 'On the Electro-chemical Polarity of Gases', *Phil. Trans.*, 142 (1852), 100. The striations appear at pressures in the range of 10^{-4} atm or 0·1 mm Hg.

[17] John Peter Gassiot, 'On the Stratification in Electrical Discharges Observed in Torricellian and Other Vacua—Second Communication', *Phil. Trans.*, 149 (1859), 137–60.

[18] Stokes to Julius Plücker, 27 May 1864 (Royal Society archive).

[19] Faraday (footnote 15), paragraph 1615.

[20] Crookes (footnote 3), 509–10.

[21] Crookes to Gimingham, 9 November 1872 (SML archive).

[22] Crookes to Gimingham, 3 July 1873 (SML archive).

[23] Crookes (footnote 3), 511.

[24] *Ibid.*, 523.

[25] *Ibid.*, 515.

[26] Thomas Andrews, 'On a Method of Obtaining a Perfect Vacuum in the Receiver of an Air Pump', *Philosophical Magazine*, 4th Series, 3 (1852), 104–8.

[27] Gassiot (footnote 17).

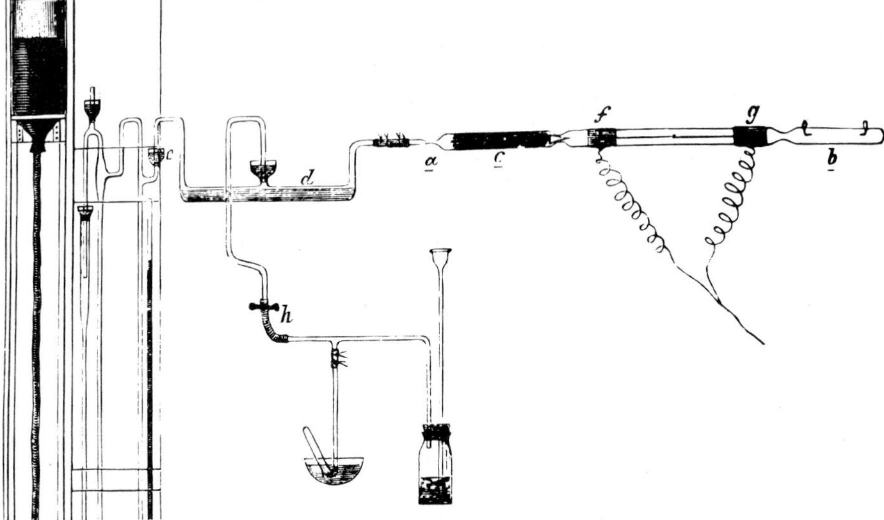

Figure 2.

potash. To the right of *c* was the tube containing pith disks, balanced on an arm, whose movement under stimulation by heat would measure the effect under study; *fg* were collars of silver foil encircling the tube that were to carry off any electricity induced in the tube by the heat source. Into tube *b* passed two platinum wires connected to an induction coil. After preliminary evacuation with the Sprengel pump, Crookes displaced the residual air with CO_2 and continued evacuating. Thereafter, he heated the caustic potash to fusion, sealed off the apparatus at *a* and allowed time for the potash to absorb the residual CO_2. By the summer of 1873, he had indeed attained a vacuum impeding the passage of a current from an induction coil. In this vacuum the pith consistently retreated from candle flame and sunlight, so his belief that residual gas molecules played no role in the phenomenon seemed justified.[28]

Crookes's rendition of this work, submitted for publication to the *Philosophical Transactions*, won approval from referees Maxwell and William Thomson in early 1874. Maxwell praised the author's 'great discovery with respect to the mode in which a body placed in air of different densities is acted on when bodies of different temperatures are placed near it'. Maxwell suggested that the repulsion was due to radiation—stressing that if so the effect depended not on temperature differences but 'on the excess of the sum of the radiations exchanged between the hot and the movable body over the sum of the radiations between the movable body and that which lies on the opposite side of it.' Maxwell was puzzled that this hypothesized repulsive force of radiation had greater vigour than he had predicted for it in his earlier *Treatise on Electricity*.[29] Thomson commended 'clear and thorough' descriptions of experiments from which Crookes 'seems to me to have made a discovery of transcendent interest and first rate importance'. But Thomson terminated his report with the challenge that 'Crookes' conclusion might be confirmed and the argument strengthened by an estimate of the absolute amount of the fluid pressure which could possibly exist.'[30]

[28] Crookes (footnote 3), 517–18.
[29] Maxwell to Stokes, 24 February 1874 (Royal Society archive).
[30] Thomson to Stokes, 18 February 1874 (Royal Society archive).

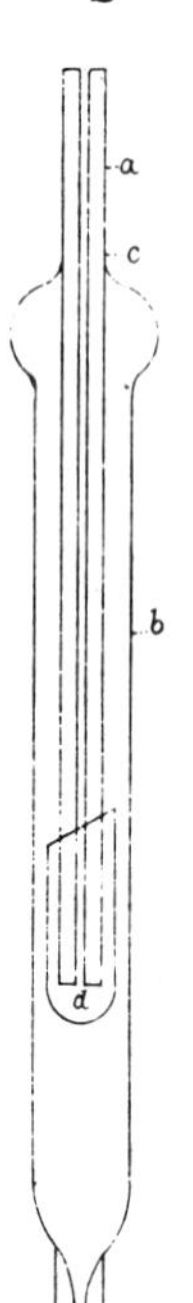

Figure 3.

Just how much gas remained in Crookes's Sprengel vacuum seemed an even more critical issue later that spring, when Osborne Reynolds attempted to explain the repulsive effect by differences in the temperature of the bodies moving in vacuo. Reynolds denied that radiation directly induced the movements. He invoked the kinetic theory of gases to argue that the temperature of a surface determined the velocity at which residual gas molecules rebounded from the surface. Therefore, differentials in surface temperature—in rebound velocities of molecules—caused the movements Crookes had observed.[31] On 18 June 1874, Crookes reiterated his belief 'that the repulsion accompanying radiation is directly due to the impact of the waves upon the surface of the moving mass, and not secondarily through the intervention of air-currents, electricity, or evaporation and condensation'. He emphasized in this response to Reynolds that the repulsive effect intensified 'as the vacuum approaches perfection, and attains its maximum when there is no air or vapour whatever present, or at all events not sufficient to permit the passage of an induction spark'.[32]

Indeed, several improvements in his vacuum apparatus had appeared by June, 1874. Crookes and Gimingham substituted glass spiral tubing for the flexible rubber joint to the receiver. In addition, Crookes had devised an air trap, shown in Figure 3, that freed mercury of air before the mercury entered the fall-tube: he

[31] Osborne Reynolds, 'On the Forces caused by Evaporation from, and Condensation at, a Surface', *Proceedings of the Royal Society*, 22 (1874), 406–7. For a rendition of Reynold's conceptual development and interaction with Maxwell in the effort to explain the radiometer effect, see S. G. Brush and C. W. F. Everitt, 'Maxwell, Osborne Reynolds, and the Radiometer', *Historical Studies in the Physical Sciences*, 1 (1969), 105–25. See also C. W. Draper, 'The Crookes Radiometer Revisited', *Journal of Chemical Education*, 53 (1976), 356–7.

[32] W. Crookes, 'Remarks on a Paper by Professor Osborne Reynolds 'On the Forces Caused by Evaporation from and Condensation at a Surface', '*Chemical News*, 30 (1874), 24–5.

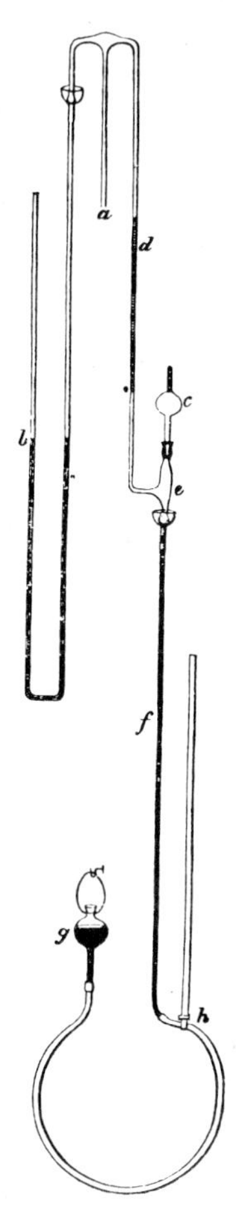

Figure 4.

blew tube *a* into *b* at point *c* such that *a* passed some distance inside, and arranged a small glass cap *d* to cover the end. Mercury flowed into *b*, and any imbibed air collected around the joint *c*, while the flow of mercury continued into cap *d* and up tube *a* toward the fall-tube.[33] Crookes still enhanced the pump action with chemical agents that absorbed residual gases and vapours. By August, following a suggestion by James Dewar, he employed charcoal for this 'getting' action as well.[34]

To indicate levels of exhaustion at pressures below 0·1 mm Hg, Crookes depended heavily upon electrical induction. He routinely introduced into a mercury joint a small induction tube with a gap of about $\frac{1}{4}$ inch. By noting how powerful an induction was necessary to force passage across the gap, he had a quantitative means to distinguish among degrees of rarefaction. He had reached exhaustions where an induction spark would not strike even across the tiny $\frac{1}{4}$ inch gap.[35] But this electrical procedure, of course, afforded no way to estimate the actual pressure within the pump and receiver.

[33] W. Crookes, 'On Attraction and Repulsion Accompanying Radiation', *Philosophical Magazine*, 4th Series, 48 (1874), 81–95. Lecture to Physical Society, 20 June 1874.

[34] Crookes to Dewar, 7 August 1874 (Royal Institution archive). P. G. Tait and J. Dewar, 'On a New Method of Obtaining very Perfect Vacua', *Proceedings of the Royal Society of Edinburgh*, 8 (1872–5), 348–9.

[35] W. Crookes, 'On Repulsion resulting from Radiation—Part II', *Phil. Trans.*, 165 (1875), 520.

On 13 June 1874, the Physical Society first learned of a new gauge that McCleod had designed for estimating extremely low pressures.[36] This soon benefited Crookes and others working in high vacua; it remains useful today.[37] The method involved condensing a known volume of the residual gas into a much smaller volume and measuring its pressure under the latter condition. As Figure 4 illustrates, the gauge connects to the Sprengel pump and receiver through *a*. A stopcock *h* permits mercury from a reservoir *g* to flow up tube *f* and into *e*—at which it cuts off communication between the gas in globe *c* and the gas in the rest of the apparatus. Ultimately, the mercury condenses all of the gas in globe *c* into a narrow 'volume tube' appended to *c* above. One ascertains the pressure in the 'volume tube' by measuring the difference between the mercury level in tube *d* and that in the 'volume tube'. Allied with knowledge of the ratio of gas volume in the 'volume tube' to the volume from the level of cutoff in *e* to the top of the 'volume tube', one can then determine the original pressure of the gas using Boyle's Law.

Aided by his modified Sprengel pump and still persuaded that repulsive effects in vacuo derived from a direct mechanical efficacy of incident radiation, Crookes devoted the remainder of 1874 and the first few months of 1875 to experiments that yielded no startling new information.[38] But in April, 1875, he exploited a discovery that light repels a black pith surface more vigorously than a white pith surface to construct the first radiometer. The instrument consisted of four arms suspended on a steel point that, in turn, rested on a cup such that the arms could revolve horizontally. At the ends of each arm Crookes attached thin pith disks, each lampblacked on one side with the black surfaces facing the same way. Light impinging in a Sprengel vacuum induced continuous rotation consistent with a retreat of the black surfaces from the source of light. The speed of revolution appeared to be proportional to the intensity of the incident light.[39]

Once again, Thomson and Maxwell refereed Crookes's paper submitted for publication in the *Philosophical Transactions*. Thomson briefly remarked as to the 'acuteness, intelligence, and inventiveness' with which Crookes had verified and illustrated his 'very neat discovery'.[40] In his more lengthy report, Maxwell rejected his own previous inclination to explain the repulsive effects by a direct action of incident radiation on the mass(es). He now accorded more importance to the temperatures of the pith surfaces.

> if the cause of the repulsion were radiation alone, there would be greater repulsion of the white surface where there is a reflected radiation than of the black where there is none. But if the result depends on the heating of the surface, we know that the black surface is heated by luminous rays more than the white one.[41]

Indeed, if the temperatures of the surfaces, rather than the 'impact' of radiation, were crucial to explaining the movements, the way opened for enhancing the importance of mechanical effects of residual gas molecules in tune with kinetic theory. Crookes was

[36] H. McLeod, 'Apparatus for Measurement of Low Pressures of Gas', *Proceedings of the Physical Society*, 1 (1874–5), 30–4.

[37] Saul Dushman, *Scientific Foundations of Vacuum Technique*, 2nd edition (New York, 1962), pp. 225–34.

[38] Crookes (footnote 35), 519–47.

[39] Crookes, 'On Attraction and Repulsion Resulting from Radiation', *Proceedings of the Royal Society*, 23 (1875), 373–8.

[40] Thomson to Stokes, 16 June 1875 (Royal Society archive).

[41] Maxwell to Stokes, ? June 1875 (Royal Society archive).

not disposed at this time either by his theoretical inclination or his faith in the capability of his improved Sprengel instrumentation[42] to take seriously the role of gas molecules. Maxwell did not yet press kinetic theory, but Reynolds continued to advocate it in the very face of Crookes's experimentally derived objections. Reynolds believed that

> it follows as a direct result of the kinetic theory of gas, that ... forces ... surrounding the surface ... will not be diminished by diminishing the tension of the gas; and consequently no amount of pumping would destroy such forces.... Whereas the smaller the tension of the gas the freer the surface will be to move, and the less its motion would be opposed by convection currents; hence ... the only effect of improving the vacuum would be to intensify the action.[43]

Despite Crookes's bravado, Reynolds' public challenge motivated him to seek even more extreme exhaustions than he had already attained. Moreover, Maxwell had cited in his report experiments of Kundt and Warburg (to which we must return) that showed 'properties of the most perfect vacuum are very different from those of a good Sprengel vacuum'. He alluded specifically to their discovery that the rate of cooling of a thermometer bulb surrounded by a thin vacuous shell is at least twice as slow for the best vacuum as for an ordinary vacuum.[44] Perhaps Crookes's desire to plunge still deeper into studies of effects in vacuo derived nourishment from Maxwell's remark that

> This fact, combined with the non-discharge of the electric spark and Mr. Crookes' own results show that the research into the properties of a real vacuum is likely to yield valuable results.[45]

Throughout the latter half of 1875, Crookes conspicuously avoided explanations for radiometric activity that invoked the kinetic theory of gases. In June, he had marked out a strategy of distinguishing between the effects of light and heat in the hope of showing that the radiometer rotation was not due to heat accompanying the light.[46] This originated from his discovery that 'dark heat'—from a bulb of warm water or warm piece of glass or metal—repelled the black and white surfaces equally, while of course radiation from the sun or a candle repelled the black surface more vigorously. By the end of the year, however, Crookes abandoned this approach. Instead he attributed the motion to the greater radiation *from* the black surface owing to its higher temperature:

> A ray of light falls on a white surface and is reflected back again; it does no work there; but if the ray falls on a black surface it is absorbed and quenched ... the ray becomes converted into thermometric heat, and ... its energy is in whole or in great part used up in raising the temperature of the dark body. But having thus become warm, the powerful radiating action of the surface for heat comes into

[42] W. Crookes, 'Attraction and Repulsion Caused by Radiation', *Nature*, 12 (1875), 125. Crookes referred to 'my experiments ... where there is no amount of gas known to be present; for example, in a chemical vacuum'.

[43] O. Reynolds, 'The Attraction and Repulsion caused by the Radiation of Heat', *Nature*, 12 (1875), 6.

[44] A. Kundt and E. Warburg, 'Über Reibung un Wärmeleitung verdünnter Gase', *Monatsberichte der Königlich Preussischen Akademie der Wissenschaften zu Berlin* (1875), pp. 160–173. Translated in *Philosophical Magazine*, 4th series, 50 (1875), 53–62.

[45] Maxwell (footnote 41).

[46] Crookes to Gimingham, 15 June 1875 (SML archive).

> play, and heat ... is quickly radiated back again. It would appear as if this radiation of heat from the surface of a body caused the latter to retreat backwards, and so produced the motion.[47]

Maxwell made short shrift of this in yet another referee's report:

> if the same radiation falls on a white and a black surface, the total radiation from the white surface will be greater than that from the black, because some of the radiation is reflected directly by the white surface, whereas what falls on the black is absorbed, and the radiation of the black is only that due to its own temperature.... It is not, therefore, because the black surface radiates more, but because it is *hotter*, that it is more repelled.[48]

2. The crucial year of 1876

Instrumentally, experimentally, and theoretically the year of 1876 marked a watershed in Crookes's high-vacua investigations. In February, Arthur Schuster conducted the well-known experiment in which he suspended a radiometer by two cocoon fibres and noted a rotation of the case opposite to that of the vanes; this, by Newton's third law, indicated that the residual gas directly caused the vane motion.[49] Crookes responded publicly at first with caution. At his request, on 28 March, Gimingham had floated a radiometer case on water, prevented rotation of the arms by an external magnet and observed rotation of the case opposite to that of the vanes when the latter moved free of magnetic constraint.[50] Before the Royal Society on 30 March, Crookes conceded that this experiment implied 'considerable internal friction', but professed his 'wish to keep free from theories'.[51] Privately, however, he remained sceptical about the efficacy of residual gas molecules. He requested Gimingham to seek out a 'molecular wind' in the radiometer by suspending an index of rice paper beside the rotating vanes and observing whether the index moved.[52] Gimingham at first failed to observe any motion of the index,[53] but by 4 April was recording a 'considerable movement' of the suspended body at a specific rotational velocity of the radiometer arms.[54]

A new dimension of high-vacua experimentation opened up for Crookes in late March. Stokes suggested that he determine the viscosities of gases at extreme exhaustions, and the two men commenced what was to be a long correspondence on the subject. No doubt Stokes had been inspired to this enquiry by Maxwell's reference to recent experiments of Kundt and Warburg that showed viscosity in a 'good vacuum' nearly equal to that at ordinary pressure.[55] Maxwell himself had actually furnished the theory and technique underlying the Kundt-Warburg determinations in his 1866

[47] W. Crookes, 'On Repulsion Resulting from Radiation—Part IV', *Phil. Trans.*, 166 (1876), 362–3.

[48] Maxwell to Stokes, 10 February 1876 (Royal Society archive). Published in *Memoirs and Scientific Correspondence of the Late Sir George Gabriel Stokes*, edited by Joseph Larmor (Cambridge, 1907), I, 435.

[49] Arthur Schuster, 'On the Nature of the Force Producing the Motion of a Body Exposed to Rays of Heat and Light', *Phil. Trans.*, 166 (1876), 715–24.

[50] Crookes, Laboratory Notebook, Vol. IV, 28 March 1876 (SML archive).

[51] W. Crookes, 'On the Movement of the Glass Case of a Radiometer', *Chemical News*, 32 (1876), 163.

[52] Crookes to Gimingham, 26 March 1876 (SML archive).

[53] Crookes to Stokes, 31 March 1876 (Cambridge University archive).

[54] Crookes, Laboratory Notebook, Vol. IV, 4 April 1876 (SML archive).

[55] Maxwell (footnote 41). To Maxwell, the Kundt–Warburg result implied that 'any currents formed in a rare gas will subside very rapidly, and any phenomenon which would be disturbed by such currents [i.e., the radiometer effect] will take place more regularly'.

article on viscosity of gases. There he had reported experiments dealing with torsional vibrations of disks placed between two parallel fixed disks at small, but measurable, distances. He found that the logarithmic decrement of swing (the decrement per swing of the logarithm of arc amplitude) remained essentially constant over a sixty-fold variation in pressure at constant temperature. This was consistent with his theoretically-derived conclusion that viscosity of a gas is independent of pressure.[56] Openly acknowledging that their experiments on viscosity followed 'Maxwell's method', Kundt and Warburg confirmed Maxwell's earlier findings by observing that the logarithmic decrement of torsional vibrations displayed by a disk between two fixed disks close to it stayed constant over a large pressure range down to about 20 mm Hg. But in 1875 Kundt and Warburg reported determinations at considerably lower pressures than Maxwell's nine years earlier. They employed a Geissler pump and constructed an apparatus with glass junctions throughout. Below 20 mm Hg the logarithmic decrement decreased faster and, at the lowest pressure reported (1 mm), sank to about 2/3 of its value at atmospheric pressure. They noted that their attempts to exhaust to yet lower pressures were frustrated by the escape of water vapour from solid parts of the instrumentation.[57] Crookes now joined in this experimental activity.

Exploiting a technique suggested by Stokes,[58] Crookes suspended a thin plate of pith at the end of a fine glass fibre in an evacuated bulb. He contrived to rotate the bulb through a small angle while holding fixed the stopper that supported the glass fibre. This induced an oscillation of the pith plate in arcs whose plane was perpendicular to the suspending fibre. As Maxwell's theory and earlier experiments had indicated, the logarithmic decrement of the oscillation diminished only slightly as the pressure in Crookes's apparatus decreased from atmospheric to a value at which the gauge barometer stood at the same height as a nearby Torricellian barometer. By mid-April, as Crookes exhausted still further with a Sprengel pump, the logarithmic decrement had not decreased significantly.

The letters to Stokes of mid-April reveal how stubbornly Crookes continued to resist the implications of Schuster's experiment. He had attached to the Sprengel pump and the viscosity apparatus a small radiometer whose revolutions in a given time by a standard candle he counted. On 15 April, he expressed

> surprise to find such an amount of viscosity in a vacuum.... I have far exceeded any of my previous attempts to get a vacuum (except the chemical method) and still there is a considerable drag between the bulb and the plate. Were it not that the repulsion from radiation is as strong as ever I should be inclined to think this action was due to the residual air. Is it possible that the viscosity may be due to the exchange of heat rays which is always going on between the plate and the bulb?[59]

Four days later, he wrote that

> I am scarcely yet satisfied that the residual gas is the cause [of the radiometer movement].... I have always found that the better the vacuum the better the [radiometer] motion.[60]

[56] J. C. Maxwell, 'On the Viscosity or Internal Friction of Air and Other Gases', *Phil. Trans.*, 156 (1866), 249–68.

[57] Kundt and Warburg (footnote 44).

[58] Crookes, Laboratory Notebook, Vol. IV, 8 April 1876 (SML archive).

[59] Crookes to Stokes, 15 April 1876 (Cambridge University archive).

[60] Crookes to Stokes, 19 April 1876 (Cambridge University archive).

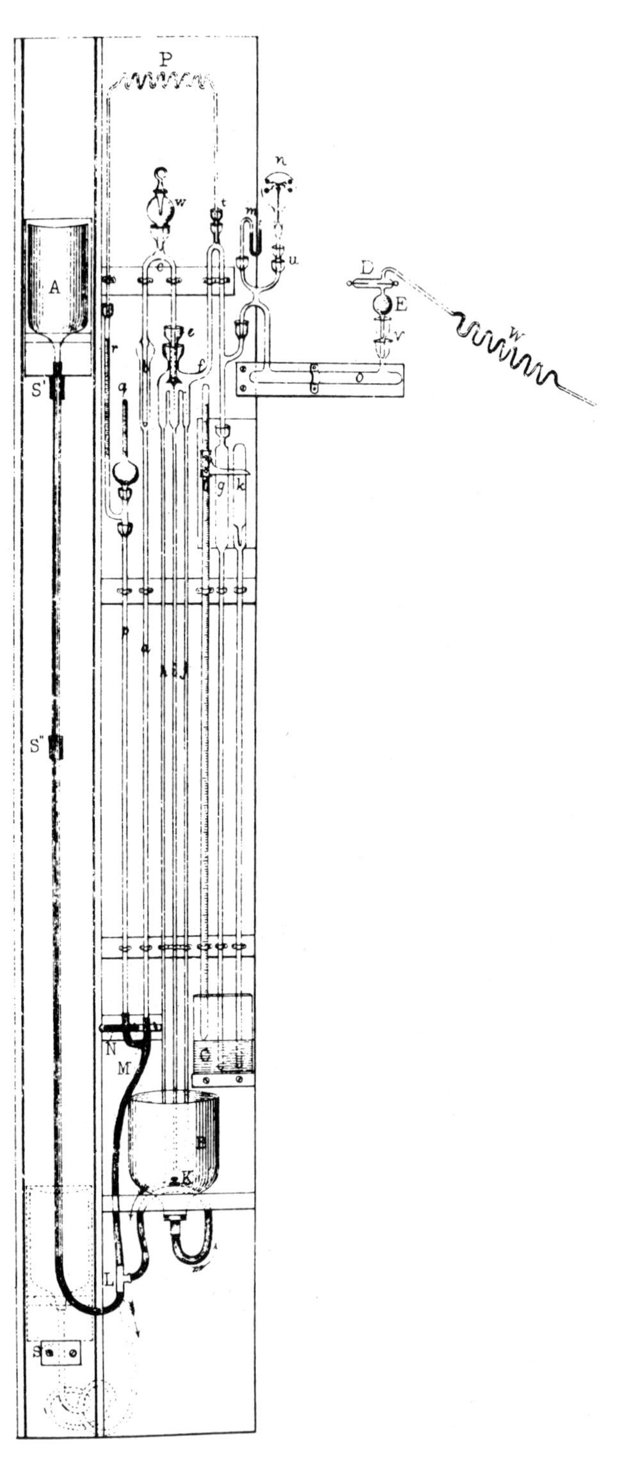

Figure 5.

By 21 April, however, Crookes had achieved exhaustions with continuous operation of his Sprengel pump in which the logarithmic decrement declined precipitously (the amplitude of swing diminished less rapidly).[61] Over the next several weeks he attained vacua in which observations of changes in radiometric activity and in viscosity convinced him once and for all that radiometric activity did indeed stem from the mechanical effects of gas molecules. By mid-June, Crookes was noting a *decrease* in the force of radiometric repulsion with increasing exhaustion after the force reached a maximum at a rarefaction hitherto unattained. His measure of radiometric force was the number of vane rotations caused by a standard candle at a given distance in a given time. This eliminated his private suspicion of the residual gas theory, because his previous assumption that 'the better the vacuum the better the [radiometer] motion' no longer held. Meanwhile, the logarithmic decrement in the viscosity determinations at his lowest pressure had sunk to about twenty times smaller than the original value. A logarithmic decrement, though smaller, indicated the presence of residual gas molecules—at exhaustions now well beyond that at which Crookes experienced the failure of an induction coil to occasion discharge. The force of radiometric action reached its maximum at a pressure lower than that at which the logarithmic decrement in the viscosity determinations commenced to fall off precipitously: the latter already had decreased to $\frac{1}{4}$ of its original value when the radiometric force maximized.[62]

These breakthroughs in experimentation at high exhaustions resulted directly from improvements in the Crookes–Gimingham vacuum apparatus. Figure 5 illustrates Gimingham's embellished Sprengel arrangement,[63] which they employed at least as early as mid-April.[64] As they pushed for yet more extreme rarefactions the frustrations of leakage were inevitable. Crookes complained on 12 June that the vacuum apparatus was 'now so large and complicated that it is almost impossible to localize a leakage or even to discover whether there is one or not'.[65] Note that Gimingham now operated simultaneously three 'fall-tubes' (*h*, *i*, *j*) down which mercury dropped after undergoing division into three columns at jet *e*. Observe as well the air trap *b* toward the top of tube *a* and the McLeod gauge (*p q r*), connected to the exhaustion arm *f* through spiral tubing *P*. At *n* was a small radiometer; *D* was a discharge tube. Bulb *E* contained a gold leaf that stopped mercury vapour from entering the receiver through spiral tubing *W*.[66] The wide tube *o* held sulphuric or anhydrous phosphoric acid to absorb water vapour.

It merits emphasis that each aspect of Crookes's vacuum investigations had two dimensions. On the one hand, radiometric, viscous and electrical effects were of interest in themselves because of what they might reveal about kinetic theory, electricity, perhaps matter and radiation. But they also served Crookes as means by which he could determine the progress of exhaustion and perhaps yield through

[61] Crookes, Laboratory Notebook, Vol. IV, 21 April 1876 (SML archive).

[62] W. Crookes, 'On Repulsion Resulting from Radiation. Influence of the Residual Gas', *Chemical News*, 34 (1876), 23–4. Read before the Royal Society, 15 June 1876.

[63] Charles H. Gimingham, 'On a New Form of the 'Sprengel' Air-pump and Vacuum-tap', *Proceedings of the Royal Society*, 25 (1876–7), 396–402.

[64] Crookes, Laboratory Notebook, Vol. IV, 11 April 1876 (SML archive).

[65] *Ibid.*, 12 June 1876.

[66] Earlier, Sprengel had striven to keep the receiver of his apparatus free of mercury vapour by filling the tube leading to the receiver with finely divided gold or charcoal to absorb mercury and by exposing that tube to a cold of $-10°C$ to condense mercury vapour. Sprengel, (footnote 4), 19. Later, Crookes found that an interposed warmed tube containing powdered sulphur absorbed mercury vapour more effectively than did gold foil. Crookes, 'On the Viscosity of Gases at High Exhaustion', *Phil. Trans.*, 172 (1881), 392–3.

theoretical connections the exact pressures at various levels of extreme rarefactions. True, he did utilize a McLeod gauge, reporting pressures measured by it regularly by midsummer 1876. But those using the McLeod gauge were haunted by the fact that the mercury within it exerted a vapour pressure, and the pressure of mercury—known today to be about 10^{-6} atm at room temperature—was an object of controversy.[67] We now realize as well that uncertain measurement of differences in mercury heights due to capillary depression in the McLeod gauge probably led to errors in pressure determination exceeding + or − 100%.[68] Crookes continued to solicit the aid of both electrical effects and speeds of radiometric rotations[69] to monitor the progress of exhaustion, as the discharge tube and radiometer attached to the Sprengel arrangement in Figure 5 indicate.

Moreover, his correspondence leaves little doubt that Crookes sought in the viscosity determinations an alternative method to determine the pressures produced by his vacuum apparatus. 'Is there any means,' he inquired of Stokes, 'of ascertaining from these [logarithmic decrement] curves how near I am to 'absolute'?' In the same letter of 19 April he asked

> At the same pressure would not the logarithmic decrement of the swing vary in different gases? Could not this be calculated before hand? If so I can ... get the pressure accurately by the swing.[70]

Later in July, he wrote to Stokes that a plot of the logarithmic decrements against the number of bottles of mercury run through the Sprengel pump yielded a straight line. The first bottle of mercury removed $\frac{3}{4}$ of the contained air. Would it be safe to assume that each successive bottle run through left $\frac{1}{4}$ of the gas present at the beginning of the cycle? If so, the fraction of gas remaining would be $1/N^4$, where N is the number of bottles of mercury run through. 'In default of a more accurate [estimate of the pressure] ... I want you to tell me if ... this rule will hold good beyond the 4th or 5th bottle?' His letter of 11 July terminated with

> If I may assume this, the 20th bottle will give me an exhaustion of 1/1,099,511,627,776th of an atmosphere, equal to 1/835,628,837,109,760th of a millimeter of mercury, and it is now conceivable that I have got a limited number of molecules in the apparatus. At this exhaustion the repulsion by radiation is at its maximum, and although I have got to the limit of the power of my present pump I can go beyond that by chemical means of absorbing the gas.[71]

Stokes, however, failed to concur with Crooke's reasoning.[72]

The McLeod gauge measured pressures as low as 0·01 mm Hg (about $1{\cdot}3\times10^{-5}$ atm) by early August in Gimingham's vacuum apparatus illustrated in Figure 5.[73] Crookes now focused more directly on chemical methods of enhancing the

[67] 'Discussion' at Society of Arts following paper read by Sylvanus Thompson, (footnote 5), 47.

[68] A. W. Porter, 'Capillary Ascent or Depression of Liquids in Cylindrical Tubes', *Transactions of the Faraday Society*, 29 (1933), 702–7.

[69] The first successful use of a quantitative measure of radiometric force as a determination of rarefaction came with Knudsen's work around 1910. Martin Knudsen, 'Thermischer Molekulardruck der Gase in Röhren und porösen Körpern', *Annalen der Physik*, 31, Ser. 4 (1910), 633–40.

[70] Crookes (footnote 60).

[71] Crookes to Stokes, 11 July 1876 (Cambridge University archive).

[72] Crookes to Stokes, 14 July 1876 (Cambridge University archive).

[73] Crookes, Laboratory Notebook, Vol. IV, 10 August 1876 (SML archive).

exhaustion process. He discarded sulphuric acid in favor of phosphoric acid in wide tube *o*, because he suspected that sulphuric acid exerted a vapor pressure too large for the extremely low pressures that they were now achieving.[74] The work proceeded well, and an optimistic Crookes wrote to Gimingham on 13 August that 'Now I think is the time to try and get a *perfect* vacuum with a bit of KHO in a tube between the apparatus and the P_2O_5 tube'.[75] His idea was to put in oil of vitriol and exhaust alternately, thus replacing all air with SO_3 vapor that the KOH then could absorb. By mid-November, he had attained a vacuum which the McLeod gauge measured as 10^{-7} atm.[76]

By mid-November as well, he had conceived the new interpretation of a fourth state of matter at extreme rarefactions that would guide his subsequent experimentation in high vacua.[77] Discussions about the character of the Crookes fourth state, the experimentation it spawned, its service to Crookes as a concept unifying various high-vacua phenomena and its weaknesses have appeared elsewhere.[78] In the context of this essay, an additional significance of the fourth state was the shift in emphasis it induced in Crookes's high-vacua experimentation. Heretofore, a major interest had been the quest for instrumental achievements of more and more extreme levels of exhaustion. Electrical, radiometric and viscous effects had served as means to indicate the extent of this accomplishment. Hereafter, the effects assumed an increasing significance in their own right as objects of study and—most importantly—served the greater end of aiding Crookes's elucidation and defense of the fourth state. Not until June 1876 had he commenced a systematic investigation of electrical effects in vacuo. At this time he began to concentrate for a brief period upon electrically-induced luminous effects at various positions relative to the poles, while correlating them to viscosity changes as measured by alterations in logarithmic decrement.[79] (Incidentally, he showed no hint of being aware that he was treading on ground already cultivated by Hittorf, Plücker and Varley.) Intensive study of electrical effects did not proceed again until September 1877,[80] motivated then by his desire to identify the newly-discovered 'dark space' about the cathode resulting from electrical induction with the thickness of his hypothesized 'molecular pressure' thought to extend from the activated surface of a radiometer vane.

This molecular pressure at extreme exhaustion, streaming off perpendicularly from energized cathodic and radiometric surfaces, was the key manifestation of his fourth state of matter. The fourth state had emerged in the autumn of 1876 in response to the events forcing upon him a molecular explanation of the radiometer effect. From October to December he studied radiometer phenomena, in particular the consequences of altering the shape of radiometer vanes.[81] The idea of the fourth state arose directly from his recognition of different responses to luminosity exhibited by concave and convex vanes.[82] The fourth state subsequently dominated his experimental design

[74] Crookes (footnote 72).

[75] Crookes to Gimingham, 13 August 1876 (SML archive).

[76] W. Crookes, 'Experimental Contributions to the Theory of the Radiometer', *Chemical News*, 34 (1876), 275. Reported to the Royal Society on 16 November.

[77] *Ibid.*, 276–7.

[78] DeKosky (footnote 1).

[79] Crookes, Laboratory Notebook, Vol. IV, 28 June–1 August 1876 (SML archive).

[80] *Ibid.*, 28 September 1877.

[81] *Ibid.*, 2 October–4 December 1876.

[82] Crookes (footnote 76), 278.

and interpretation of high-vacua electrical and radiometric investigations—before[83] and after the fruitful interlude of the 1880s, when his attention turned to the spectroscopic effects produced by cathode rays upon rare earths.[84] The theme of the fourth state governed Crookes's attempt in 1891 to affiliate his theory of the chemical elements with electrical and radiometric effects in high vacua;[85] it dictated the course of several experimental investigations of cathode-ray phenomena in the 1890s.[86] Throughout what became an immensely productive period of experimental studies on electrical effects in high vacua, the fourth state ossified Crookes's perspective. Its influence perhaps explains in part why, despite his experimental skill and experience, he missed two prize discoveries that emerged from that field of research—X-rays and the electron. In this vein as well, Maxwell's premature death at the age of 48 in 1879 may have been particularly unfortunate for Crookes, tragic as it was for physics on several levels. With his referee's reports, Maxwell had begun to influence the course of Crookes's experimentation on cathode rays, showing particular interest in possible incompatibilities between electromagnetic theory and the notion that cathode rays were electrically-charged molecules.[87] Through Stokes, Crookes expressed a wish to meet Maxwell in April 1879, and acknowledged the value of Maxwell's comments on his work.[88] Maxwell died in November 1879. Crookes's major scientific correspondent in ensuing years, Stokes, was not an active participant in the growth of electromagnetic theory and rarely displayed interest in that area of physics.

3. Conclusion: post-1876

Not that the transition of 1876 was absolute. Prior to the period demanding a molecular interpretation of the radiometer effect, Crookes had demonstrated interest in the effect itself and its cause. In the period after 1876, he and Gimingham continued their efforts to achieve a vacuum approaching perfection. But their optimism in the autumn of 1876 proved premature in this regard. Vacuum technology was not to ascend much above the plateau of the 1870s for several decades. True, Gimingham introduced further alterations: he increased the number of fall-tubes from three to five by February 1877, and demonstrated his incredible glassblowing talent with the complete elimination of mercury joints by fusing all connections directly.[89] Later, he devised an apparatus with seven fall tubes.[90] But the maximum exhaustion Gimingham ever claimed to reach was 10^{-8} atm. Ogden Rood, Professor of Physics at Columbia University, proclaimed in 1881 an enhancement of evacuation to the range of 10^{-9} atm with an improved Sprengel arrangement in which he could heat all parts of

[83] W. Crookes, 'On Repulsion Resulting from Radiation—Part V', *Phil. Trans.*, 169 (1878), 243–318. 'On Repulsion Resulting from Radiation—Part VI', *Phil. Trans.*, 170 (1879), 87–134. 'On the Illumination of Lines of Molecular Pressure, and the Trajectory of Molecules', *Phil. Trans.*, 170 (1879), 135–64.

[84] W. H. Brock, 'William Crookes', *Dictionary of Scientific Biography*, III (New York, 1971), 478–9. Robert K. DeKosky, 'Spectroscopy and the Elements in the Late Nineteenth Century: The Work of Sir William Crookes', *British Journal for the History of Science*, 6 (1973), 400–23.

[85] W. Crookes, 'Electricity in Transitu: From Plenum to Vacuum', *Chemical News*, 63 (1891), 53–6, 68–70, 77–80, 89–93, 98–100, 112–14.

[86] Crookes, Laboratory Notebook, Vol. XII, October–November, 1890. Laboratory Notebook, Vol. XV, July, 1896 (Royal Institution archive).

[87] Maxwell to Stokes, 31 December 1878 (Royal Society archive). Discussed in DeKosky (footnote 1), 59.

[88] Crookes to Stokes, 26 April 1879 (Cambridge University archive).

[89] Crookes, Laboratory Notebook, Vol. IV, 8 February 1877 (SML archive).

[90] C. Gimingham, 'Contributions to the Development of the Sprengel Air Pump', *Journal of the Society of Chemical Industry*, 3 (1884), 83–9.

the pump with a Bunsen burner,[91] and thus remove films of air otherwise stubbornly remaining between the surfaces of mercury and glass. Yet even at these exhaustions, as Crookes later noted, billions of molecules existed in the receiver. By the mid-1880s, he had despaired of the possibility that any available means could produce a perfect vacuum.[92]

Indeed, well-founded doubts arose as to the veracity of late nineteenth-century estimates in the range of 10^{-8} and 10^{-9} atm because of uncertain influence of mercurial vapour pressure in the McLeod gauge. Yet Crookes and others certainly obtained pressures of at least 10^{-7} atm and perhaps lower in their receivers. Though their techniques were tedious and delicate, they permitted the discovery and investigation of cathode rays. This stage of vacuum technology enabled J. J. Thomson, using an improved modification of a Töpler pump in the 1890s, to deflect cathode rays with an electrical field—a critical step leading to the identification of the electron. The innovations of the 1860s and 1870s dominated vacuum technique until Gaede invented his rotary pump in 1905.[93] A decade later, Gaede's and Langmuir's vapour pumps decreased pressures by several orders of magnitude beyond the capabilities of the late nineteenth-century instrumentation.[94]

Acknowledgements

I am pleased to acknowledge the aid of the National Science Foundation and the graduate Research Fund of the University of Kansas in the preparation of this essay. I thank as well the staffs of the Royal Society, the Royal Institution, Cambridge University Library and the Science Museum, London, for their cooperation.

[91] O. Rood, 'On an Improvement in the Sprengel Pump', *American Journal of Science*, 20 (1880), 57–8. 'On a Method of Obtaining and Measuring very High Vacua with a Modified Form of Sprengel Pump', ibid., 22 (1881), 90–102.

[92] Footnote 67, 46–9.

[93] W. Gaede, 'Demonstration einer rotierenden Quecksilberluftpumpe', *Physikalische Zeitschrift*, 6 (1905), 758–60.

[94] W. Gaede, 'Die Diffusion der Gase durch Quecksilberdampf bei niederen Drucken und die Diffusionsluftpumpe', *Annalen der Physik*, 4th series, 46 (1915), 357–92. Irving Langmuir, 'The Condensation Pump: An Improved Form of High Vacuum Pump', *Journal of the Franklin Institute*, 182 (1916), 719–43.

XVI. *Apparatus for Measurement of Low Pressures of Gas.* By Professor M'LEOD, *Indian Civil-Engineering College, Cooper's Hill**.

THIS apparatus was devised for estimating the pressure of a gas when its tension is so low that the indications of the barometer cannot safely be relied on, unless indeed a very wide barometer and an accurate cathetometer be employed. The method consists in condensing a known volume of the gas into a smaller space and measuring its tension under the new conditions.

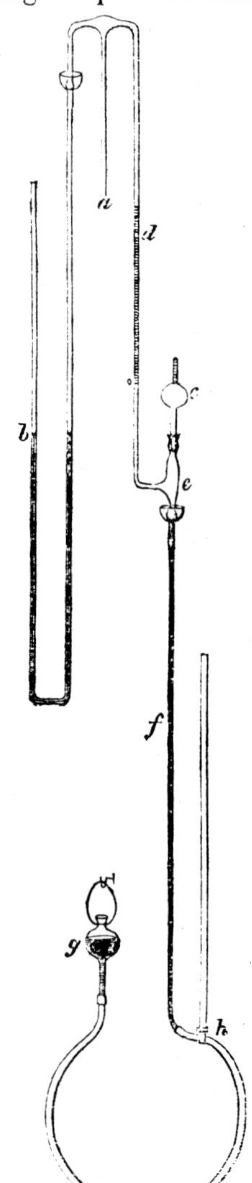

The form of the apparatus is the following:—The tube *a* communicates with the Sprengel, and with the apparatus to be exhausted; *b* is a siphon-barometer with a tube about 5 millimetres in diameter; and the principal parts of the measuring-apparatus consist of *c*, a globe of about 48 cubic centims. capacity with the volume-tube at the top, and *d* the pressure-tube; these two are exactly of the same diameter, to avoid error from capillarity. The tube at the bottom of the globe is ground into a funnel-shaped portion at the top of the wide tube *e*; and to the side of the latter the pressure-tube *d* is joined. The volume-tube at the top of the globe is graduated in millimetres from above downwards, the lowest division in this particular apparatus being 45; the pressure-tube *d* is also graduated in millimetres, the 0 being placed at the level of the 45th division on the volume-tube. A ball-and-socket joint connects the bottom of *e* with a vertical tube *f* about 800 millims. long, which is connected at its lower extremity by means of a flexible tube with the mercury-reservoir *g*; a stopcock at *h* permits the regulation of the flow of mercury into the apparatus: this may be conveniently turned by a rod, so that the operator may watch the rise of

* Read before the Physical Society, June 13, 1874. Communicated by the Society.

Originally published in Philos. Mag., 110–113 (1874).

the mercury through a telescope and have the stopcock at the same time at command.

The volume-tube was calibrated in the usual way, by introducing weighed quantities of mercury into it, and making the necessary corrections for the meniscus. The capacity of the volume-tube, the globe, and upper part of the tube *e* was determined by inverting the apparatus and introducing mercury through *e* until the mercury flowed down the pressure-tube; the weight of this quantity of mercury, divided by the weight of that contained in the volume-tube, gives the ratio between the volumes; in the present case it is 1 to 54·495. While the apparatus is being exhausted, the reservoir *g* is lowered so as to prevent the mercury rising out of the tube *f*; but when it is desired to make a measurement of the pressure, the reservoir is raised and the mercury allowed to pass through the stopcock *h*. On the mercury rising into the tube *e* it cuts off the communication between the gas in the globe and that in the rest of the apparatus. Ultimately the whole of the gas in the globe is condensed into the volume-tube; and its tension is then found by measuring the difference of level between the columns of mercury in the volume- and pressure-tubes. On dividing this difference by the ratio between the capacities of the globe and volume-tube, a number is obtained which is approximately the original pressure of the gas; this number must now be added to the difference between the columns, since it is obvious that the column in the pressure-tube is depressed by the tension of the gas in the remaining part of the apparatus; on dividing this new number once more by the ratio between the volumes the exact original tension is found.

An example will best illustrate this. A quantity of gas was compressed into the volume-tube, and the flow of mercury was arrested when its surface reached the lowermost division on the tube. The volume was then $\frac{1}{54 \cdot 495}$ of its original volume, and the difference between the levels of the mercury in the volume- and pressure-tubes was 66·9 millims.; this number, divided by 54·495, gives 1·228 as the approximate pressure. 1·2 must therefore be added to the observed column, which thus becomes 68·1, and on dividing by 54·495, the number 1·2497 is obtained as the actual pressure.

The relations existing between the contents of the other divisions of the volume-tube and the total contents of the globe were determined by measuring the tensions of the same quantity of gas when compressed into the different volumes. By this means the values of the divisions 40, 35, 30, 25, 20, 15, 10, 9, 8, 7, 6, 5, 4, 3, and 2 have been found; the experimenter is thus enabled to employ a division suitable to the quantity of gas with which he has to deal. The smallest division contains only $\frac{1}{1492 \cdot 35}$ of the globe; consequently when a quantity of gas has been condensed into this space, its original tension will be multiplied 1492·35 times. In one case an amount of gas, which originally filled the globe, exhibited a pressure of only ·5 millim. when it had been compressed into the smallest division of the volume-tube; this indicates an original pressure of only ·00033 millim.

When measuring a tension, it is advisable to make two readings under different condensations, and to take the mean of the results. The following will give some notion of the precision attainable:—

I. At division 5 ·0225, „ 2 ·0235 — Mean ·0230.

Remeasured.

At division 5 ·0228, „ 2 ·0236 — Mean ·0232.

II. Barometer 0 millim.:—

At division 10 ·1985, „ 5 ·1980 — Mean ·1982.

Remeasured.

At division 10 ·1953, „ 5 ·1967 — Mean ·1960.

III. Barometer 0·6 millim.:—

At division 15 ·5488, „ 10 ·5488, „ 6 ·5501 — Mean ·5492.

Remeasured.

At division 15 ·5464, „ 10 ·5464, „ 6 ·5480 — Mean ·5469.

IV. Barometer 1 millim.:—

At division 20 1·2042, „ 15 1·2069 — Mean 1·2055.

Remeasured.

At division 20 1·2082, „ 15 1·2099 — Mean 1·2090.

V. Barometer 1·5 millim.: —

At division 30 1·9139, „ 25 1·9080 — Mean 1·9109.

Remeasured.

At division 30 1·9041, „ 25 1·9039 — Mean 1·9040.

VI. Barometer 2·1 millims.:—

At division 35 2·6017 } Mean 2·6045.
,, 30 2·6073 }

Remeasured.

At division 35 2·6160 } Mean 2·6190.
,, 30 2·6220 }

It may be mentioned incidentally that connexions for apparatus may be conveniently made by means of ball-and-socket joints of glass. The ball is made by thickening a piece of tube in the blowpipe-flame, and the socket by cutting in half a thick bulb blown on a glass tube. The ball is then ground into the socket by means of emery and solution of soda, and afterwards polished with rouge and soda solution. When slightly greased and with a small quantity of mercury in the cup, a joint is obtained which is perfectly air-tight and flexible*.

of his show-rooms and his personal belongings, and after that date he appears neither to have done business in Paris nor to have visited it. Five years later the invasion of Neuwied led to the closing of his workshops; prosperity never returned, and he died half ruined at Wiesbaden on the 12th of February 1807.

Röntgen was not a great cabinet-maker. His forms were often clumsy, ungraceful and commonplace; his furniture lacked the artistry of the French and the English cabinet-makers of the great period which came to an end about 1790. His bronzes were poor in design and coarse in execution—his work, in short, is tainted by commercialism. As a *marqueteur*, however, he holds a position of high distinction. His marquetry is bolder and more vigorous than that of Riesener, who in other respects soared far above him. As an adroit deviser of mechanism he fully earned a reputation which former generations rated more highly than the modern critic, with his facilities for comparison, is prepared to accept. On the mechanical side he produced, with the help of Kintzing, many long-cased and other clocks with ingenious indicating and registering apparatus. Röntgen delighted in architectural forms, and his marquetry more often than not represents those scenes from classical mythology which were the dear delight of the 18th century. He is well represented at South Kensington.

RÖNTGEN, WILHELM KONRAD (1845–), German physicist, was born at Lennep on the 27th of March 1845. He received his early education in Holland, and then went to study at Zürich, where he took his doctor's degree in 1869. He then became assistant to Kundt at Würzburg and afterwards at Strassburg, becoming *privat-docent* at the latter university in 1874. Next year he was appointed professor of mathematics and physics at the Agricultural Academy of Hohenheim, and in 1876 he returned to Strassburg as extra-ordinary professor. In 1879 he was chosen ordinary professor of physics and director of the Physical Institute at Giessen, whence in 1885 he removed in the same capacity to Würzburg. It was at the latter place that he made the discovery for which his name is chiefly known, the Röntgen rays. In 1895, while experimenting with a highly exhausted vacuum tube on the conduction of electricity through gases, he noticed that a paper screen covered with barium platinocyanide, which happened to be lying near, became fluorescent under the action of some radiation emitted from the tube, which at the time was enclosed in a box of black cardboard. Further investigation showed that this radiation had the power of passing through various substances which are opaque to ordinary light, and also of affecting a photographic plate. Its behaviour being curious in several respects, particularly in regard to reflection and refraction, doubt arose in his mind whether it was to be looked upon as light or not, and he was led to put forward the hypothesis that it was due to longitudinal vibrations in the ether, not to transverse ones like ordinary light; but in view of the uncertainty existing as to its nature, he called it X-rays. For this discovery he received the Rumford medal of the Royal Society in 1896, jointly with Philip Lenard, who had already shown, as also had Hertz, that a portion of the cathode rays could pass through a thin film of a metal such as aluminium. Röntgen also conducted researches in various other branches of physics, including elasticity, capillarity, the conduction of heat in crystals, the absorption of heat-rays by different gases, piezo-electricity, the electromagnetic rotation of polarized light, &c.

RÖNTGEN RAYS, W. K. Röntgen discovered in 1895 (*Wied. Ann.* 64, p. 1) that when the electric discharge passes through a tube exhausted so that the glass of the tube is brightly phosphorescent, phosphorescent substances such as potassium platinocyanide became luminous when brought near to the tube. He found that if a thick piece of metal, a coin for example, were placed between the tube and a plate covered with the phosphorescent substance a sharp shadow of the metal was cast upon the plate; pieces of wood or thin plates of aluminium cast, however, only partial shadows, thus showing that the agent which produced the phosphorescence could traverse with considerable freedom bodies opaque to ordinary light. He found that as a general rule the greater the density of the substance the greater its opacity to this agent. Thus while this effect could pass through the flesh it was stopped by the bones, so that if the hand were held between the discharge tube and a phosphorescent screen the outline of the bones was distinctly visible as a shadow cast upon the screen, or if a purse containing coins were placed between the tube and the screen the purse itself cast but little shadow while the coins cast a very dark one. Röntgen showed that the cause of the phosphorescence, now called Röntgen rays, is propagated in straight lines starting from places where the cathode rays strike against a solid obstacle, and the direction of propagation is not bent when the rays pass from one medium to another, *i.e.* there is no refraction of the rays. These rays, unlike cathode rays or *Canalstrahlen*, are not deflected by magnetic force; Röntgen could not detect any deflection with the strongest magnets at his disposal, and later experiments made with stronger magnetic fields have failed to reveal any effect of the magnet on the rays. The rays affect a photographic plate as well as a phosphorescent screen, and shadow photographs can be readily taken. The time of exposure required depends upon the intensity of the rays, and this depends upon the state of the tube, and the electric current going through it, as well as upon the substances traversed by the rays on their journey to the photographic plate. In some cases an exposure of a few seconds is sufficient, in others hours may be required. The rays coming from different discharge tubes have very different powers of penetration. If the pressure in the tube is fairly high, so that the potential difference between its electrodes is small, and the velocity of the cathode rays in consequence small, the Röntgen rays coming from the tube will be very easily absorbed; such rays are called "soft rays." If the exhaustion of the tube is carried further, so that there is a considerable increase in the potential differences between the cathode and the anode in the tube and therefore in the velocity of the cathode rays, the Röntgen rays have much greater penetrating power and are called "hard rays." With a highly exhausted tube and a powerful induction coil it is possible to get appreciable effects from rays which have passed through sheets of brass or iron several millimetres thick. The penetrating power of the rays thus varies with the pressure in the tube; as this pressure gradually diminishes when the discharge is kept running through the tube, the type of Röntgen ray coming from the tube is continually changing. The lowering of pressure due to the current through the tube finally leads to such a high degree of exhaustion that the discharge has great difficulty in passing, and the emission of the rays becomes very irregular. Heating the walls of the tube causes some gas to come off the sides, and by thus increasing the pressure creates a temporary improvement. A thin-walled platinum tube is sometimes fused on to the discharge tube to remedy this defect; red-hot platinum allows hydrogen to pass through it, so that if the platinum tube is heated, hydrogen from the flame will pass into the discharge tube and increase the pressure. In this way hydrogen may be introduced into the tube when the pressure gets too low. When liquid air is available the pressure in the tube may be kept constant by fusing on to the discharge tube a tube containing charcoal; this dips into a vessel containing liquid air, and the charcoal is saturated with air at the pressure which it is desired to maintain in the tube. Not only do bulbs emit different types of rays at different times, but the same bulb emits at the same time rays of different kinds. The property by which it is most convenient to identify a ray is the absorption it suffers when it passes through a given thickness of aluminium or tin-foil. Experiments made by McClelland and Sir J. J. Thomson on the absorption of the rays produced by sheets of tin-foil showed that the absorption by the first sheets of tin-foil traversed by the rays was much greater than that by the same number of sheets when the rays had already passed through several sheets of the foil. The effect is just what would occur if some of the rays were much more readily absorbed by the tin-foil than others, for the first few layers would stop all the easily absorbable rays while the ones left would be those that were but little absorbed by tin-foil.

Originally published in Encycl. Brit., 694–696 (1911).

The fact that the rays when they pass through a gas ionize it and make it a conductor of electricity furnishes the best means of measuring their intensity, as the measurement of the amount of conductivity they produce in a gas is both more accurate and more convenient than measurements of photographic or phosphorescent effects. Röntgen rays when they pass through matter produce—as Perrin (*Comptes rendus*, 124, p. 455), Sagnac (*Jour. de Phys.*, 1899, (3), 8, and J. Townsend (*Proc. Camb. Phil. Soc.*, 1899, 10, p. 217, have shown—secondary Röntgen rays as well as cathodic rays. A very complete investigation of this subject has been made by Barkla and Sadler (Barkla, *Phil. Mag.*, June 1906, pp. 812–828; Barkla and Sadler, *Phil. Mag.*, October 1908, pp. 550–584; Sadler, *Phil. Mag.*, July 1909, p. 107; Sadler, *Phil. Mag.*, March 1910, p. 337). They have shown that the secondary Röntgen rays are of two kinds: one kind is of the same type as the primary incident ray and may be regarded as scattered primary rays, the other kind depends only on the matter struck by the rays—their quality is independent of that of the incident ray. When the atomic weight of the element exposed to the primary rays was less than that of calcium, Barkla and Sadler could only detect the first type of ray, *i.e.* the secondary radiation consisted entirely of scattered primary radiation; elements with atomic weights greater than that of calcium gave out, in addition to the scattered primary radiation, Röntgen rays characteristic of the element and independent of the quality of the primary rays. The higher the atomic weight of the metal the more penetrating are the characteristic rays it gives out. This is shown in the table, which gives for the different elements the reciprocal of the distance, measured in centimetres, through which the rays from the element can pass through aluminium before their energy sinks to 1/2·7 of the value it had when entering the aluminium; this quantity is denoted in the table by λ.

Element.	Atomic weight.	λ.
Chromium	52	367
Iron	55·9	239
Cobalt	59·0	193·2
Nickel	58·7? (61·3)	159·5
Copper	63·6	128·9
Zinc	65·4	106·3
Arsenic	75·0	60·7
Selenium	79·2	51·0
Strontium	87·6	35·2
Molybdenum	96·0	12·7
Rhodium	103·0	8·44
Silver	107·9	6·75
Tin	119·0	4·33

The radiation from chromium cannot pass through more than a few centimetres of air without being absorbed, while that from tin is as penetrating as that given out by a fairly efficient Röntgen tube. Barkla and Sadler found that the radiation characteristic of the metal is not excited unless the primary radiation is more penetrating than the characteristic radiation. Thus the characteristic radiation from silver can excite the characteristic radiation from iron, but the characteristic radiation from iron cannot excite that from silver. We may compare this result with Stokes's rule for phosphorescence, that the phosphorescent light is of longer wave-length than the light which excites it.

The discovery that each element gives out a characteristic radiation (or, as still more recent work indicates, a line spectrum of characteristic radiation) is one of the utmost importance. It gives us, for example, the means of getting homogeneous Röntgen radiation of a perfectly definite type; it is also of fundamental importance in connexion with any theory of the Röntgen rays. We have seen that there is no evidence of refraction of the Röntgen rays; it would be interesting to try if this were the case when the rays passing through the refracting substance are those characteristic of the substance.

Secondary Cathodic Rays.—The incidence of Röntgen rays on matter causes the matter to emit cathodic rays. The velocity of these rays is independent of the intensity of the primary Röntgen rays, but depends upon the "hardness" of the rays; it seems also to be independent of the nature of the matter exposed to the primary rays. The velocity of the cathodic rays increases as the hardness of the primary Röntgen rays increases. Innes (*Proc. Roy. Soc.* 79, p. 442) measured the velocity of the cathodic radiation excited by the rays from Röntgen tubes, and found velocities varying from $6{\cdot}2\times10^9$cm./sec. to $8{\cdot}3\times10^9$ cm./sec. according to the hardness of the rays given out by the tube. The cathodic rays given out under the action of the homogeneous secondary Röntgen radiation characteristic of the different elements have been studied by Sadler (*Phil. Mag.*, March 1910) and Beatty (*Phil. Mag.*, August 1910). The following table giving the properties of the cathode rays excited by the radiation from various elements is taken from Beatty's paper; t_1 is the thickness of air at atmospheric pressure and temperature required to absorb one-half of the energy of the cathode particles, t_2 is the corresponding quantity for hydrogen.

Radiator.	t_1	t_2
Iron	·00804	·0410
Copper	·0135	·0733
Zinc	·0164	·0909
Arsenic	·0255	
Tin	·1672	1·37

The properties of the cathode rays excited by the radiation from tin correspond very closely with those produced in a discharge tube when the potential difference between the anode and cathode is about 30,000 volts. When Röntgen rays pass through a thin plate the cathodic radiation on the side the rays emerge is more intense than on the side they enter. Kaye (*Phil. Trans.* 209, p. 123) has shown that when cathode rays fall upon a metal two kinds of Röntgen rays are excited, one being the characteristic radiation of the metal and the other a kind independent of the nature of the metal and dependent only upon the velocity of the cathode rays. The faster the cathode rays the harder the Röntgen rays they produce. It would be interesting to see if there is any connexion between the velocity of the cathode rays required to excite Röntgen rays as hard as those given out say by tin and the velocity of the cathode rays which the radiation from tin produces when it falls upon any metal. Sadler has shown that metals can give off cathodic radiation even when the incident Röntgen rays are too soft to excite the characteristic Röntgen radiation of the metal, but that there is a large increase in the cathodic radiation as soon as the characteristic Röntgen radiation is excited. It is possible that the shock produced by the emission of these cathode particles starts the vibrations which give rise to the characteristic rays; the cathode particles emitted when the incident rays are too soft to excite the characteristic radiation coming from a different source from those tapped by the hard rays.

Absorption of Röntgen Rays.—The wide variations in the penetrating power of Röntgen rays from different sources is shown by the above table of the penetrating power of the characteristic rays of the different elements. Many experiments have been made on the penetration of the same rays for different substances. It is a rûle to which there is no well-established exception that the greater the density of the substance the greater is its power of absorbing the rays. The connexion, however, between the absorption and the density of the substance is not in general a simple one, though there is evidence that for exceedingly hard rays the absorption is proportional to the density.

The power of any material to absorb rays is usually measured by a coefficient λ, the definition of which is that a plate 1/λ centimetres thick reduces the energy of the rays when they pass through it normally to $1/e$ of their original value, where e is the base of the Napierian logarithms and equal to 2·7128. It has been shown that however the physical state of a substance may alter,—if, for example, it changes from the liquid to the gaseous,—λ/D, where D is the density of the substance, remains constant. It has also been shown that if we have a mass M made up of masses $M_1, M_2, M_3, \ldots$ of substances having coefficients of absorption $\lambda_1, \lambda_2, \lambda_3, \ldots$ and densities $D_1, D_2, D_3, \ldots$ then if λ/D for the mixture is given by the equation

$$M\lambda/D = M_1\lambda_1/D_1 + M_2\lambda_2/D_2 + M_3\lambda_3/D_3 + \ldots$$

this equation is true whether the substances are chemically combined or chemically mixed. From this equation, when we know λ/D for a binary compound and for one of its constituents, we can find the value of λ/D for the other constituent. By the use of this principle we can find the value of λ/D for the elements which cannot be obtained in a free state. Benoist (*Jour. de Phys.* (7), 28, p. 289) has shown that if the values of λ/D are plotted against the atomic weight we get a smooth curve; if we draw this curve it is evident that we have the means of determining the atomic weight of an element by measuring its transparency to Röntgen rays when in combination with elements whose transparency is known. Benoist has applied this method to determine the atomic weight of indium.

The value of λ/D for any one substance depends upon the type of ray used, and the ratio of the values of λ/D for two substances may vary very greatly with the type of ray; this is especially the case when one of the substances is hydrogen. Thus Crowther (*Proc. Roy. Soc.*, March 1909) has shown that the ratio of λ for air to λ for hydrogen varied from 100 for rays given out by a Röntgen tube at a comparatively high pressure when the rays were very soft to 5·56 when the pressure in the bulb was very low and the rays very hard. Beatty (*Phil. Mag.*, August 1910) found that this ratio was as large as 175 for the characteristic rays given out by iron, copper, zinc and arsenic, but fell to 25·0 for the rays from tin.

Polarization of Röntgen Rays.—A great deal of attention has been paid to a phenomenon called the polarization of the

Röntgen rays. The nature of this effect may be illustrated by fig. 1. Suppose that AB is a stream of cathode rays striking against a solid obstacle B and giving rise to Röntgen rays, let these rays impinge on a small body P, P under these conditions will emit secondary rays in all directions. Barkla (*Phil. Trans.*, 1905, A, 204, p. 467; *Proc. Roy. Soc.* 77, p. 247) found that the intensity of the secondary rays, tested by the ionization they produced in air, was less intense in the plane ABP than in a plane through PB at right angles to this plane, the distances from P being the same in the two cases; the difference in the intensities amounting to about 15%. Haga (*Ann. d. Phys.* 28, p. 439), who tried a similar experiment but used a photographic method to measure the intensity of the secondary rays, could not detect any difference of intensity in the two planes, but experiments by Bassler (*Ann. der Phys.* 28, p. 808) and Vegard (*Proc. Roy. Soc.* 83, p. 379) have confirmed Barkla's original observations.

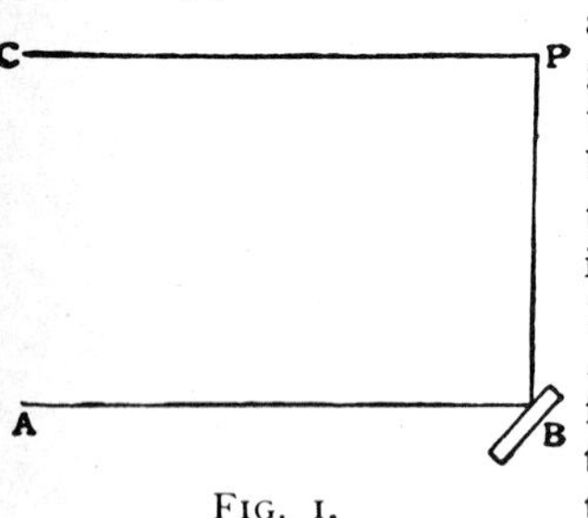

FIG. 1.

The "polarization" is much more marked if instead of exciting the secondary radiation in P by the Röntgen rays from a discharge tube we do so by means of secondary rays. If, for example, in the case illustrated by fig. 1 we allow a beam of Röntgen rays to fall upon B instead of the cathode rays, the difference between the intensities in the plane ABP and in the plane at right angles to it are very much increased. It is only the scattered secondary radiation which shows this "polarization"; the characteristic secondary radiation emitted by the body at P is quite unpolarized. The existence of this effect has a very important bearing on the nature of Röntgen rays. Whether Röntgen rays are or are not a form of light, *i.e.* are some form of electromagnetic disturbance propagated through the aether, is a question on which opinion is not unanimous. They resemble light in their rectilinear propagation; they affect a photographic plate and, Brandes and Dorn have shown, they produce an effect, though a small one, on the retina, giving rise to a very faint illumination of the whole field of view. They resemble light in not being deflected by either electric or magnetic forces, while the characteristic secondary radiation may be compared with the phosphorescence produced by ultra-violet light, and the cathodic secondary rays with the photo-electric effect. The absence of refraction is not an argument against the rays being a kind of light, for all theories of refraction make this property depend upon the relation between the natural time of vibration T of the refracting substance and the period t of the light vibrations, the refraction vanishing when t/T is very small. Thus there would be no refraction for light of a very small period, and this would also be true if instead of regular periodic undulations we had a pulse of electromagnetic disturbance, provided the time taken by the light to travel over the thickness of the pulse is small compared with the periods of vibration of the molecules of the refracting substance. Experiments on the diffraction of Röntgen rays are very difficult, for, in addition to the difficulties caused by the smallness of the wave-length or the thinness of the pulse, the secondary radiation produced when the rays strike against a photographic plate or pass through air might give rise to what might easily be mistaken for diffraction effects. Röntgen has never succeeded in observing effects which prove the existence of diffraction. Fomm (*Wied. Ann.* 59, p. 50) observed in the photograph of a narrow slit light and dark bands which looked like diffraction bands; but observation with slits of different sizes showed that they were not of this nature, and Haga and Wind (*Wied. Ann.* 68, p. 884) have explained them as contrast effects. These observers, however, noticed with a very narrow wedge-shaped slit a broadening of the image of the narrow part which they are satisfied could not be explained by the causes. Walter and Pohl (*Ann. der Phys.* 29, p. 331) could not observe any diffraction effects, though their arrangement would have enabled them to do so if the wave-length had not been smaller than $1 \cdot 5 \times 10^{-9}$cm. Sir George Stokes (*Proc. Manchester Lit. and Phil. Soc.*, 1898) put forward the view that the disturbances which constitute the rays are not regular periodic undulations but very thin pulses. Thomson (*Phil. Mag.* 45, p. 172) has shown that when charged particles are suddenly stopped, pulses of very intense electric and magnetic disturbances are started. As the cathode rays consist of negatively electrified particles, the impact of these on a solid would give rise to these intense pulses. The electromagnetic theory therefore shows that effects resembling light, inasmuch as they are electromagnetic disturbances propagated through the aether, must be produced when the cathode rays strike against an obstacle. Since under these circumstances Röntgen rays are produced, it seems natural, unless direct evidence to the contrary is obtained, to connect the Röntgen rays with these pulses. This view explains very simply the "polarization" of the rays; for, suppose the cathode particle moving from A to B were stopped at its first impact with the plate B (fig. 1), the electric force transmitted along BP would be in the plane ABP at right angles to BP. When this electric force reached the body at P it would accelerate any electrified particles in that body, the acceleration being parallel to AB. Each of these accelerated particles would start electric waves. The theory of such waves shows that their intensity vanishes along a line through the particle parallel to the direction of acceleration, while it is a maximum at right angles to this line; thus the intensity of the rays along a horizontal line through P would vanish, while it would be a maximum in the plane at right angles to this line. In this case there would be complete polarization. In reality the cathode particle is not stopped at its first encounter, but makes many collisions, changing its direction between each; and these collisions will send out electric disturbances which when they fall on P are able to excite waves which send some energy along PC. The polarization will therefore be only partial and will be of the kind found by Barkla.

The velocity with which the waves travel has not yet been definitely settled. Marx (*Ann. der Phys.* 20, p. 677) by an ingenious but elaborate method came to the conclusion that they travelled with the velocity of light; his interpretation of his experiments has, however, been criticized by Franck and Pohl (*Verh. d. D. Physik Ges.* 10, p. 489).

Another view of the nature of Röntgen rays has been advocated by Bragg (*Phil. Mag.* 14, p. 429); he regards them as neutral electric doublets consisting of a negative and a positive charge of electricity which are usually held together by the attraction between them, but which may be knocked asunder when the rays strike against matter and turned into cathodic rays. On this view when the rays pass through a gas only a few of the molecules of the gas are struck by the rays and so we can easily understand why so few of the molecules are ionized. On the ordinary view of an electric wave all the molecules would be affected by the wave when it passed through a gas, and to explain the small fraction ionized we must either suppose that systems sensitive to the Röntgen rays are at any time present only in a very small fraction of the molecule or else that the front of an electric or light wave is not continuous but that the energy is concentrated in patches which only occupy a fraction of the wave front.

Apparatus for producing Röntgen Rays.—The tube now used most frequently for producing Röntgen rays is of the kind introduced by Porter and known as a focus tube (fig. 2). The cathode is a portion of a hollow sphere, and the cathode rays come to a point on or near a metal plate A, called the anti-cathode, connected with the anode; this plate is the source of the rays. This ought to be made of a very unfusible metal such as platinum or, still better, tantalum, and kept cool by a water-cooling arrangement. The anti-cathode is generally set at an angle of 45° to the rays; it is probable that the action of the tube would be improved by putting the anti-cathode at right angles to the cathode rays. The walls of the tube get strongly electrified. This electrification affects the working of the tube, and the production of rays can often be improved by having an earth-connected piece of tin-foil on the outside of the bulb, and moving it about until the best position is attained. To produce the discharge an induction coil is generally employed with a mercury interrupter. Excellent results have been obtained by using an electrostatic induction machine to produce the current, the emission of rays is more uniform than when an induction coil is used. The rays are emitted pretty uniformly in all directions until the plane of the anti-cathode is approached; in the neighbourhood of this plane there is a rapid falling off in the intensity of the rays. After long use the glass of the bulb often becomes distinctly purple. This is believed to be due to the presence of manganese compounds in the glass. (J. J. T.)

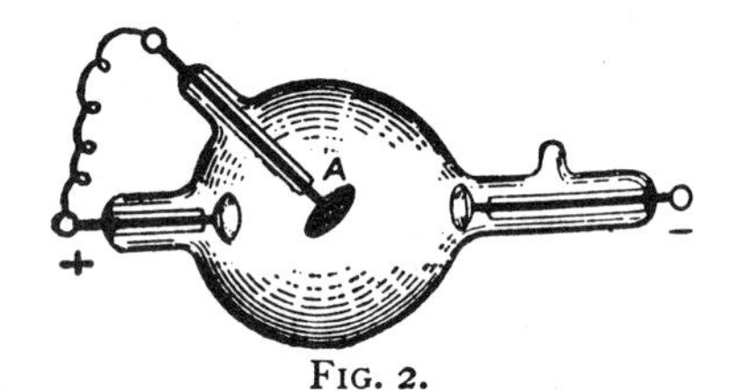

FIG. 2.

ROOD (O.E. *rōd*, a stick, another form of "rod," O.E. *rodd*, possibly cognate with Lat. *rudis*, a staff), properly a rod or pole, and so used as the name of a surface measure of land. The rood varies locally but is generally taken as = 40 square rods, poles or perches; 4 roods = 1 acre. The term was, however, particularly applied, in O.E., to a gallows or cross, especially to the Holy Cross on which Christ was crucified, the sense in which the word survives. A crucifix, often accompanied by figures of St John and the Virgin Mary, was usually placed in churches above the screen, hence known as "rood screen" public vaccinator was now required to visit the homes of children for the purpose of offering vaccination with glycerinated calf lymph, "or such other lymph as may be issued by the Local Government Board." The operative procedure in public vaccinations was formerly based on the necessity of carrying on a weekly series of transferences of vaccine lymph from arm to arm; and for the purposes of such arm-to-arm vaccination the provision of stations, to which children were brought first for the performance of the operation, and again, after a week's interval, for inspection of the results, was an essential. The occasional hardships to the mothers, and a somewhat remote possibility of danger to the children, involved in being taken long journeys to a vaccination station in bad weather, or arising from the collecting together in one room of a number of children and adults, one or more of whom might happen to be suffering at the time from some infectious disorder, are a few of the reasons which appeared to render a change in this regulation desirable; as a matter of fact, it would appear that nothing but good has arisen from the substitution of domiciliary for stational vaccination. There have naturally been some curious discussions before the magistrates as to what is "conscientious" or not, but the working of the so-called "conscience clause" by no means justified the somewhat gloomy forebodings expressed, both in Parliament and elsewhere, at the time of its incorporation in the act of 1898. On the contrary, its operation appeared to tend to the more harmonious working of the Vaccination Acts, by affording a legal method of relief to such parents and guardians as were prepared to affirm that they had a conscientious belief that the performance of the operation might, in any particular instance, be prejudicial to the health of the child.

AUTHORITIES.—Acland, "Vaccinia," Allbutt and Rolleston, *System of Medicine* (1906); Baron, *Life of Jenner*; Henry Colburn (London, 1838); Copeman, *Vaccination: Its Natural History and Pathology* (Milroy Lectures) (Macmillan, London, 1899); "Modern Methods of Vaccination and their Scientific Basis," *Trans. Royal Med. and Chir. Society* (1901–2); M'Vail, "Criticism of the Dissentient Commissioners' Report," *Trans. Epidemiological Society* (1897); Reports of the Royal Commission on Vaccination (1889–1896); "The History and Effects of Vaccination," *Edinburgh Review*, No. 388 (1899); *Vaccination Law of German Empire* (Berlin, 1904).

VACHEROT, ÉTIENNE (1809–1897), French philosophical writer, was born of peasant parentage at Torcenay, near Langres, on the 29th of July 1809. He was educated at the École Normale, and returned thither as director of studies in 1838, after some years spent in provincial schoolmasterships. In 1839 he succeeded his master Cousin as professor of philosophy at the Sorbonne. His *Histoire critique de l'école d'Alexandrie* (3 vols. 1846–51), his first and best-known work, drew on him attacks from the Clerical party which led to his suspension in 1851. Shortly afterwards he refused to swear allegiance to the new imperial government, and was dismissed the service. His work *Démocratie* (1859) led to a political prosecution and imprisonment. In 1868 he was elected to the French Academy. On the fall of the Empire he took an active part in politics, was *maire* of a district of Paris during the siege, and in 1871 was in the National Assembly, voting as a Moderate Liberal. In 1873 he drew nearer the Conservatives, after which he was never again successful as a parliamentary candidate, though he maintained his principles vigorously in the press. He died on the 28th of July 1897. Vacherot was a man of high character and adhered strictly to his principles, which were generally opposed to those of the party in power. His chief philosophical importance consists in the fact that he was a leader in the attempt to revivify French philosophy by the new thought of Germany, to which he had been introduced by Cousin, but of which he never had more than a second-hand knowledge. Metaphysics he held to be based on psychology. He maintains the unity and freedom of the soul, and the absolute obligation of the moral law. In religion, which was his main interest, he was much influenced by Hegel, and appears somewhat in the ambiguous position of a sceptic anxious to believe. He sees insoluble contradictions in every mode of conceiving God as real, yet he advocates religious belief, though the object of that belief have but an abstract or imaginary existence.

His other works are: *La Métaphysique et la science* (1858), *Essais de philosophie critique* (1864), *La Religion* (1869), *La Science et la conscience* (1870), *Le Nouveau Spiritualisme* (1884), *La Démocratie libérale* (1892).

See Ollé Laprune, *Étienne Vacherot* (Paris, 1898).

VACQUERIE, AUGUSTE (1819–1895), French journalist and man of letters, was born at Villequier (Seine Inférieure) on the 19th of November 1819. He was from his earliest days an admirer of Victor Hugo, with whom he was connected by the marriage of his brother Charles with Léopoldine Hugo. His earlier romantic productions include a volume of poems, *L'Enfer de l'esprit* (1840); a translation of the *Antigone* (1844) in collaboration with Paul Meurice; and *Tragaldabas* (1848), a melodrama. He was one of the principal contributors to the *Événement* and followed Hugo into his exile in Jersey. In 1869 he returned to Paris, and with Paul Meurice and others founded the anti-imperial *Rappel*. His articles in this paper were more than once the occasion of legal proceedings. After 1870 he became editor. Other of his works are *Souvent homme varie* (1859), a comedy in verse; *Jean Baudry* (1863), the most successful of his plays; *Aujourd'hui et demain* (1875); *Futura* (1900), poems on philosophical and humanitarian subjects. Vacquerie died in Paris on the 19th of February 1895. He published a collected edition of his plays in 1879.

VACUUM-CLEANER, an appliance for removing dust from carpets, curtains, &c., by suction, and consisting essentially of some form of air-pump drawing air through a nozzle which is passed over the material that has to be cleaned. The dust is carried away with the air-stream and is separated by filtration through screens of muslin or other suitable fabric, sometimes with the aid of a series of baffle-plates which cause the heavier particles to fall to the bottom of the collecting receptacle by gravity. In the last decade of the 19th century compressed air came into use, especially in America, for cleaning railway carriages, but it was found difficult to arrange for the collection of the dust that was blown out by the jets of air, and in consequence recourse was had to working by suction. From this beginning several types of vacuum cleaner have developed.

In the first instance the plants were portable, consisting of a pump driven by a petrol engine or electric motor, and were periodically taken round to houses, offices &c., when cleaning was required. The second stage was represented by the permanent installation of central plants in large buildings, with a system of pipes running to all floors, like gas or water pipes, and provided at convenient points with valves to which could be attached flexible hose terminating in the actual cleaning tools. The vacuum thus rendered available is in some cases utilized for washing the floors in combination with another system of piping connected to a tank containing soap and water, which having been sprayed over the floor by compressed air is removed with the dirt it contains and discharged into the sewers; or in a simpler arrangement the soap and water is contained in a portable tank from which it is distributed, to be sucked up by means of the vacuum as before. In their third stage vacuum cleaners have become ordinary household implements, in substitution for, or in addition to the broom and duster, and small machines are now made in a variety of forms, driven by hand, by foot, or by an electric motor attached to the lighting circuit. In addition to their domestic uses, other applications have been found for them, as for instance in removing dust from printers' type-cases.

VACUUM TUBE. The phenomena associated with the passage of electricity through gases at low pressures have attracted the attention of physicists ever since the invention of the frictional electrical machine first placed at their disposal a means of producing a more or less continuous flow of electricity through vessels from which the air had been partially exhausted. In recent years the importance of the subject in connexion with the theory of electricity has been fully realized; indeed, the modern theory of electricity is based upon ideas which have been obtained from the study of the electric discharge through gases. Most of the important principles deduced from these investigations are given in the article CONDUCTION, ELECTRIC (*Through Gases*); here we shall confine ourselves to the consideration of the more striking features of the luminous phenomena observed when electricity passes through a luminous gas.

Originally published in Encycl. Brit., 834–837 (1911).

Methods of producing the Discharge.—To send the current through the gas it is necessary to produce between electrodes in the gas a large difference of potential. Unless the electrodes are of the very special type known as Wehnelt electrodes, this difference of potential is never less than 200 or 300 volts and may rise to almost any value, as it depends on the pressure of the gas and the size of the tube. In very many cases by far the most convenient method of producing this difference of potential is by means of an induction coil; there are some cases, however, when the induction coil is not suitable, the discharge from a coil being intermittent, so that at some times there is a large current going through the tube, while at others there is none at all, and certain kinds of measurement cannot be made under these conditions. Not only is the current intermittent, but it is apt with the coil to be sometimes in one direction and sometimes in the opposite; there is a tendency to send a discharge through the tube not only when the current through the primary is started but also when it is stopped. These discharges are in opposite directions, and though that produced by stopping the current is more intense than that due to starting it, the latter may be quite appreciable. The reversal of the current may be remedied by inserting in series with the discharge tube a piece of apparatus known as a "rectifier" which allows a current to pass through it in one direction but not in the opposite. A common type of rectifier is another tube containing gas at a low pressure and having one of its electrodes very large and the other very small; a current passes much more easily through such a tube from the small to the large electrode than in the opposite direction. Sometimes an air-break inserted in the circuit with a point for one electrode and a disk for the other is sufficient to prevent the reversal of the current without the aid of any other rectifier.

There are cases, however, when the inevitable intermittence of the discharge produced by an induction coil is a fatal objection. When this is so, the potential difference may be produced by a battery of a large number of voltaic cells, of which the most convenient type, where more than a few milli-ampères of current are required, are small storage cells. As each of these cells only produces a potential difference of two volts, a very large number of cells are required when potential differences of thousands of volts have to be produced, and the expense of this method becomes prohibitive. When continuous currents at these high potential differences are rquired, electrostatic induction machines are most generally used. By means of Wimshurst machines, with many plates, or the more recent Wehrsen machines, considerable currents can be produced and maintained at a very constant value.

The exhaustion of the tubes can, by the aid of modern mercury pumps, such as the Töpler pump or the very convenient automatic Gaede pump, be carried to such a point that the pressure of the residual gas is less than a millionth of the atmospheric pressure. For very high exhaustions, however, the best and quickest method is that introduced by Sir James Dewar. In this method a tube containing small pieces of dense charcoal (that made from the shells of coco-nuts does very well) is fused on to the tube to be exhausted. The preliminary exhaustion is done by means of a water-pump which reduces the pressure to that due to a few millimetres of mercury and the charcoal strongly heated at this low pressure to drive off any gases it may have absorbed. The tube is then disconnected from the water-pump and the charcoal tube surrounded by liquid air; the cold charcoal greedily absorbs most gases and removes them from the tube. In this way much higher exhaustions can be obtained than is possible by means of mercury pumps; it has the advantage, too, of getting rid of the mercury vapour which is always present when the exhaustion is produced by mercury pumps. Charcoal does not absorb much helium even when cooled to the temperature of liquid air, so that the method fails in the case of this gas; the absorption of hydrogen, too, is slower than that of other gases. Both helium and hydrogen are vigorously absorbed when the charcoal is cooled to the temperature of liquid hydrogen.

When first the discharge is sent through an exhausted tube, a considerable amount of gas (chiefly hydrogen and carbon monoxide) is liberated from the electrodes and the walls of the tube, so that to obtain permanent high vacua the exhaustion must be continued until the discharge has been going through the tube for a considerable time. One of the greatest difficulties experienced in getting these high vacua is that even when all the joints are carefully made there may be very small holes in the tube through which the air is continually leaking from outside, and when the hole is very small it is sometimes very difficult to locate the leak. The writer has found that a method due to Goldstein is of the greatest service for this purpose. In this method one of the electrodes in the tube and one of the terminals of the induction coil are put to earth, and the pressure of the gas in the tube is reduced so that a discharge would pass through the tube with a small potential difference. The point of an insulated wire attached to the other terminal of the induction coil is then passed over the *outside* of the tube. When it comes to the hole, a very bright white spark may be seen passing through the glass, and in this way the leak located. The appearance of the discharge when the exhaustion is going on is a very good indication as to whether there is any leakage in the tube or not. If the colour of the discharge remains persistently red in spite of continued pumping, there is pretty surely a leak in the tube, as the red colour is probably due to the continued influx of air into the tube. Platinum is the only metal which can be fused through the glass with any certainty that the contact between the glass and the metal will be close enough to prevent air leaking into the tube. Platinum, however, when used as a cathode at low pressures "sputters," and the walls of the tube get covered with a thin deposit of the metal: to avoid this, the platinum is often fastened to a piece of aluminium, which does not sputter nearly so much. Tantalum is also said to possess this property, and it has the advantage of being much less fusible than aluminium. This sputtering depends to some extent on the kind of gases present in the tube, as in monatomic gases, such as mercury vapour, even aluminium sputters badly.

Electrodeless Tubes.—As some gases, such as chlorine and bromine, attack all metals, it is impossible to use metallic electrodes when the discharge through these gases has to be investigated. In these cases "electrodeless" tubes are sometimes used. These are of two kinds. The more usual one is when tin-foil is placed at the ends of the tube on the outside, and the terminals of the induction coil connected with these pieces of foil; the glass under the foil virtually acts as an electrode. A more interesting form of the electrodeless discharge is what is known as the "ring" discharge. The tube in this case is placed inside a wire solenoid which forms a part of a circuit connecting the outside coatings of two Leyden jars, the inside coatings of these jars being connected with the terminals of an induction coil or electrical machine; the jars are charged up by the machine, and are discharged when sparks pass between its terminals. As the discharge of the jars is oscillatory (see ELECTRIC WAVES), electric currents surge through the solenoid surrounding the discharge tube, and these currents reverse their direction hundreds of thousands of times per second. We may compare the solenoid with the primary coil of an induction coil, and the exhausted bulb with the secondary; the rapidly alternating currents in the primary induce currents in the secondary which show themselves as a luminous ring inside the tube. Very bright discharges may be obtained in this way, and the method is especially suitable for spectroscopic purposes (see *Phil. Mag.* [5], 32, pp. 321, 445).

Appearance of the Discharge in Vacuum Tubes.—Fig. 15 b of the article CONDUCTION, ELECTRIC (*Through Gases*) represents the appearance of the discharge when the pressure in the tube is comparable with that due to a millimetre of mercury and for a particular intensity of current. With variations in the pressure or the current some of these features may disappear or be modified. Beginning at the negative electrode *k*, we meet with the following phenomena: A velvety glow runs, often in irregular patches, over the surface of the cathode; this glow is often called the first negative layer. The spectrum of this layer is a bright line spectrum, and Stark has shown that it shows the Döppler effect due to the rapid motion of the luminous particles towards the cathode. Next to this there is a comparatively dark region known as the "Crookes' dark space," or the second negative layer. The luminous boundary of this dark space is approximately such as would be got by tracing the locus of the extremities of normals of constant length drawn from the negative electrode, thus if the electrode is a disk, the luminous boundary of the dark sphere is nearly plain over a part of its surface as in fig. 1, while if the electrode is a ring of wire (fig. 2) the luminous boundary resembles that

FIG. 1.

shown in fig. 17 of the article CONDUCTION, ELECTRIC (*Through Gases*). The length of the dark space depends on the pressure of the gas and on the intensity of the current passing through it. The width of the dark space increases as the pressure diminishes, and may, according to the experiments of Aston (*Pro. Roy. Soc.* 79, p. 81), be represented with considerable accuracy by the expression $a+b/p$ or $a+c\lambda$, where a, b, c are constants, p the pressure and λ the mean free path of a corpuscle through the gas. The thickness of the dark space is larger than this free path; for hydrogen, for example, the value of c is about 4.

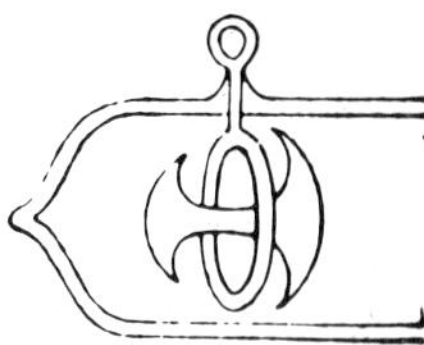

FIG. 2.

When the current is so large that the whole of the cathode is covered with glow the width of the dark space depends upon the current decreasing as the current increases. In helium and hydrogen Aston (*Pro. Roy. Soc.* 80 A., p. 45) has detected the existence of another thin dark space quite close to the cathode whose thickness is independent of the pressure. The farther boundary of the Crookes' dark space is luminous and is known as the negative glow or the third negative layer. Until the current gets so large that the glow next the cathode covers the whole of its surface the potential difference between the cathode and the negative glow is independent of the pressure of the gas and the current passing through it; it depends only on the kind of gas and the metal of which the cathode is made. This difference of potential is known as the cathode fall of potential; the values of it in volts for some gases and electrodes as determined by Mey (*Verh. deuts. Phys. Ges.*, 1903, v. p. 72) are given in the table.

CATHODE FALL

GAS	ELECTRODE										
	Pt	Hg	Ag	Cu	Fe	Zn	Al	Mg	Na	Na-K	K
O_2	369	..	..	..	..	..	..	..	..	..	—
H_2	300	..	295	280	230	213	190	168	185	169	172
N_2	232	226	..	..	..	..	..	207	178	125	170
He	226	..	..	..	..	..	..	..	80	78·5	69
Arg	167	..	..	..	..	..	100	..	..	..	—

The cathode fall of potential measures the smallest difference of potential which can produce a spark through the gas. Thus, for example, it is not possible to produce a spark through nitrogen with platinum electrodes with a potential difference of less than 232 volts, except when the electrodes are placed so close together that with a smaller potential difference the electric force between the terminals amounts to more than a million volts per centimetre; for this to be the case the distance between the electrodes must be comparable with the wave-length of sodium light.

When the current is small the glow next the cathode does not cover the whole of the surface, and when this occurs an increase in the current causes the glow to cover a greater area, but does not increase the current density nor the cathode fall. When the current is so much increased that the glow covers the whole of the cathode an increase in current must result in an increase of the current density over the cathode, and this is accomplished by a rapid increase in the cathode fall of potential. The cathode fall in this case has been investigated by Stark (*Phys. Zeit.* III, p. 274), who finds that its value K can be represented by the equation

$$K = K_n + k(C - xpf)^{\frac{1}{2}}/pf^{\frac{1}{2}},$$

where K_n is the normal cathode fall, f the area of the cathode, C the current through the tube, p the pressure of the gas and k and x constants.

The increase in the potential fall is much more marked in small tubes than in large ones, as with small tubes the formation of the negative glow is restricted; this gives rise to a greater concentration of the current at the cathode and an increase in the cathode fall. The intensity of the electric field in the dark space has been measured by many observers. Aston used very large plain cathodes and measured the electric force by observing the deflection of a small pencil of cathode rays sent across the dark space at different distances from the cathode. He found that the magnitude of the force at a point in the dark space was proportional to the distance of the point from the junction of the negative glow and the dark space. This law of force shows that positive electricity must be in excess in the dark space, and that the density of the electrification must be constant throughout that space. The force inside the negative glow if not absolutely zero is so small that no one has as yet succeeded in measuring it; thus the surface of this glow must be very approximately an equi-potential surface. In the dark space there is a stream of positively electrified particles moving towards the cathode and of negatively electrified corpuscles moving away from it, these streams being mutually dependent; the impact of the positive particles against the cathode gives rise to the emission of corpuscles from the cathode; these, after acquiring kinetic energy in the dark space, ionize the gas and produce the positive ions which are attracted by the cathode and give rise to a fresh supply of corpuscles. The corpuscles which carry the negative electricity are very different from the carriers of the positive; the former have a mass of only $\frac{1}{1700}$ of the atom of hydrogen, while the mass of the latter is never less than that of this atom.

The stream of positive particles towards the cathode is often called the *Canalstrahlen*, and may be investigated by allowing the stream to flow through a hole in the cathode and then measuring, by the methods described in CONDUCTION, ELECTRIC (*Through Gases*), the velocity and the value of e/m when e is the charge on a carrier and m its mass. It has been found that this stream is somewhat complex and consists of—

α. A stream of neutral particles.

β. A stream of positively electrified particles moving with a constant velocity of 2×10^8 cm./sec., and having $e/m=10^4$. This is a secondary stream produced by the passage of α through the gas, and it is very small when the pressure of the gas is low.

γ. Streams of positively electrified atoms and perhaps molecules of the gases in the tube. The velocity of these depends upon the cathode fall of potential.

The streams of negative corpuscles and positive particles produce different kinds of phosphorescence when they strike against a solid obstacle. The difference is especially marked when they strike against lithium chloride. The corpuscles make it phosphoresce with a steely blue light giving a continuous spectrum; the positive particles, on the other hand, make it shine with a bright red light giving in the spectroscope the red lithium line. This affords a convenient method of investigating the rays; for example, the distribution of the positive stream over the cathode is readily studied by covering the cathode with fused lithium chloride and observing the distribution of the red glow. Goldstein has observed that the film of metal which is deposited on the sides of the tube through the sputtering of the cathode is quickly dissipated when the positive stream impinges on it. This suggests that the sputtering of the cathode is caused by the impact against it of the positive stream. This view is supported by the fact that the sputtering is not very copious until the increase in the current produces a large increase in the cathode fall of potential. The magnitude of the potential fall and the length of the dark space are determined by the condition that the positive particles when they strike against the cathode must give to it sufficient energy to liberate the number of cathode particles which produce, when they ionize the gas, sufficient positive particles to carry this amount of energy. Thus the cathode fall may be regarded as existing to make the cathode emit negative corpuscles. If the cathode can be made to emit corpuscles by other means the cathode fall of potential is not required and may disappear. Now Wehnelt (*Ann. Phys.*, 1904, 14, p. 425), found that when lime or barium oxide is heated to redness large quantities of negative corpuscles are emitted; hence if a cathode is covered with one of these substances and made red hot it can emit corpuscles without the assistance of an electric field, and we find that in this case the cathode fall of potential disappears, and current can be sent through the gas with very much smaller differences of potential than with cold cathodes. With these hot cathodes a luminous current can under favourable circumstances be sent through a gas with a potential difference as small as 18 volts.

The dimensions of the parts of the discharge we have been considering—the dark space and the negative glow—depend essentially upon the pressure of the gas and the shape of the cathode, and do not increase when the distance between the anode and cathode is increased. The dimensions of the other part of the discharge which reaches to the anode and is called the positive column depends upon the length of the tube, and in long tubes constitutes by far the greater part of the discharge. This positive column is separated from the negative glow by a dark interval generally known as the Faraday dark space; the dimensions of this dark interval are very variable—it is sometimes altogether absent.

The positive column assumes a considerable variety of forms as the current through the gas and the pressure are varied: sometimes it is a column of uniform luminosity, at others it breaks up into a series of bright and dark patches known as striations. Some examples of these are given in fig. 17 of CONDUCTION, ELECTRIC (*Through Gases*). The distance between the striations varies with the pressure of the gas and the diameter of the tube, the bright parts being more widely separated when the pressure is low and the diameter of the tube large, than when the pressure is high and the tube small. The striations are especially brilliant and steady when a Wehnelt cathode covered with hot lime is used and the discharge produced by a number of storage cells; by this means large currents can be sent through the tube, resulting in very brilliant striations. When the current is increased the positive column shortens, retreating backwards towards the anode, and may, by using very low currents, be reduced to a glow over the surface of the anode. The electric force in the positive column has been measured by many observers. It is small compared with the forces which exist in the dark space; when the luminosity in the positive column is uniform, the force there is uniform; when the positive column is striated there are periodic variations in the electric force, the force being greater in the bright parts of the striation than in the dark.

Anode Drop of Potential.—Skinner (*Wied. Ann.* 68, p. 752; *Phil. Mag.* [6], 8, p. 387) has shown that there is a sudden change in potential between the anode itself and a point in the gas close to the anode. This change amounts to about 20 volts in air; it is thus much smaller than the cathode fall of potential, and it is also much more abrupt. There does not seem to be any region at the anode comparable in dimensions with the Crookes' dark space in which the drop of potential occurs.

The highly differentiated structure we have described is not the only way in which the current can pass through the tube. If a large Leyden jar is suddenly discharged through the tube the discharge passes as a uniform, continuous column stretching without interruption from anode to cathode; Goldstein has shown (*Verh. deutsch. phys. Ges.* 9, p. 321) that the spectrum of this discharge shows very interesting characteristics. (J. J. T.)

VÁCZ (Ger. *Waitzen*), a town of Hungary, in the county of Pest-Pilis-Solt-Kis-Kun, 20 m. N. of Budapest by rail. Pop. (1900) 16,563. It is situated on the left bank of the Danube, at the point where this river takes its southern course, and at the foot of the Nagyszál (Ger. *Waitzenberg*), on the outskirts of the Carpathians. It is the seat of a Roman Catholic bishopric, founded in the 11th century, and contains a beautiful cathedral, built in 1761–1777, after the model of St Peter's at Rome. Amongst other buildings are the episcopal palace, with a museum of Roman and medieval antiquities, several convents, and the principal deaf and dumb institute in the country. There are large vineyards in the neighbouring hilly district, and the exportation of grapes is extensively carried on. Vácz was the scene of two victories gained by the Austrians against the Turks, one in 1597 and the other in 1684.

VADE-MECUM, a Latin phrase meaning literally "come with me" (*vade*, imperative of *vadere*, to go or come; *cum*, with; *me*, abl. of *ego*, I), and used in French, Spanish and English for something that a person is in the habit of constantly taking about with him, especially a book of the nature of a handy guide or work of reference.

VAGRANCY (formed from "vagrant," wandering, unsettled; this word appears in Anglo-Fr. as *wakerant* and O.Fr. as *wancrant*, and is probably of Teut. origin, cf. M.L.G. *welkern*, to walk about; it is allied to Eng. "walk," and is not to be directly referred to Lat. *vagari*), the state of wandering without any settled home; in a wider sense the term is applied in England and the United States to a great number of offences against the good order of society. An English statute of 1547 contains the first mention of the word "vagrant," using it synonymously with "vagabond" or "loiterer." Ancient statutes quoted by Blackstone define vagrants to be "such as wake on the night and sleep on the day and haunt customable taverns and alehouses and routs about; and no man wot from whence they come ne whither they go." The word vagrant now usually includes idle and disorderly persons, rogues, vagabonds, tramps, unlicensed pedlars, beggars, &c.

The social problem of vagrancy is one that in 1910 had not yet been satisfactorily dealt with, so far as the United Kingdom is concerned. Indeed, the legislation of the early 19th century remained still in force in England and Wales. In early times, legislation affecting the deserving poor and vagrants was blended. It was only very gradually that the former were allowed to run a freer course, but provisions as to vagrancy and mendicity, including stringent laws in relation to constructive "sturdy beggars," "rogues" and "vagabonds," formed, until well on in the 19th century, a prominent feature of Poor Law legislation. In 1713 an act was passed for reducing the laws relating to rogues, vagabonds, sturdy beggars and vagrants into one act, and for more effectually punishing them and sending them to their homes, the manner of conveying them including whipping in every county through which they passed. This act was in turn repealed in 1740; the substituted consolidation act (13 Geo. II. c. 24), embracing a variety of provisions, made a distinction between idle and disorderly persons, rogues and vagabonds and incorrigible rogues. Four years later was passed another statute which continued the rough classification already mentioned. The laws relating to idle and disorderly persons, rogues and vagabonds, incorrigible rogues and other vagrants in England were again consolidated and amended in 1822, but the act was superseded two years later by the Vagrancy Act (5 Geo. IV. c. 83), which in 1910 was the operative statute.

The offences dealt with under the act of 1824 may be classified as follows: (1) offences committed by persons of a disreputable mode of life, such as begging, trading as a pedlar without a licence, telling fortunes, or sleeping in outhouses, unoccupied buildings, &c., without visible means of subsistence; (2) offences against the poor law, such as leaving a wife and family chargeable to the poor rate, returning to and becoming chargeable to a parish after being removed therefrom by an order of the justices, refusing or neglecting to perform the task of work in a workhouse, or damaging clothes or other property belonging to the guardians; (3) offences committed by professional criminals, such as being found in possession of house-breaking implements or a gun or other offensive weapon with a felonious intent, or being found on any enclosed premises for an unlawful purpose, or frequenting public places for the purpose of felony.

Offences specially characteristic of vagrancy are begging, sleeping out, and certain offences in casual wards, such as refusal to perform a task of work and destroying clothes. Persons committing these last-mentioned offences are classed as "idle and disorderly persons" and are liable on summary conviction to imprisonment with hard labour for fourteen days or on conviction by a petty sessional court to a fine of £5 or a month's imprisonment with or without hard labour. A second conviction makes a person a "rogue and vagabond" liable on summary conviction to imprisonment for fourteen days or on conviction by a petty sessional court to a fine of £25 or imprisonment for three months with or without hard labour. Any person sleeping out without visible means of subsistence is a rogue and vagabond, or on second conviction an incorrigible rogue, while an ordinary beggar is an idle and disorderly person. Under the poor law as reformed in 1834 the primary duty of boards of guardians was to relieve destitute persons within their district, but legislation and administration gradually widened that duty, so that eventually they came to administer relief to vagrants also, or casual paupers, as they are officially termed.

Casual Wards.

Within the limits prescribed by the local government board the treatment in English casual wards varies in a striking degree. Before admission to a casual ward a vagrant requires an order, obtained either from a relieving officer or his assistant. In cases of sudden or urgent necessity, however, the master of the workhouse has power to admit without an order. Generally speaking, vagrants are not admitted to the casual wards before 4 p.m. in winter or 6 p.m. in summer. On admission, they are supposed to be searched, but this is not usually done with much thoroughness; broken food found on them is sometimes allowed to be eaten in the ward; money, pipe, tobacco, &c., are restored to them on discharge. As soon as practicable after admission vagrants are required to be cleansed in a bath with water of suitable temperature. Their clothes are taken away and disinfected and a night-shirt provided. Sleeping accommodation is provided either on the cellular system or in associated wards, the proportion of workhouses providing the former being 2 to 1. Vagrants are, as a general rule, supposed to be detained two nights and are required to perform a task of work. This consists of stone-breaking, wood-sawing, wood-chopping, pumping, digging or oakum picking, and should represent nine hours' work.

The supervising authority has endeavoured, in prescribing the work to which vagrants are put, to make it as deterrent as possible, but in practice the work presents little difficulty to the habitual vagrant, and very few workhouses enforce the two nights' detention rule. The fare provided for vagrants is a welcome relief, too, from their usual scanty fare, and in many of the modern workhouses the wards are almost luxurious in their style and equipment.

AN IONIZATION MANOMETER

By O. E. Buckley

RESEARCH LABORATORY, AMERICAN TELEPHONE AND TELEGRAPH COMPANY AND WESTERN ELECTRIC COMPANY

Received by the Academy, November 13, 1916

Heretofore the only manometers available for measuring extreme vacua have been the Knudsen manometer and the Langmuir molecular gauge. Both of these have serious disadvantages due to their delicate construction and slowness of action. A new manometer free from these objections and with a greater range of presssure than either has been developed. This manometer makes use of the ionization of gas by an electron discharge.

The manometer consists of three electrodes sealed in a glass bulb which serve as cathode, anode, and collector of positive ions. The cathode may be any source of pure electron discharge such as a Wehnelt cathode or a heated tungsten or other metallic filament. The exact forms of the electrodes are not of great importance. The collector is preferably situated between the other two electrodes and of such form as not to entirely block the electron current to the anode. A milliammeter is used to measure the current to the anode and a sensitive galvanometer to measure the current from the collector which is maintained negative with respect to the cathode so as to pick up only the positive ions.

If there were no gas at all in the space between the electrodes a pure electron current would flow from cathode to anode and no current would flow to the collector. However, if gas is present positive ions are formed by collision in amount proportional to the electron current and the number of gas molecules in the space. Since the collector is negative with respect to the cathode a certain proportion of the positive ions, depending on the form, dimensions, and potentials of the electrodes, will flow to the collector. Hence the ratio of the collector current to the anode current is proportional to the pressure and may be used to measure the pressure when the constant of proportionality has been determined.

This relation has been tested experimentally with air over a pressure range from 10^{-3} to 4×10^{-6} mm. of mercury by comparison with McLeod and Knudsen manometers. The actual apparatus used consisted of a glass bulb 6 cm. in diameter enclosing three parallel, V shaped filaments of thin platinum strip, each about 3.5 cm. long, placed 5 mm. apart, the collector being between the other two. Leads from both ends of each filament were brought through the glass. This arrange-

Originally published in Proc. Natl. Acad. Sci. **2**, 683–685 (1916).

ment permits glowing the electrodes to free them from occluded gases. An oxide coated filament was used for the cathode. The bulb was sealed to a large glass reservoir which was connected to a high vacuum pump and either the Knudsen or McLeod manometers. When the latter was used a liquid-air trap served to keep the mercury vapor of the McLeod manometer out of the ionization manometer.

Currents from 0.2 to 2.0 milliamperes were used with from 100 to 250 volts between cathode and anode. The collector was held at 10 volts negative with respect to the cathode. The resulting current to the collector at a pressure of 10^{-3} mm. was about one-thousandth the current to the anode and at lower pressures was proportionately less. Hence at a pressure of 10^{-6} mm. with a current of 2.0 milliamperes to the anode a collector current of 2×10^{-9} amperes could be obtained. With a sensitive galvanometer much lower pressures could easily be measured.

Experiments with hydrogen and with mercury vapor in place of air gave constants of proportionality nearly the same as with air.

The advantages of this type of manometer are readily apparent. Its range compared to that of other high vacuum gauges is very large, extending from more than 10^{-3} mm. to as low pressures as can be obtained, without any change of apparatus. On account of its simplicity of construction it is inexpensive and exactly reproducible. Since there are no moving parts there are no difficulties due to vibration. The pressures of vapors which would not be registered on the McLeod gauge are measured by the ionization manometer. One of the greatest advantages is the rapidity and ease with which measurements of a varying pressure may be made since only the reading of a galvanometer need be followed.

Many applications for which other manometers cannot readily be used at once suggest themselves, such as the measurement of vapor pressures of metals, etc. Since the device may be made with extremely small volume the pressure of very small quantities of gas may be measured. It would also be useful to measure pressure changes over a long period of time for which more expensive manometers could not well be employed.

A number of interesting physical measurements other than the measurement of pressure can be made with devices operating on the principle of this manometer, among which is that of the removal of occluded gases by electron bombardment. It is also hoped that experiments with various gases will give some information as to the relative cross sectional areas which different kinds of molecules present to the electron discharge, for although the constant of the manometer was found approximately the same for hydrogen, air, and mercury vapor, more exact measurements might show differences due to different molecular diameters.

PROCEEDINGS

OF THE

NATIONAL ACADEMY OF SCIENCES

Volume 3 MARCH 15, 1917 Number 3

THE CONDENSATION AND EVAPORATION OF GAS MOLECULES

By Irving Langmuir

RESEARCH LABORATORY, GENERAL ELECTRIC COMPANY, SCHENECTADY, N. Y.

Communicated by A. A. Noyes, January 11, 1916

Several years ago,[1] I gave evidence that atoms of tungsten, molybdenum, or platinum vapors, striking a clean, dry glass surface in high vacuum, are condensed as solids at the first collision with the surface. Subsequently, similar evidence[2] was obtained in connection with a study of chemical reactions in gases at low pressures. It was concluded that in general, when gas molecules strike a surface, the majority of them "do not rebound from the surface by elastic collisions, but are held by cohesive forces until they evaporate from the surface." In this way a theory of adsorption was developed[3] which has been thoroughly confirmed by later experiments. It was stated: "The amount of material adsorbed depends on a kinetic equilibrium between the rate of condensation and the rate of evaporation from the surface. Practically every molecule striking the surface condenses (independently of the temperature). The rate of evaporation depends on the temperature (van't Hoff's equation) and is proportional to the *fraction* of the surface covered by the adsorbed material."

R. W. Wood[4] described some remarkable experiments in which a stream of mercury atoms impinges upon a plate of glass held at a definite temperature. With the plate cooled by liquid air, all the mercury atoms condense on the plate, but at room temperature all the atoms appear to be diffusely reflected.

The whole question of the evaporation, condensation, and possible reflection of gas molecules has been discussed at some length in two recent papers[5,6]. It was pointed out that, in Wood's experiments, there are excellent reasons for believing that the mercury vapor actually condenses on the glass at room temperature, but evaporates so rapidly

Originally published in Proc. Natl. Acad. Sci. **3**, 141–147 (1917).

that no visible deposit of mercury is formed. Further evidence of the absence of reflection is furnished by the operation of the 'Condensation Pump.'[7]

In a second paper, Wood[8] gives an account of some still more striking experiments. A stream of cadmium atoms, striking the walls of a well exhausted glass bulb, does not form a visible deposit unless the glass is at a temperature below about −90°C. If, by cooling the bulb for a moment with liquid air, a deposit is started, this continues to grow in thickness even after it is warmed to room temperature. From these and similar observations, Wood concludes that:

1. Cadmium atoms all condense on cadmium surfaces at any temperature.

2. Cadmium atoms condense on glass only if it is at a temperature below about −90°C. At higher temperatures, nearly all the atoms are reflected.

This viewpoint leads to no explanation of the changes in the reflection coefficient. The results of Wood's experiments may, however, be explained by the theory that all the atoms, striking either the glass or the cadmium surface, condense, and that subsequent evaporation accounts for the apparent reflection.

Cadmium atoms on a glass surface are acted on by totally different forces from those holding cadmium atoms on a cadmium surface. When a thick deposit of cadmium which has been distilled onto glass in vacuum, is heated quickly above its melting-point, the molten cadmium gathers together into little drops on the surface of the glass. In other words, molten cadmium does not wet glass. Therefore cadmium atoms have a greater attractive force for each other than they have for glass. Thus, single cadmium atoms on a glass surface evaporate off at a lower temperature than that at which they evaporate from a cadmium surface. It is not unreasonable to assume that in Wood's experiments, even at −90°C., the cadmium evaporated off of the glass as fast as it condensed upon it.

This theory possesses the advantage that it automatically explains the apparent reflection of cadmium atoms from a glass surface at room temperature, and indicates why this effect should be absent at low temperatures.

We shall see, moreover, that this condensation-evaporation theory explains many other facts incompatible with the reflection theory.

Let us examine for a moment the essential differences between these two theories. Wood describes his remarkable experimental results, but he has not attempted to discuss the mechanism of the underlying processes. It is clear that Wood uses the term 'reflection' merely to express the fact that under certain conditions no visible deposit is formed when the atoms strike a surface. From this point of view, condensation followed by evaporation is the same as reflection. In considering the possible mechanisms of the process, however, we must sharply distinguish between the two theories.

When an atom strikes a surface and rebounds elastically from it, we are justified in speaking of this process as a reflection. Even if the collision is only partially elastic, we may still use this term. The idea that should be expressed in the word 'reflection' is that the atom leaves the surface by a process which is the direct result of the collision of the atom against the surface.

On the other hand, according to the condensation-evaporation theory, there is no direct connection between the condensation and subsequent evaporation. The chance that a given atom on a surface will evaporate in a given time is not dependent on the length of time that has elapsed since the condensation of that atom. Atoms striking a surface have a certain average 'life' on the surface, depending on the temperature of the surface and the intensity of the forces holding the atom. According to the 'reflection' theory, the life of an atom on the surface is simply the duration of a collision, a time practically independent of temperature and of the magnitude of the surface forces.

To determine definitely which of the two theories corresponds best with the facts, I have repeated Wood's experiments under somewhat modified conditions. A small spherical bulb, together with an appendix containing cocoanut charcoal, was heated to 600°C. for about four hours while being exhausted by a condensation pump. A liquid-air trap was placed between the pump and the bulb. Some cadmium was purified by distillation and was distilled into the bulb, which was then sealed off from the pump. The cadmium in the bulb was then all distilled into the lower hemisphere. By heating this lower half of the bulb to about 140°C., the upper half remained clear, but by applying a wad of cotton, wet with liquid air, to a portion of the upper hemisphere, a uniform deposit formed within less than a minute and continued to grow, even after the liquid air was removed.

When the liquid air was applied only long enough to start a deposit, it was found in the first experiments that the deposit did not grow uniformly, but became mottled, or showed concentric rings. The outer edges of the deposit were usually much darker than the central portions. By cooling the cocoanut charcoal in liquid air, this effect disappeared entirely and the cadmium deposits became remarkably uniform in density. It is thus evident that traces of residual gas may prevent the growth of the deposit, particularly in those places which have been the most effectively cooled. This is probably due to the adsorption of the gas by the cooled metal deposit. This gas is apparently retained by the metal, even after it has warmed up to room temperature, so that vapor condensing on the surface evaporates off again at room temperature.

These results indicate how enormously sensitive such metal films are to the presence of gas. However, by using liquid air and charcoal continually during the experiments, most of these complicating factors were eliminated.

If all the cadmium is distilled to the lower half of the bulb and this is then heated to 220° in an oil bath while the upper half is at room temperature, a fog-like deposit is formed on the upper part of the bulb in about fifteen seconds. This deposit is very different from that obtained by cooling the bulb in liquid air. Microscopic examination shows that it consists of myriads of small crystals. According to the condensation-evaporation theory, the formation of this fog is readily understood. Each atom of cadmium, striking the glass at room temperature, remains on the surface for a certain length of time before evaporating off. If the pressure is very low, the chance is small that another atom will be deposited, adjacent to the first, before this has had time to evaporate. But at higher pressures this frequently happens. Now if two atoms are placed side by side on a surface of glass, a larger amount of work must be done to evaporate one of these atoms than if the atoms were not in contact. Not only does the attractive force between the cadmium atom and the glass have to be overcome, but also that between the two cadmium atoms. Therefore the rate of evaporation of atoms from pairs will be much less than that of single atoms. Groups of three and four atoms will be still more stable. Groups of two, three, four, etc., atoms will thus serve as nuclei on which crystals can grow. The tendency to form groups of two atoms increases with the square of the pressure, while groups of three form at a rate proportional to the cube of the pressure. Therefore the tendency for a foggy deposit to be formed increases rapidly as the pressure is raised or the temperature of the condensing surface is lowered.

On the other hand, according to the reflection theory, there seems to be no satisfactory way of explaining why the foggy deposit should form under these conditions.

Experiments show clearly that when a beam of cadmium vapor at very low pressure strikes a given glass surface at room temperature, no foggy deposit is formed, although when the *same quantity* of cadmium is made to impinge against the surface in a shorter time (and therefore at higher pressure) a foggy deposit results. This fact constitutes strong proof of the condensation evaporation theory.

A deposit of cadmium of extraordinary small thickness will serve as a nucleus for the condensation of more cadmium at room temperature. Let all the cadmium be distilled to the lower half of the bulb. Now heat the lower half to 60°C. Apply a wad of cotton, wet with liquid air, to a portion of the upper half for one minute, and then allow the bulb to warm up to room temperature. Now heat the lower half of the bulb to 170°C. In about thirty seconds a deposit of cadmium appears which rapidly grows to a silver-like mirror. This deposit only occurs where the bulb was previously cooled by liquid air.

The question arises: how much cadmium could have condensed on the bulb in one minute while the lower part of the bulb was at 60°C.?

The vapor pressure of cadmium has been determined by Barus[9] between the temperatures 549° and 770°C. If the logarithms of the pressures are plotted against the reciprocals of the temperature, a straight line is obtained from which the following equation for the vapor pressure (in bars) is obtained as a function of absolute temperature

$$\log p = 11.77 - \frac{6060}{T} \qquad (1)$$

At 60°C. the vapor pressure of cadmium is of the order of magnitude of 4×10^{-7} bars. Now the number of molecules of gas which strike a square centimeter of surface per second is

$$n = 2.65 \times 10^{19}\, p/\sqrt{MT} \qquad (2)$$

Substituting $M = 112$, $T = 333°$, and $p = 4 \times 10^{-7}$, we find that with saturated cadmium vapor at 60°C., $n = 5 \times 10^{10}$ atoms per second per square centimeter.

The maximum number of atoms of cadmium which can condense in *one minute* on a spot cooled in liquid air when the lower part of the bulb is at 60°C. is therefore 3.0×10^{12} atoms per square centimeter. The diameter of a cadmium atom is approximately 3.1×10^{-8} cm., so that it would require 1.0×10^{15} atoms to cover 1 square centimeter with a single layer of atoms.

Therefore the deposit which forms in one minute with the vapor from cadmium at 60°, contains only enough cadmium atoms to cover 3/1000 of the surface of the glass. Yet this deposit serves as an effective nucleus for the formation of a visible deposit.

If the lower part of the bulb is heated to 78° instead of 60°, the nucleus formed by applying liquid air for one minute causes a visible deposit to grow more rapidly (with the lower part of the bulb at 170°). But the nucleus obtained with temperatures above about 78° are not any more effective than those formed at 78°.

A calculation similar to that above shows that the deposit formed in one minute at 78° contains 2.5×10^{13} atoms per square centimeter, or enough to cover 25/1000 of the surface. If we consider that the surface of the glass contains elementary spaces each capable of holding one cadmium atom, the chance that any given cadmium atom will be adjacent to another is $1 - (1 - 0.025)^7$, or 0.16. When the surface is allowed to warm up, the single atoms evaporate, but the pairs remain. The surface is then covered to the extent of 16% of 25/1000, or 4/1000. About 2% of the atoms striking such a surface will fall in positions adjacent to those atoms already on the surface. With cadmium vapor at 170°, 1.4×10^{15} atoms per square centimeter strike the surface each second, so that 2.8×10^{13} would condense in the first second around the 4×10^{12} atoms remaining on the surface. Thus in only a few seconds the whole surface becomes covered with a layer of cadmium atoms. This explains why a surface only partially covered with cadmium atoms can serve so effectively as a nucleus. If a much smaller fraction than 0.025 of the surface is covered, however, there is a long delay in completing the first layer of atoms, so that the visible deposit is formed much more slowly.

The above experiments prove that the range of atomic forces is very small and that they act only between atoms practically in contact with each other. Thus a surface covered by a single layer of cadmium atoms behaves, as far as condensation and evaporation are concerned, like a surface of massive cadmium. This absence of transition layer is in accord with my theory of heterogeneous reactions.[10]

One of the best proofs of the correctness of the condensation-evaporation theory was obtained in experiments in which nuclei formed at liquid air temperature, were not allowed to warm up to room temperature, but only to −40°C. In this case the nuclei were formed in one minute from cadmium vapor at 54°C. The nuclei which were kept at −40°C. developed rapidly into cadmium mirrors in cadmium vapor at 170°, while those at room temperature developed extremely slowly. A still more striking demonstration of the theory was obtained when one of the nuclei was allowed to warm up to room temperature and then cooled to −40° before exposure to cadmium vapor at 170°. This nucleus did not develop nearly as rapidly as that which had not been allowed to warm up to room temperature.

These experiments prove that single cadmium atoms actually evaporate off of a glass surface at temperatures below room temperature, although they do not do so at an appreciable rate from a cadmium surface.

This theory affords a very satisfactory explanation of Moser's breath figures on glass and the peculiar effects observed in the formation of frost crystals on window panes. In fact, the theory appears capable of extension to the whole subject of nucleus formation, including, for example, the crystallization of supercooled liquids.

The final paper will be submitted to the *Physical Review* for publication.

[1] Langmuir, I., *Physic. Rev., Ithaca, N. Y.*, (Ser. 2), 1913, **2,** (329–342); *Physik. Zs., Leipzig*, **14,** 1913, (1273); Langmuir and Mackay, *Physic. Rev., Ithaca, N. Y.*, (Ser. 2), **4,** 1914, (377–386).

[2] Langmuir, I., *J. Amer. Chem. Soc., Easton, Pa.*, **37,** 1915, (1139–1167); *J. Ind. Eng. Chem., Easton, Pa.*, **7,** 1915, (349–351).

[3] Langmuir, I., *Physic. Rev., Ithaca, N. Y.*, (Ser. 2), **6,** 1915, (79–80).

[4] Wood, R. W., *Phil. Mag., London*, **30,** 1915, (300–304).

[5] Langmuir, I., *Physic. Rev., Ithaca, N. Y.*, (Ser. 2), **8,** 1916, (149).

[6] Langmuir, I., *J. Amer. Chem. Soc., Easton, Pa.*, **38,** 1916, (2250–2263).

[7] Langmuir, I., *Gen. Electric Rev., Schenectady, N. Y.*, **19,** 1916, (1060); *Philadelphia, J. Frank Inst.*, **182,** 1916, (719).

[8] Wood, R. W., *Phil. Mag., London*, **32,** 1916, (364–369).

[9] Barus, C., *Ibid.*, **29,** 1890, (150).

[10] Langmuir, I., *J. Amer. Chem. Soc., Easton, Pa.*, **38,** 1916, (2286).

The pressures on the fine side were measured with an extremely sensitive type of McLeod gauge except in the case of the first result given in the table which was estimated. The writer's own experiments[19] with the Gaede molecular pump at 8000 r.p.m. have shown that with a rough pump pressure of 20 mm. the fine side pressure was 0.0004 mm., so that the ratio of the pressures was 50,000—a result which is in accord with figures given by Gaede.

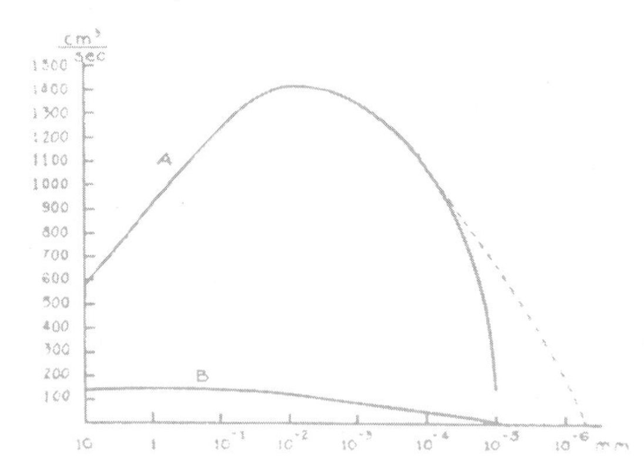

Fig. 19. Effect of Rough-pump Pressure on Speed of Gaede Molecular Pump

The speed of the pump as defined by equation (37) has been found by Gaede to vary with the magnitude of the rough-pump pressure. The curve A in Fig. 19 shows that the maximum speed is about 1400 cm.³ per second with a fore-vacuum of 0.01 mm. For comparison Gaede also shows the curve B for his rotary mercury pump, which has a speed of about 130 cm.³ per sec., at the maximum.

MERCURY VAPOR PUMPS[20]

The fact that a reduction in pressure can be obtained by a blast of steam or air has been known and applied in the industry for a long time. In steam aspirators or ejectors such as are used for producing the low pressure required in the condenser of a

[19] S. Dushman, Phys. Rev. *5*, 224 (1915).

[20] The introductory remarks are based largely upon the discussion of this subject by I. Langmuir in his paper on, "The Condensation Pump, an Improved Form of High Vacuum Pump," General Electric Review, 1916, p. 1060, also Journ. Franklin Inst., *182*, 719 (1916), Phys. Rev. *8*, 48 (1916). An excellent discussion of the mercury vapor pumps described in this section has also been written by A. Gehrts, Naturwissenschaften, 7, 983 (1919).

Originally published in *Production and Measurement of High Vacuum* (General Electric Review, 1922), pp. 57–71.

steam turbine, "the high velocity of the jet of steam causes, according to hydrodynamical principles, a lowering of pressure, so that the air to be exhausted is sucked directly into the jet." An analysis of the action of the aspirator shows, according to Langmuir, that in its action two separate processes are involved.

"1. The process by which the air is drawn into the jet.

"2. The action of the jet in carrying the admixed air along into the condensing chamber.

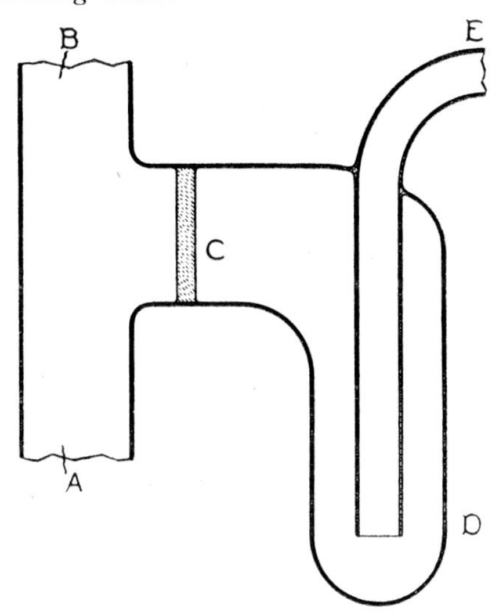

Fig. 20. **Diagram Illustrating Principle of Diffusion Pump**

"The aspirators cease operating at low pressures because of the failure of the first of these processes. If air at low pressure could be made to enter the jet, and if gas escaping from the jet could be prevented from passing back into the vessel to be exhausted, then it should be possible to construct a jet pump which would operate even at the lowest pressures."

This problem has been solved in two different ways by Gaede and Langmuir. In the pumps devised by each of these, a blast of mercury carries along the gas to be exhausted into the condenser (process 2). In order to introduce gas into the blast of mercury, Gaede has used diffusion through a narrow opening. On the other hand, Langmuir has made use of the fact that the mercury atoms on colliding with the gas molecules must impart to the latter a portion of the momentum which they possess in virtue of their high average kinetic energy, while the mercury atoms themselves are removed rapidly from the stream of mixed gases by condensation on the cooled walls.

Gaede's Diffusion Pump[21]

The action of Gaede's "diffusion" pump can best be illustrated by referring to Fig. 20. A blast of steam is blown through the tube A B, in which is fixed a porous diaphragm C. The vessel to be exhausted is attached at E. Water vapor diffuses through the capillaries in the diaphragm into the trap D where it is condensed by some refrigerating agent, while air diffuses through the diaphragm in the opposite direction into the tube A B, from where it is drawn away rapidly by the blast of steam. The result is that the pressure in E decreases and finally reaches a very low value.

A study of the phenomena of diffusion of gases through mercury vapor in narrow tubes led Gaede to the conclusion that at sufficiently low pressures, where the mean free path L of the air molecules in mercury vapor is comparable with the diameter d of the tube, the volume of air, V, diffusing through a tube of length l, per unit time, is given by relation of the form:

$$V = k\pi d^3/l \quad (40)$$

where k is a constant for any given gas A. In other words, the speed of exhaust is independent of the actual pressure of the gas in the vessel to be exhausted, and the relative decrease in the pressure per unit time therefore remains constant as the pressure in the system is decreased. This result obtained empirically by Gaede is evidently in agreement with Knudsen's deductions as stated in equations (28) and (29), Chapter I.

The actual construction of the diffusion pump is shown in Fig. 21. The porous diaphragm is replaced by a steel cylinder C with a narrow slit S whose width can be altered by means of the set screws H. The cylinder is set in the mercury trough G, which forms a seal between the low and high pressure parts. The mercury at A is heated and the stream of vapor passes over the slit in the steel cylinder in the directions indicated by the arrows. The air or other gas from the system to be exhausted (connected at F) diffuses into this mercury stream at S, and then passes out through E into the fore-pump which is connected at V. Any mercury vapor passing out through S is condensed on the

[21] Ann. Phys. *46*, 357-392 (1915).

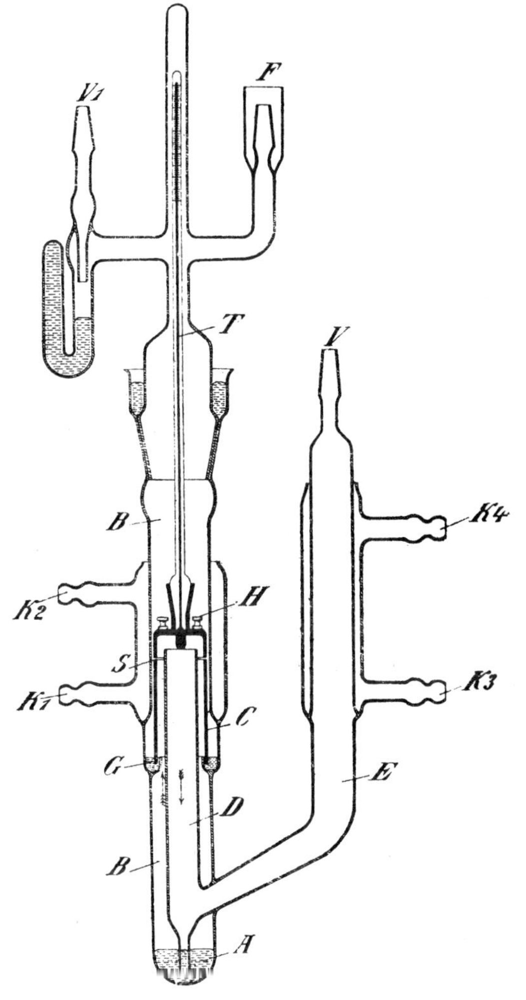

Fig. 21. Gaede Diffusion Pump

glass in the immediate neighborhood, by means of the water cooling jacket K_1 K_2. The opening V connects with the fore-pump or other source of rough vacuum and is used for exhausting the system until the pressure gets low enough for the operation of the diffusion pump to become effective. As soon as this stage is reached the mercury in the trap automatically closes this opening and the exhaust then continues by means of the diffusion pump.

According to Gaede's theory the maximum speed of the pump is attained when the width of the slit S is of the same order of magnitude as the mean free path of the gas molecules in the slit

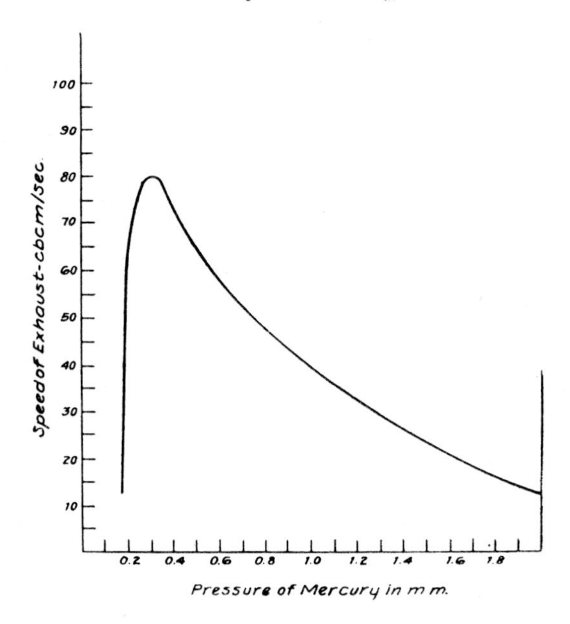

Fig. 22. Effect of Mercury Vapor Pressure on Speed of Diffusion Pump

and when the vapor pressure of the mercury is only slightly in excess of the pressure in the fore-vacuum (at V). Consequently the temperature of the mercury vapor has to be maintained at a fairly constant value. For this purpose a thermometer, T, is placed inside the tube B.

The effect of varying the temperature of the vapor (and consequently its pressure) on the speed of exhaust is shown by the data (given by Gaede) in Table XI and the plot of these in Fig. 22. These results were obtained with a slit width of 0.012 cm. The maximum speed of 80 cm.^{3} per sec. was attained at a temperature of the mercury vapor of 99° C. At this temperature the pressure of the mercury vapor is 0.27 mm., while the mean free path for air in mercury at this pressure according to Gaede's calculation is about 0.023 cm.

Table XII shows the effect of varying the width of the slit. The noteworthy fact is that the speed of the pump remains constant as the pressure in the exhausted system is decreased, a result which Gaede previously deduced from theoretical consideration, as mentioned already.

The great advantage of the diffusion pump over all the previous types of pump consists therefore in the fact that there is theoretically no limit to the degree of vacuum which can be attained by its operation. In the case of the Gaede rotary pump and all mechanical pumps the speed of exhaust decreases with decrease in pressure. In the case of the Gaede molecular pump the minimum pressure attainable depends upon the pressure in the fore-vacuum as the ratio of the pressures is constant for the pump. There is thus with all these pumps a fixed lower limit to the lowest pressure attainable in the exhaust system. While there is no such limitation with the diffusion pump, it does have the double disadvantage of low exhaust speed and the necessity of carefully regulating the temperature of the mercury vapor.

TABLE XI

EFFECT OF PRESSURE OF MERCURY VAPOR ON SPEED OF EXHAUST WITH DIFFUSION PUMP

T	P (mm.)	S	T	P	S
90°C.	0.165	13.4	118.5	0.72	51
94	0.20	60	127.5	1.10	38
97	0.24	70	134	1.51	23
99	0.27	80	139	1.84	15
113	0.55	62	143.5	2.2	11

TABLE XII

EFFECT OF WIDTH OF SLIT ON SPEED OF EXHAUST WITH DIFFUSION PUMP

WIDTH OF SLIT = 0.025 CM.		WIDTH OF SLIT = 0.004 CM.	
p	*S*	*p*	*S*
0.025 mm.	77	0.07 mm.	52
0.009	72	0.028	48
0.0025	67	0.006	40
0.0008	72	0.0015	38
0.0002	73	0.0004	41
0.00006	70	0.00007	40

Langmuir's Condensation Pump

Both these disadvantages are removed in the type of mercury vapor pump designed by Langmuir, while the advantages of the diffusion pump are retained. In constructing and operating a pump of this type it occurred to Langmuir that "the limitation of speed could be removed if some other way could be found to bring the gas to be exhausted into the stream of mercury vapor."[20] As stated previously, Langmuir comes to the conclusion that the ejector pump must become inoperative at low pressure, since at these pressures, "according to the kinetic theory of gases, the molecules in a jet of gas, passing out into a high vacuum must spread laterally, so that there would be no tendency for a gas at low pressures to be *drawn* into such a blast."

Furthermore, under these conditions, the mercury atoms condense on the walls of the inner tube near the inlet and owing to the latent heat of evaporation raise the temperature of the walls so that condensation ceases, and the mercury atoms are merely reflected from the walls in *all* directions. Consequently there is just as much tendency for the mercury to diffuse back towards the exhaust system as away from it, and the air molecules are thus prevented from entering into the mercury blast at the nozzle.

These considerations and the results of his previous investigations on the mechanism of condensation of gas molecules on solid surfaces* led Langmuir to the conclusion that the mercury atoms could readily be prevented from diffusing back in the direction from which the gas molecules are diffusing by simply cooling the walls of the tube near the mercury vapor outlet. Under these conditions the mercury atoms ought to be rapidly condensed as they strike the walls. At the same time the gas molecules diffusing in from the system to be exhausted would collide with the high speed mercury atoms at the jet and thus acquire a velocity component from the latter which would remove them rapidly from the space around the jet opening. The whole action of the pump constructed on the basis of this reasoning thus rests on the fundamental principle that the mercury vapor is rapidly condensed as it leaves the jet and the temperature is maintained so low that the mercury does not re-evaporate to any measurable extent. Langmuir has therefore suggested that pumps based on this principle should be designated as "*Condensation*" pumps.

* These investigations are discussed in Chapter VI.

A convenient type of glass condensation pump constructed on this principle is shown in Fig. 23.

"In order that the pump may function properly it is essential that the end of the nozzle L shall be located below the level at which the water stands in the condenser J. In other words,

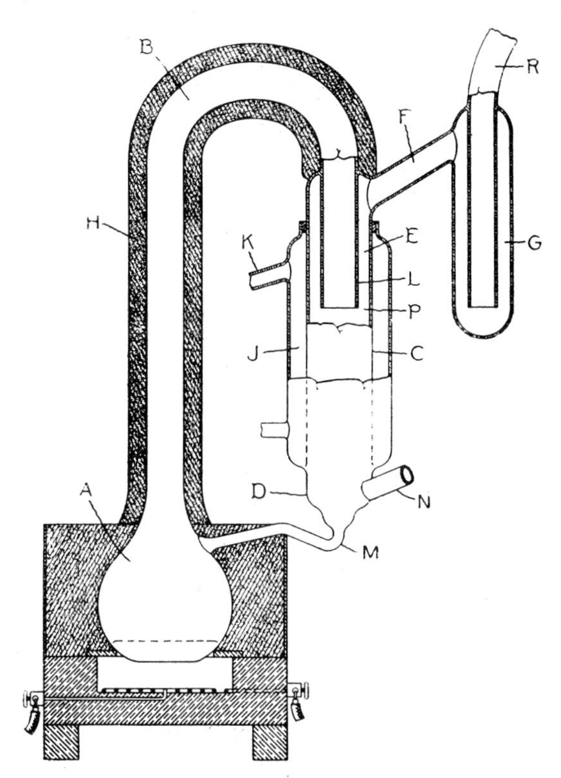

Fig. 23. Langmuir Condensation Pump, Glass Form

the overflow tube K must be placed at a somewhat higher level than the lower end of the nozzle as is indicated in the figure. The other dimensions of the pump are of relative unimportance. The distance between L and D must be sufficiently great so that no perceptible quantity of gas can diffuse back against the blast of mercury vapor, and so that a large enough condensing area is furnished.

"The pump may be made in any suitable size. Some have been constructed in which the tube B and the nozzle L were one and a quarter inches in diameter while in the other pumps this tube was only one quarter of an inch in diameter and the length of the whole pump was only about four inches. The larger the pump the greater is the speed of exhaustion that may be obtained.

"In the operation of the pump the mercury boiler A is heated by either gas or electric heating so that the mercury evaporates at a moderate rate. A thermometer placed in contact with the tube B, under the heat insulation, usually reads between 100 and 120 deg. C. when the pump is operating satisfactorily. Under these conditions the mercury in the boiler A evaporates quietly from its surface. No bubbles are formed, so there is never any tendency to bumping.

"Unlike Gaede's diffusion pump, there is nothing critical about the adjustment of the temperature. With an electrically heated pump in which the nozzle L was 7/8 in. in diameter, the pump began to operate satisfactorily when the heating unit delivered 220 watts. The speed of exhaustion remains practically unchanged when the heating current is increased even to a point where about 550 watts is applied.

"The back pressure against which the pump will operate depends, however, upon the amount and velocity of the mercury vapor escaping from the nozzle. Thus in the case above cited, with 220 watts, the pump would not operate with a back pressure exceeding about 50 bars, whereas with 550 watts back pressures as high as 800 bars did not affect the operation of the pump."

Condensation Pumps Built of Metal

For most practical purposes a glass pump has many disadvantages. Langmuir has therefore applied the same principles to the construction of a metal pump.

One such type of pump which has proved relatively simple in construction and efficient in operation is shown diagrammatically in Fig. 24. "A metal cylinder A is provided with two openings, B and C, of which B is connected to the backing pump and C is connected to the vessel to be exhausted. Inside of the cylinder is a funnel-shaped tube F which rests on the bottom of the cylinder A. Suspended from the top of the cylinder is a cup E inverted over the upper end of F. A water jacket, J, surrounds the walls of the cylinder A from the level of B to a point somewhat above the lower edge of the cup E.

"Mercury is placed in the cylinder as indicated at D. By applying heat to the bottom of the cylinder the mercury is caused to evaporate. The vapor passes up through F and is deflected by E and is thus directed downward and outward against the water-cooled walls of A. The gas entering at C passes down between A and E and at P meets the mercury vapor

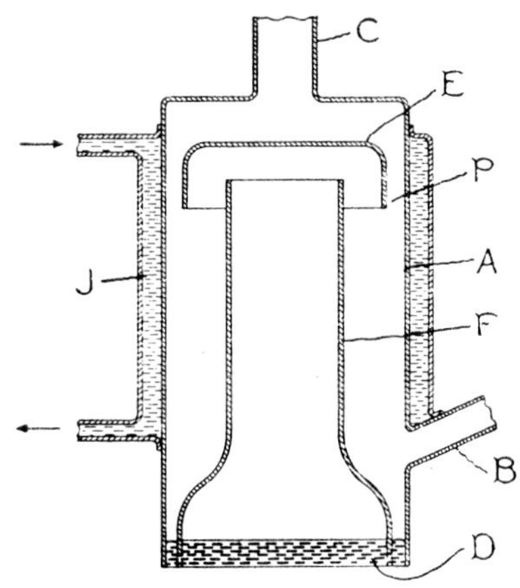

Fig. 24. Diagram of Construction of Condensation Pump, Metal Form

blast and is thus forced down along the walls of A and out of the tube B. The mercury which condenses on the walls of A falls down along the lower part of the funnel F and returns again to D through small openings provided where the funnel rests upon the bottom of the cylinder. A more detailed drawing of the pump as actually constructed is shown in Fig. 25."

A pump in which the funnel F is 3 cm. in diameter and the cylinder A is 7 cm. in diameter gives a speed of exhaustion for air of about 3000–4000 cm.^{3} per second. It operates best against an exhaust pressure of 10 bars or less and requires about 300 watts energy consumption in the heater circuit.

Degree of Vacuum Obtainable

"The condensation pump resembles Gaede's diffusion pump in that there is no definite lower limit (other than zero) below which the pressure cannot be reduced. This is readily seen from its method of operation. A lower limit could only be caused by diffusion of gas from the exhaust side (N in Fig. 23) back against the blast of mercury vapor passing down from L. The mean free path of the atoms in this blast is of the order of magnitude of a millimeter or less and the blast is moving downward with a velocity at least as great as the average molecular velocity (100 meters per second for mercury).*

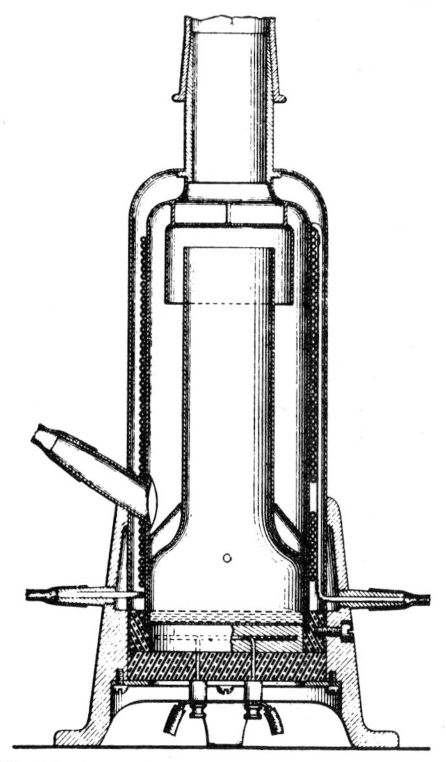

Fig. 25. Langmuir Condensation Pump, Metal Form

"The chance of a molecule of gas moving a distance about 4.6 times the mean free path without collision is only one in a hundred. To move twice this distance the chance is only 1 in 100^2, etc. If the mean free path were one millimeter the chance of a molecule moving a distance of 4.6 cm. against the blast without collision would be 1 in 10^{20}. In other words, an entirely negligible chance."

* This is apparent when we consider that no appreciable number of atoms pass up into the space E.

Actual observations with the ionization gauge* in this laboratory have shown that it is possible with the Langmuir condensation pump to obtain pressures which are of the order of 10^{-4} bar or less. The limiting factor which ordinarily makes it impossible to obtain pressures as low as this, is the continuous liberation of gas from the glass walls or metal parts, so

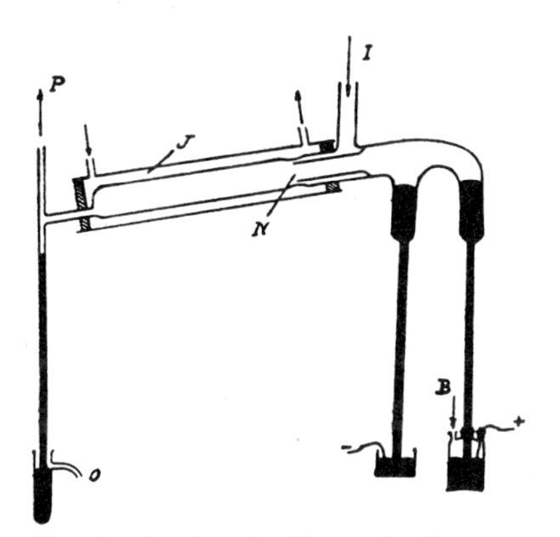

Fig. 26. Condensation Pump, Arc Type

that it becomes extremely difficult to obtain vacua in excess of the above order of magnitude. The necessary precautions in using the pump are discussed at greater length in Appendix I.

Other Forms of Mercury Vapor Pumps

Other forms of mercury vapor pumps have been described by H. B. Williams,[22] Chas. T. Knipp,[23] and L. T. Jones and N. O. Russell.[24] The construction used by the latter is shown in Fig. 26. The advantage of this form is that it "permits using the pump as a mercury still at the same time that it is being used for exhaustion purposes. Two barometer columns introduce the mercury into the arc, the arc being started by blowing in one neck of the Woulff bottle. As shown at B the mercury vapor is driven through the nozzle N, and condenses in the chamber surrounded by the water jacket, J. The condensed clean mercury is then drawn off at O." With a current of 10–15 amps., a speed of exhaust of 400 cm.^{3} per sec. was obtained.

* See Chapter III for description of this gauge.
[22] Phys. Rev. *7*, 583 (1916).
[23] Phys. Rev. *9*, 311 (1917), and *12*, 492 (1918).
[24] Phys. Rev. *10*, 301 (1916).

A simple construction for a condensation pump has also been described by W. C. Baker.[25] In this form as well as Knipp's the main object of the design is to simplify the glass blowing.

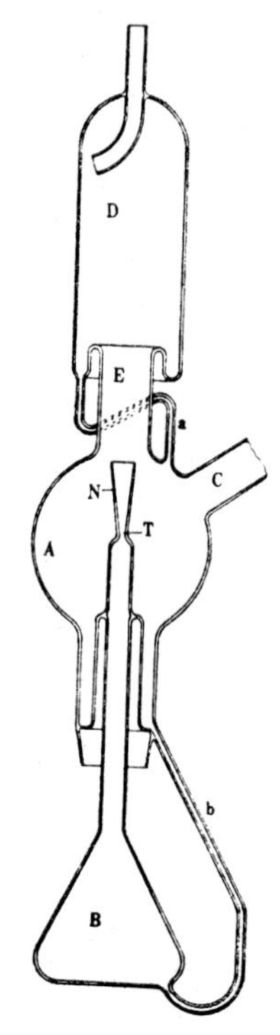

Fig. 27. Crawford's Form of Condensation Pump, Vertical Type

J. E. Shrader and R. G. Sherwood[26] have used a modified form of Langmuir condensation pump made of pyrex. Full details with all necessary dimensions are given in the original paper. The speed of the pump was measured for different amounts of energy input into the mercury heater and was found to be a maximum at about 400–500 watts input. With

[25] Phys. Rev. *10*, 642 (1916).
[26] Phys. Rev. *12*, 70 (1918).

the speed of exhaustion purposely cut down by a special constriction, the maximum speed observed was around 225 cm.^{3} per sec. and pressures as low as 2×10^{-8} mm. Hg were obtained after care had been taken to heat up all the glass parts to a temperature of 500° C. for a long time.

An interesting form of mercury vapor pump is that devised by W. W. Crawford[27] and shown in Figs. 27 and 28. "The mercury vapor generated in the boiler B at a pressure of 10 mm. of mercury or more, escapes through the narrow throat T

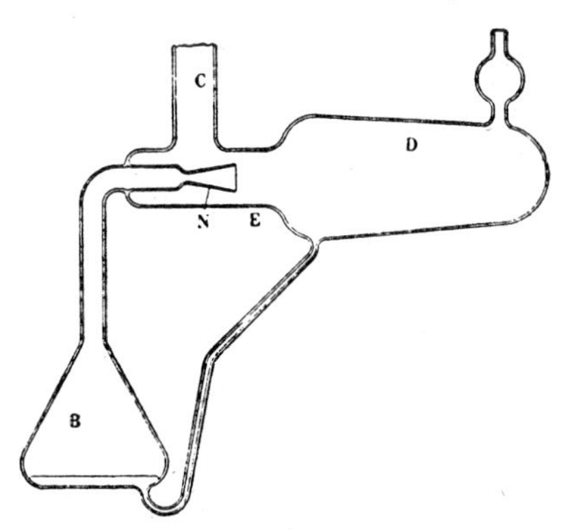

Fig. 28. Crawford's Form of Condensation Pump, Horizontal Type

(Fig. 27), ahead of the point of entrainment. The vapor expands in the diverging nozzle N, and the issuing jet passes through the tube E, which it fills, and condenses in D, mostly, where it is found at the upper end. A slight amount of vapor escapes into the chamber A and condenses there. The condensed vapor drains back through the tubes *a* and *b*, to the boiler." The vessel to be exhausted is connected at *c*, while D connects with the rough pump. The speed of the pump in series with 10 cm. of tubing, 1.9 cm. in diameter was observed to be around 1300 cm.^{3} per sec. at a boiler pressure of 10 mm. of Hg.

A two-stage mercury vapor pump to work against a primary vacuum of 2 cm. given by a water aspirator, has been described by C. A. Kraus.[28] It consists essentially of two Langmuir condensation pumps in series. The pump is very rapid and is capable of exhausting 1500 cm.^{3} to less than 10^{-4} mm. in 10 min.

[27] Phys. Rev. *10*, 557 (1917).

[28] J. Am. Chem. Sec. 39, 2183 (1917).

H. F. Stimson has also constructed a two-stage pump along the same principles,[29] which is illustrated in Fig. 29. "The operation of the pump is as follows: Cooling water entering at tube A flows up through the water jacket B above the lower end of nozzle F, up through the water jacket C above nozzle G, and out tube D. Mercury vapor from the boiler entering through tubes E flows through the nozzles F and G, is liquefied in the condensation chambers H and I, falls into the tubes K, and returns to the boiler through tube L. Gas from the vessel to be exhausted enters at M, flows past nozzle F, is compressed by the jet of mercury vapor in the condensation chamber H, and flows up through N to the intermediate pump. From here it flows past the nozzle G and is compressed through O into the chamber I to a pressure measured by the attached manometer, then out by tube P to the water aspirator."

The speed of the pump as defined by Gaede's equation was observed to be about 250 ccm. per sec.[30]

General Remarks Regarding Exhaust Procedure

As has been pointed out in a previous section, the vacuum actually attained by the use of any pump is dependent, first, on the type of pump used, and second, on the rate at which gases are given off from the walls of the vessel to be exhausted and metal parts inside it. In the case of the Gaede molecular pump, as stated previously, the degree of vacuum attainable (P_1) is dependent upon the exhaust pressure (P_2) produced by the rough pump. As the value of the ratio P_2/P_1 is about 50,000, it is evident that even with a rough pump pressure of one bar, the pressure attainable with this pump is less than 10^{-4} bar. In the case of the mercury vapor pumps there is theoretically no lower limit pressure, and the only limitation is therefore that due to the second cause mentioned above.

The gases occluded on the walls of the glass vessels consist for the most part of water vapor and carbon dioxide gas along with slight amounts of carbon monoxide and other gases which are not condensible at the temperature of liquid air. Metal parts usually contain carbon monoxide and hydrogen gases.* In order to eliminate these gases it is essential to heat the glass walls and metal parts to as high a temperature as practicable.

[29] J. Washington Acad. Sciences 7, 477 (1917).

[30] M. Volmer, Ber. *52* (b), 804 (1919), has also constructed a similar form of two-stage condensation pump, which is described briefly in an abstract in J. Chem. Soc. *116*, ii, 225 (1919).

* See Chapter IV for discussion of the absorption of gases by glass and metals.

Electron diffraction: fifty years ago

A look back at the experiment that established the wave nature of the electron, at the events that led up to the discovery, and at the principal investigators, Clinton Davisson and Lester Germer.

Richard K. Gehrenbeck

An article that appeared in the December 1927 issue of *Physical Review*, "Diffraction of Electrons by a Crystal of Nickel," has been referred to in countless articles, monographs and textbooks as having established the wave nature of the electron—in principle, of all matter.[1] Now, fifty years later, it is fitting to look back at the events that led up to this historical discovery and at the discoverers, Clinton Davisson and Lester Germer. Figure 1 shows them in their lab in 1927, together with their assistant Chester Calbick.

A shy midwesterner

Clinton Joseph Davisson, the senior investigator, was born in Bloomington, Illinois, on 22 October 1881, the first of two children. His father, Joseph, who had settled in Bloomington after serving in the Civil War, was a contract painter and paperhanger by trade. His mother, Mary, occasionally taught in the Bloomington school system. Their home was, as Davisson's sister, Carrie, characterized it, "a happy congenial one—plenty of love but short on money."

Davisson, slight of frame and frail throughout his life, graduated from high school at age 20. For his proficiency in mathematics and physics he received a one-year scholarship to the University of Chicago; his six-year career there was interrupted several times for lack of funds. He acquired his love and respect for physics from Robert Millikan; Davisson was "delighted to find that physics was the concise, orderly science [he] had imagined it to be, and that a physicist [Millikan] could be so openly and earnestly concerned about such matters as colliding bodies."

Richard K. Gehrenbeck is an assistant professor of physics and astronomy at Rhode Island College, Providence, Rhode Island.

Before finishing his undergraduate degree at Chicago, he became a part-time instructor in physics at Princeton University, where he came under the influence of the British physicist Owen Richardson, who was directing electronic research there. Davisson's PhD thesis at Princeton, in 1911, extended Richardson's research on the positive ions emitted from salts of alkaline metals. Davisson later credited his own success to having caught "the physicist's point of view—his habit of mind—his way of looking at things" from such men as Millikan and Richardson.

After completing his degree, Davisson married Richardson's sister, Charlotte, who had come from England to visit her brother. After a honeymoon in Maine Davisson joined the Carnegie Institute of Technology in Pittsburgh as an instructor in physics. The 18-hour-per-week teaching load left little time for research, and in six years there he published only three short theoretical notes. One notable break during this period was the summer of 1913, when Davisson worked with J. J. Thomson at the Cavendish laboratory in England.

In 1917, after he was refused enlistment in the military service because of his frailty, Davisson obtained a leave of absence from Carnegie Tech to do war-related research at the Western Electric Company, the manufacturing arm of the American Telephone and Telegraph Company, in New York City. His work was to develop and test oxide-coated nickel filaments to serve as substitutes for the oxide-coated platinum filaments then in use. At the end of World War I he turned down an offered promotion at Carnegie Tech to accept a permanent position at Western Electric. It was at this time that he began the sequence of investigations that ultimately led to the discovery of electron diffraction; it was also at this time that he was joined by a young colleague, Lester Halbert Germer, just discharged from active service.

An adventurous New Yorker

Germer was born on 10 October 1896, the first of two children of Hermann Gustav and Marcia Halbert Germer, in Chicago, where Dr Germer was practicing medicine. In 1898 the family moved to Canastota in upper New York state, the childhood home of Mrs Germer. Germer's father became a prominent citizen in the little town on the Erie canal, serving as mayor, president of the board of education and elder in the Presbyterian church.

Germer attended school in Canastota and won a four-year scholarship to Cornell University, graduating from there in the spring of 1917, six weeks early because of the outbreak of the war. The local newspaper, after applauding 18-year-old Lester for working as a laborer for the local paving contractors during his summer vacation, proceeded to ridicule his lazier contemporaries for sitting "day after day in the lounging places of the village," saying there is "nothin' doin'" and that "a young feller has no chanst in this durn town." (Lester, must have taken a bit of ribbing from the "idle boys" after this appeared!) Germer's studies at Cornell were partly self-directed; in their junior year he and two classmates, finding themselves "unsatisfied with the course in electricity and magnetism given ... bought a more advanced text and met regularly in the vacant class room ... and really learned something."

Upon graduation from Cornell, Germer obtained a research position at Western Electric, which he held for about two months before volunteering for the Army (aviation section of the signal corps). He

Originally published in Phys. Today **31**(1), 34–41 (1978).

apparently made no contact with Davisson then. Lieutenant Germer, among those piloting the first group of airplanes on the Western Front, was officially credited with having brought down four German warplanes. Discharged on 5 February 1919, Germer was treated in New York City for severe headache, nervousness, restlessness and loss of sleep, conditions attributed to his military campaigns, but he refused to file for compensation because "others were worse off." After three weeks of rest, he was re-hired by Western Electric—and had as his first assignment the preparation of an annotated bibliography for a new project being directed by his new supervisor, Davisson.

That fall Germer married his Cornell sweetheart, Ruth Woodard of Glens Falls, New York.

Electron emission—in court

The assignment that engaged Davisson and Germer in their first joint effort reflects one of the chief interests of the parent company, AT&T, at this time: to conduct a fundamental investigation into the role of positive-ion bombardment in electron emission from oxide-coated cathodes. Although Germer later remembered this project as having been directly related to the famous Arnold–Langmuir patent suit, that occupied Western Electric (Harold Arnold) and General Electric (Irving Langmuir) from 1916 until it was finally settled[2] by the US Supreme Court (in favor of Western Electric) in 1931, a careful examination of the documents makes it clear that Davisson and Germer's project could have related to it only in a very indirect way. The patent case concerned improvements to the earliest deForest triode tubes with metallic (tungsten or tantalum) cathodes; it dealt with evidence obtained in the years 1913 to 1916, before Davisson and Germer appeared on the scene. Nevertheless, because AT&T was deeply concerned about the efficiency and effectiveness of its triode amplifiers—key components in its recently constructed transcontinental telephone lines—Arnold assigned Davisson and Germer the task of conducting tests on oxide-coated cathodes. They published their results in the *Physical Review* in 1920, concluding that positive-ion bombardment has a negligible effect on the electron emission from oxide-coated cathodes.[3]

With this problem settled, a related question came up: What is the nature of secondary electron emission from grids and plates subjected to electron bombardment? Davisson was assigned this new task and given an assistant, Charles H. Kunsman, a new PhD from the University of California. For this work they were able to convert the positive-ion apparatus to an electron-beam apparatus. Meanwhile Germer was shifted to a project on the measurement of the thermionic properties of tungsten, a topic he pursued for about four years, both under Davisson's direction and as part of a graduate program he undertook at nearby Columbia University part time.

A startling observation

Soon after Davisson and Kunsman began their secondary electron emission studies, they observed an unexpected phenomenon that was to have crucial importance for their future experimental program: A small percentage (about 1%) of the incident electron beam was being scattered back toward the electron gun with virtually no loss of energy—the electrons were being scattered elastically. Figure 2 reconstructs this phenomenon. Previous observers had noticed this effect for low-energy electrons (about 10 eV), but none had reported it for electrons of energies over 100 eV.

Although this discovery undoubtedly had no immediate impact on the stockholders of AT&T, it affected Davisson profoundly. To him these elastically scattered electrons appeared as ideal probes with which to examine the extranuclear structure of the atom. Ernest Rutherford announced his nuclear model of the atom in 1911, the year Davisson completed his PhD; Hans Geiger and Ernest Marsden completed their definitive experimental tests of Rutherford's theory and Niels Bohr announced his planetary model of the atom in 1913, when Davisson worked with Thomson at Cambridge. So it is not surprising that Davisson was enthusiastic about the prospect of using these electrons for basic research on the structure of the atom. In Davisson's own words,

> "The mechanism of scattering, as we pictured it, was similar to that of alpha ray scattering. There was a certain probability that an incident electron would be caught in the field of the atom, turned through a large angle, and sent on its way without loss of energy. If this were the nature of electron scattering it would be possible, we thought, to deduce from a statistical study of the deflections some information in regard to the field of the deflecting atom . . . What we were attempting . . . were atomic explorations similar to those of Sir Ernest Rutherford . . . in which the probe should be an electron instead of an alpha particle."

In fact, Davisson was so enthusiastic about a full-scale assault on the atom that he was able to convince his superiors to let him and Kunsman devote a large fraction of their time to it, and to give them the necessary shop backup.

BELL LABORATORIES

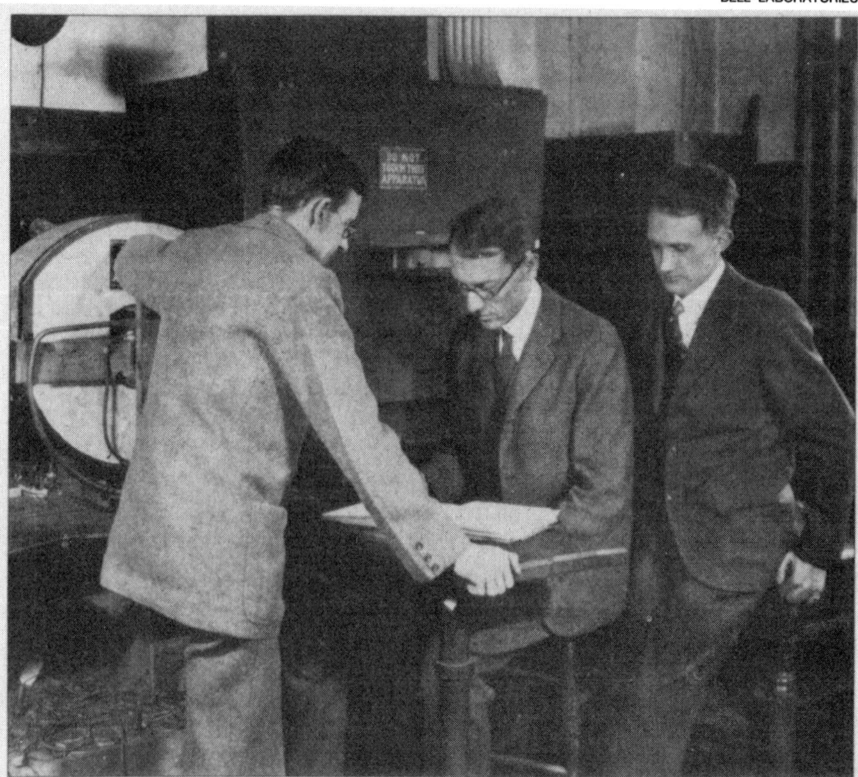

Davisson, Germer and Calbick in 1927, the year they demonstrated electron diffraction. In their New York City laboratory are Clinton Davisson, age 46; Lester Germer, age 31, and their assistant Chester Calbick, age 23. Germer, seated at the observer's desk, appears ready to read and record electron current from the galvanometer (seen beside his head); the banks of dry cells behind Davisson supplied the current for the experiments. Figure 1

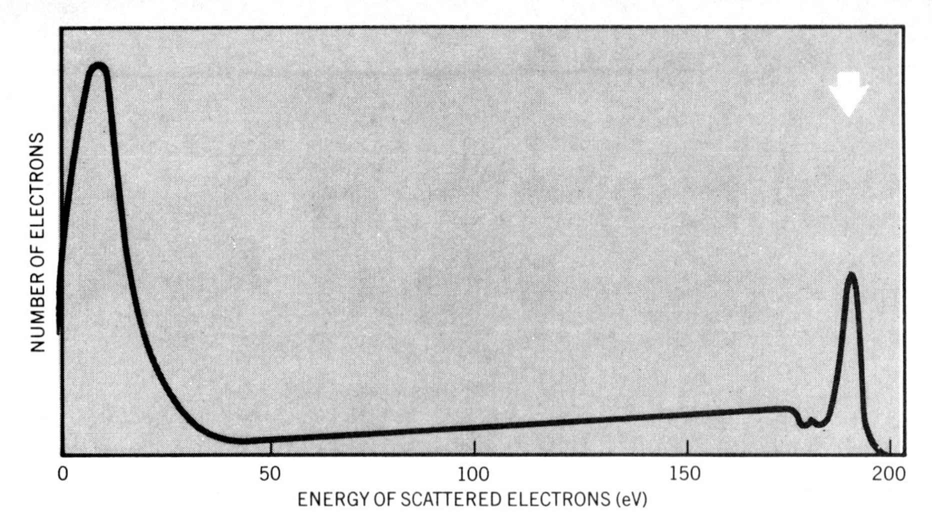

Electron-scattering peak. The energy of the scattered electrons varies from almost zero to that of the incident beam (indicated by the arrow). This is a reconstruction of the type of observation that led Davisson and Charles Kunsman to conclude that some electrons were being scattered elastically. Davisson saw these as possible probes of the electronic structure of the atom, in analogy to Rutherford's use of alpha particles to explore the nucleus. Figure 2

The basic piece of apparatus, built to order by a talented machinist and glass-blower, Geroge Reitter, was a vacuum tube with an electron gun, a nickel target inclined at an angle of 45° to the incident electron beam and a Faraday-box collector, which could move through the entire 135° range of possible scattered electron paths; it is diagrammed in figure 3. The Faraday box was set at a voltage to accept electrons that were within 10% of the incident electron energy.

After two months of experimentation, Davisson and Kunsman submitted a two-column paper to *Science,* in which they sketched the main features of their scattering program, presented a typical curve of their data, proposed a shell model of the atom for interpreting these results, and offered a formula for the quantitative prediction of the implications of the model.[4] Unfortunately their attempts to link together their data, the model and the predictions were anything but definite—quite out of keeping with the Rutherford-Geiger-Marsden tradition.

Although Davisson (and Kunsman) must have been somewhat disappointed at the limited success of their initial venture, they pressed on with additional experiments. In the next two years they built several new tubes, tried five other metals (in addition to nickel) as targets, developed rather sophisticated experimental techniques at high vacuum ("the pressure became less than could be measured, i.e., less than 10^{-8} mm Hg,") and made valiant theoretical attempts to account for the observed scattering intensities. The results were uniformly unimpressive; several of the studies were not even published. In fact, the generally disheartened atmosphere that seems to have prevailed by the end of 1923 is indicated by the fact that Kunsman left the company and Davisson abandoned the scattering project.

A year later, however, Davisson was ready to have another try at electron scattering. Was this change of heart prompted by Davisson's strong attraction to the project? Was it his eagerness to obtain additional information about the extranuclear structure of the atom? In any case, in October 1924 Germer was put back on the scattering project in place of the departed Kunsman. Germer, who had already completed several thermionic-emission studies, was returning to Western Electric after a 15-month illness. Regarding his development as a physicist by this time, Germer later recollected:

"I learned relatively little at Columbia . . . but was nevertheless fortunate in working . . . with Dr C. J. Davisson. I learned a simply enormous amount from him. This included how to do experiments, how to think about them, how to write them up, how even to learn what other people had previously done in the field. I am quite certain that I do really owe to Dr Davisson much the best part of my education, and I am not really convinced that it is so inferior to that obtained in more conventional ways. It is certainly different."

A "lucky break" and a new model

So the scattering experiments were finally resumed. One can easily imagine, then, the feelings of disappointment and frustration that Davisson and Germer must have shared when, soon after the project had been restarted, they discovered a cracked trap and badly oxidized target on the afternoon of 5 February 1925, as the notebook entry in figure 4 shows. What it meant in simple terms was that the experiments with the specially polished nickel target, discontinued for almost a year, were to be delayed again. Apparently Germer's attempts to revitalize the tube after its long period of disuse by repumping and baking (outgassing) were to be for nought; an additional delay for repairs was necessary.

This was not the only time that a tube had broken during a scattering experiment, nor was it to be the last. Nor was the method of repair unique, for the method of reducing the oxide on the nickel target by prolonged heating in vacuum and hydrogen had been used once before (unsuccessfully; that time it had led to the formation of a "black precipitate" and "no apparent cleaning up of the nickel"). This particular break and the subsequent method of repair, however, had a crucial role to play in the later discovery of electron diffraction.

By 6 April 1925 the repairs had been completed and the tube put back into operation. During the following weeks, as the tube was run through the usual series of tests, results very similar to those obtained four years earlier were obtained. Then suddenly, in the middle of May, unprecedented results began to appear, as shown in figure 5. These so puzzled Davisson and Germer that they halted the experiments a few days later, cut open the tube, and examined the target (with the assistance of the microscopist F. F. Lucas) to see if they could detect the cause of the new observations.

What they found was this: The polycrystalline form of the nickel target had been changed by the extreme heating until it had formed about ten crystal facets in the area from which the incident electron beam was scattered. Davisson and Germer surmised that the new scattering pattern must have been caused by the new crystal arrangement of the target. In other words, they concluded that it was the arrangement of the atoms in the crystals, not the structure of the atoms, that was responsible for the new intensity pattern of the scattered electrons.

Thinking that the new scattering patterns were too complicated to yield any useful information about crystal structure, Davisson and Germer decided that a large single crystal oriented in a known direction would make a more suitable target than a collection of some ten small facets randomly arranged. Because neither Davisson nor Germer knew much about crystals, they, assisted by Richard Bozorth, spent several months examining the damaged target and various other nickel surfaces until they were thoroughly familiar with the x-ray diffraction patterns (note!) obtained from nickel crystals in various states of preparation and orientation.

By April 1926 they had obtained a suitable single crystal from the company's metallurgist, Howard Reeve, and cut, etched and mounted it in a new tube that allowed for an additional degree of freedom of measurement; the collector could now rotate in azimuth (the 360° angle circling the beam axis) as well as in colatitude. The design of the new tube reflected their expectation of finding certain "transparent directions" in the crystal along which the electrons would move with least resistance. They expected these special directions to coincide with the unoccupied lattice directions.

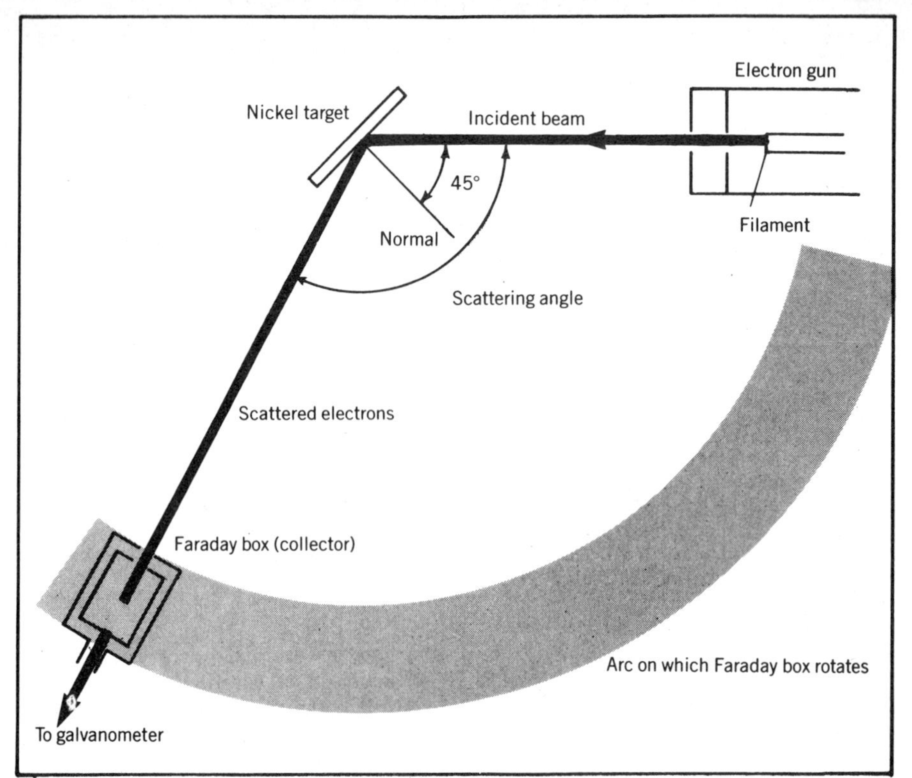

Scheme of the first scattering tube, which served as a prototype for the group's later models. Davisson and Germer later included mechanisms for rotating the target azimuthally 360° about the beam axis and for changing the angle of the incident beam with respect to the normal to the target. In their 1926–27 work the incident beam was perpendicular to the target face, and the scattering angle was called the "colatitude angle." Figure 3

More than a "second honeymoon"

Having suffered disappointment with the results of the original scattering experiments performed with Kunsman, Davisson must have been doubly disheartened by the meager returns he and Germer obtained with the new tube. After an entire year spent in preparation, and with a new tube and a new theory in hand, they obtained experimental results that were even less interesting than those from the earliest experiments. The new colatitude curves showed essentially nothing, and even the new azimuth curves gave at best only a weak indication of the expected three-fold symmetry of the nickel crystal about the incident beam.

Davisson must have been quite pleased with the prospect of getting away for a few months during the summer of 1926, when he and his wife had planned a vacation trip to relax and visit relatives in England. Mrs Davisson recalled that this summer had been chosen for the trip because her sister, May, and brother-in-law, Oswald Veblen of Princeton University, were available to stay with the Davisson children at that time. As Davisson wrote to his wife, then at the Maine cottage making arrangements for the children: "It seems impossible that we will be in Oxford a month from today—doesn't it? We should have a lovely time—Lottie darling—It will be a second honeymoon—and should be sweeter even than the first." Something was to happen on this particular trip, however, to turn it into more than the "second honeymoon" Davisson envisioned.

Theoretical physics was undergoing fundamental changes at this time. In the early months of 1926 Erwin Schrödinger's remarkable series of papers on wave mechanics appeared, following Louis de Broglie's papers of 1923–24 and Albert Einstein's quantum-gas paper of 1925. These papers, along with the new matrix mechanics of Werner Heisenberg, Max Born and Pascual Jordan, were the subject of lively discussions at the Oxford meeting of the British Association for the Advancement of Science. Davisson, who generally kept abreast of recent developments in his field but appears to have been largely unaware of these recent developments in quantum mechanics, attended this meeting. Imagine his surprise, then, when he heard a lecture by Born in which his own and Kunsman's (platinum-target) curves of 1923 were cited as confirmatory evidence for de Broglie's electron waves![5]

After the meeting Davisson met with some of the participants, including Born and possibly P.M.S. Blackett, James Franck and Douglas Hartree, and showed them some of the recent results that he and Germer had obtained with the single crystal. There was, according to Davisson, "much discussion of them." All this attention might seem strange in light of the relatively feeble peaks Davisson and Germer had obtained, but even these may have been exciting to physicists already convinced of the basic correctness of the new quantum theory. It may also reflect the fact that several European physicists, Walter Elsasser (Göttingen), E.G. Dymond (Cambridge, formerly Göttingen and Princeton), and Blackett, James Chadwick and Charles Ellis of Cambridge[6] had attempted similar experiments and abandoned them because of the difficulties of producing the required high vacuum and detecting the low-intensity electron beams. Apparently they were encouraged by these results, which appeared so unimpressive to Davisson. At any rate, Davisson spent "the whole of the westward transatlantic voyage . . . trying to understand Schrödinger's papers, as he then had an inkling . . . that the explanation might reside in them"—no doubt to the detriment of the "second honeymoon" in progress.

Back at Bell Labs (as the engineering arm of Western Electric has been called since 1925), Davisson and Germer examined several new curves that Germer had obtained during Davisson's absence. They found a discrepancy of several degrees between the observed electron intensity peaks and the angles they expected from the de Broglie–Schrödinger theory. To pursue this matter further they cut the tube open and carefully examined the target and its mounting. After finding that most of the discrepancy could be accounted for by an accidental displacement of the collector-box opening, they "laid out a program of thorough search" to pursue the quest of diffracted electron beams. In typical Davisson fashion, however, this quest was preceded by a period of careful preparation, including an important change in the experimental tube. As Davisson wrote to Richardson in November,

"I am still working at Schrödinger and others and believe that I am beginning to get some idea of what it is all about. In particular I think that I know the sort of experiment we should make with our scattering apparatus to test the theory."

Found—a "quantum bump"

It was three weeks before the "thorough search" was begun. The importance that Davisson (and Bell Labs) had come to attach to this project can be surmised from the addition to it of a new assistant, Chester Calbick, a recently graduated electrical engineer. After about a month of experimenting, during which time Calbick took charge of operating the experiment, they gave the newly prepared tube a thorough set of consistency tests. During one attempt by Germer to reactivate the tube in late November the tube broke, but with little damage. (Strangely, little damage can be considered "lucky" in this case, whereas it would have been "unlucky" in the case of the 1925 break!)

The first experiments with the new tube yielded no significant results; the colatitude and azimuth curves looked much as before, and the new experiments added by Davisson "to test the theory" were uninformative as well. These tests consisted of varying the accelerating voltage, and hence electron energy E, for fixed colatitude and azimuth settings, and were designed to see if any effect could be discerned for a changed electron wavelength λ, according to the de Broglie relationship, $\lambda = h/(2\,mE)^{1/2}$.

A concerted search for "quantum peaks" (voltage-dependent scattered electron beams) was launched by late December. These attempts revealed only "very feeble" peaks. The situation changed dramatically on 6 January 1927, however; the data for that day are accompanied by the remark, in Calbick's neat handwriting: "Attempt to show 'quantum bump' at an intermediate [colatitude] angle. Bump develops at 65 V, compared with calculated value for 'quantum bump' of V = 78 V." Then, stretched across the bottom of the page in Germer's unmistakable bold strokes, is the additional remark: "First Appearance of Electron Beam." A portion of the notebook page is reproduced in figure 6.

The data for this curve are extremely interesting. Noting from the figure that the readings were taken in one-volt intervals on either side of 79 volts, whereas the steps are 2, 5 and then 10 volts elsewhere, we see that a peak was expected at about 78 volts. But the experiment yielded a single large current at 65 volts. The experimenters took immediate notice of this spike, making a second run in one-volt steps around 65 volts, which on a graph shows a clear peak centered on 65 volts. It is easy to imagine the excitement that must have accompanied this sudden turn of events, moving Germer to sprawl his glad tidings across the bottom of the page!

With this single critical result in hand, the experimental situation changed suddenly. The next day, 7 January, they ran several additional voltage curves, one for each of four different colatitude positions. A voltage peak appeared at a colatitude angle of 45° that was even greater than that at 40°, where the collector had been set the previous day. On the eighth, a new colatitude curve was run at a voltage of 65 volts, and the first true and unmistakable colatitude peak was observed—this was what Davisson had been looking for since 1920! Skipping Sunday, they next ran an azimuth curve at 65 volts and a colatitude of 45°. This time the threefold azimuthal symmetry was immediately apparent. Figure 7 shows these curves.

The notebook entry for 5 February 1925 records, in Germer's handwriting, the discovery of the broken tube that interrupted the scattering experiments once again. It was this break, however, which initiated a chain of events that eventually led to the preparation of a single crystal of nickel as the target, and to a shift of Davisson's interest from atomic structure to crystal structure. Reproduced by courtesy of Bell Laboratories. Figure 4

The experiments that were carried out during the next two months show that Davisson, Germer and Calbick, having finally found and positively identified one set of electron beams, could now find and identify others quickly. This block of experiments continued through 3 March, when Calbick left for a month on family business. Comparing this with earlier periods of Davisson's long contact with electron scattering, we see that not since the early days of the original Davisson-Kunsman experiments had there been such intense and concentrated effort in a single well defined direction. The presence of a clear, unambiguous goal certainly must have been a major factor in the two cases, an ingredient lacking at other times.

Another factor undoubtedly urging Davisson on to rapid (but careful) experimentation and possible early publication was his feeling that others might be pursuing similar investigations at that time. Recalling his conversations at Oxford and the comments that had been made about the interest of others in this matter, he sent off an article to Richardson in March with the accompanying note:

"I hope you will be willing, if you think it at all desirable, to get in touch with the editor of *Nature* with the idea of securing early publication. We know of three other attempts that have been made to do this same job, and naturally we are somewhat fearful that someone may cut in ahead of us." As it turned out these efforts had long been abandoned, but he had no way of knowing that. Nevertheless, another investigator, unknown to Davisson at that time, was indeed making progress at revealing the phenomena of electron diffraction with high-voltage electrons and thin metal foils. This was J.J.'s son, G.P. Thomson; his and Andrew Reid's first note was published in *Nature* just one month after Davisson and Germer's.[7]

A conservative note and a bold one

Davisson and Germer's *Nature* article was an extremely conservative expression of the new experimental evidence for electron diffraction.[8] Its title, "The Scattering of Electrons by a Single Crystal of Nickel," bears a closer connection to the early work of Davisson and Kunsman than it does to the new wave mechanics. Although the paper included a table linking the scattered electron peaks to the corresponding de Broglie wavelengths, it was not until the last two paragraphs that a tentative suggestion was made about the important implications of the work: The results were "highly suggestive . . . of the ideas underlying the theory of wave mechanics."

This cautious attitude may have been due to the problem that Davisson and Germer had in making the proper correlation between their data points and the theory; they found it necessary to hypothesize an *ad hoc* "contraction factor" of about 0.7 for the nickel-crystal spacing to get approximate correspondence between the de Broglie wavelengths and their data. Even at that, only eight of the thirteen beams described were clearly amenable to this analysis.

This cautious attitude appears to have been abandoned in a concurrent article by Davisson alone for an in-house publication, the *Bell Labs Record.*[9] The very title, "Are Electrons Waves?" suggests this difference. After reviewing the evidence that led Max von Laue to think of x rays as being wave-like, he cited his and Germer's recent work with electrons, urging a similar conclusion in this case. Although this article gave its readers no actual data on the experimental evidence for electron waves, it clearly indicates that Davisson's thoughts (and certainly Germer's as well) on the subject were not nearly as reserved as the *Nature* article suggests.

One other public announcement of the recent discoveries was made at this time. In a paper presented at the Washington meeting of The American Physical Society on 22–23 April 1927 and abstracted in the *Physical Review* in June,[10] Davisson and Germer basically repeated what they had stated in their *Nature* article, and then added an intriguing final paragraph. Referring to the three anomalous beams that could not be fitted into the analysis in the *Nature* article, they suggested that these "offer strong evidence that there exists in this crystal a structure which has not been hitherto observed for nickel." This statement implies Davisson and Germer had already gone beyond the point of using the "known" structure of the nickel crystal to find out about the possibility of the wave properties of the electron; they were now using the "known" electron waves to learn new facts about the nickel crystal. Between March, when the *Nature* article was submitted, and April, when the *Phys. Rev.* abstract was prepared, results that had been embarrassing to the theory had become a potential new application of that very theory!

True to form, however, Davisson and Germer did not sit back and rest on a "job well done"; they recognized the considerable work necessary to resolve a number of questions still outstanding. Among these were:

- the problem of the "anomalous" beams mentioned above,
- the *ad hoc* "contraction factor" that they had found necessary to attribute to the nickel crystal and
- extension of their electron energies over a greater range, and sharpening and refining their diffraction peaks.

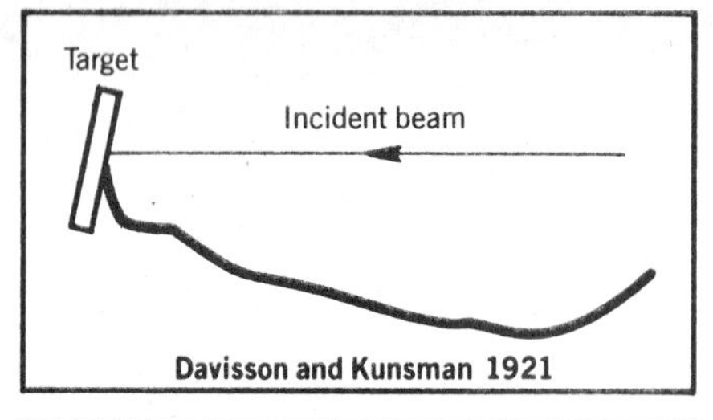

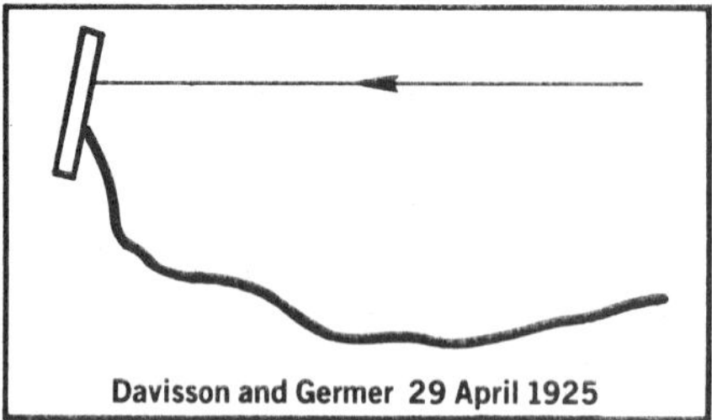

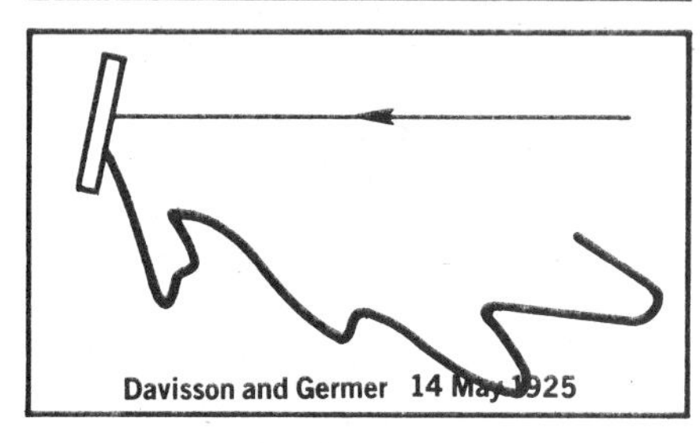

Before and after the accident of 5 February 1925. Although the first scattering curves after the repair of the broken tube (middle curve) resembled the 1921 results of Davisson and Kunsman (top curve), striking peaks soon made a sudden appearance (bottom). This development led Davisson and Germer to make a major change in their program. Figure 5

Instant acclaim

Toward this end they initiated an extensive experimental and theoretical attack that lasted from 6 April (when Calbick returned from his month's absence) until 4 August. At that time the tube was cut open for a final careful examination of the target and the other tube components. As it turned out, this intention was foiled when, in the process of being brought back to room temperature, the tube "blew up and [was] partially ruined . . . the leads being broken, filament also, and a large part of the nickel oxidized." A broken tube had served to initiate the decisive experiments on 5 February 1925, and a broken tube ended them on 4 August 1927, two and a half years later. The cover of this issue of PHYSICS TODAY shows the Davisson–Germer tube as it appears today.

The most interesting of this last group of experiments was a series designed to investigate "the anomalous peaks after bombardment," which appeared for a restricted period of time after the target had been heated by bombardment. The experiments showed that the nature—even the existence—of certain beams was not static but varied with temperature and time (and hence conditions of the target in terms of occluded gases). The notebook entries include a great variety of different terms, diagrams and calculations designed to try to make sense out of these data. Davisson and Germer found a "gas crystal" model, in which "gas atoms fit into the crystal," to be the most effective.

The task of welding data and interpretation into a comprehensive report for publication was begun in mid June, well before the experiments were completed. It appears that Davisson was responsible for most, of not all, of the writing; in a letter to his family at the summer cottage he wrote:

> "I'm busy these days writing up our experiment—It's an awful job for me. I didn't get much done yesterday as Prof. Epstein from Pasadena turned up and had to be entertained and shown things—and today I'm too sleepy [after having spent last evening at the theater with Karl Darrow]. However, I must keep at it."

More than three weeks later (23 July) he was at last able to exclaim,

> "I finished the first draft of our paper this morning. It is going to take a lot of going over and revising . . . I will leave [the drawings] to Lester—and also the thing is full of blanks in which he will have to stick in the right numbers."

A week later he made his final changes before departing for Maine. Germer, too, needed a break, and after finishing his tasks he left on 14 August for a canoe trip with several friends. The final copy was sent to the *Physical Review* in August and the article appeared in December.

The paper itself was a detailed, comprehensive report on experiments performed, conclusions reached and questions left unanswered. One of the significant features of the paper was its thoughtful examination of the possible ways of interpreting the systematic differences between observed and calculated electron wavelengths (either the suggested "contraction factor" or an "index of refraction" proposed by Carl Eckart, A.L. Patterson, and Fritz Zwicky in independent responses to the *Nature* article).[11] Summarizing the evidence, the paper concluded that of the 30 beams that had been observed, 29 were adequately accounted for by attributing wave properties to free electrons. It acknowledged, however, that the wave assumption implied the existence of eight additional beams, which had not been observed.

The discrepancies between theory and experiment, apparently fairly minor, that Davisson and Germer recorded, evidently did not reduce their fundamental belief that free electrons behave like waves. The physics community appears to have concurred, for I have not found a single voice raised in opposition. This may well have been due as much to the success of the earlier theory of wave mechanics and the acceptance of a wave–particle duality for light as to the force of the evidence inherent in the paper itself.

This may be illustrated by some remarks made by prominent physicists prior to the publication of the *Phys. Rev.*

article. In the reports and discussions of the fifth Solvay Conference held in Brussels in October 1927, Niels Bohr, de Broglie, Born, Heisenberg, Langmuir and Schrödinger all hailed the experiments of Davisson and Germer (as described in the *Nature* article) as being, in the words of de Broglie, "very important results which [appear] to confirm the general provisions and even the formulas of wave mechanics."[12] Bohr, speaking before the International Congress of Physics assembled in Como, Italy, on 16 September 1927, drew upon these experiments in establishing his views on complementarity:

> ". . . the discovery of the selective reflection of electrons from metal crystals . . . requires the use of the wave theory superposition principle . . . Just as in the case of light . . . we are not dealing with contradictory but with *complementary* pictures of phenomena."[13]

Planck, addressing the Franklin Institute on 18 May 1927, even *before* he had heard of the Davisson–Germer results, stated about the electron: "[Its] motion [in the atom] resembles . . . the vibrations of a standing wave . . . [Thanks] to the ideas introduced into science by L. de Broglie and E. Schrödinger, these principles have already established a solid foundation."[14] Yet in the same address Planck stated that he was *still* (in 1927, four years *after* the decisive Compton experiments) reluctant to accept the corpuscular implications for electromagnetic radiation inherent in his own quantum hypothesis! It appears that physicists were willing to accept the experimental evidence for electron waves almost before those experiments were performed!

BELL LABORATORIES

The sixth of January 1927 might well be regarded as the birthday of electron waves, for it was the day that data directly supporting the de Broglie hypothesis of electron waves were first observed. Note the peak deflection at 65 volts, and the detailed study of the region directly below. Calbick's handwriting is neat and cautious; Germer's is bold and expansive. Davisson made no entries in any of the research notebooks kept in the Bell Labs files. Figure 6

The world that is physics

Davisson and Germer succeeded where others had failed. In fact, the others mentioned above (Elsasser, Dymond, Blackett, Chadwick and Ellis), who had the *idea* of electron diffraction considerably *ahead* of Davisson and Germer, were not able to produce the desired experimental evidence for it. G.P. Thomson, who did find that evidence by a very different method, testified to the magnitude of the technical achievement as follows:

> "[Davisson and Germer's work] was indeed a triumph of experimental skill. The relatively slow electrons [they] used are most difficult to handle. If the results are to be of any value the vacuum has to be quite outstandingly good. Even now [1961] . . . it would be a very difficult experiment. In those days it was a veritable triumph. It is a tribute to Davisson's experimental skill that only two or three other workers have used slow electrons successfully for this purpose."[15]

Davisson and Thomson shared in the Nobel Prize for physics in 1937 for their accomplishments. Germer and Reid, as junior partners to Davisson and Thomson, did not share in the prize. Reid was tragically killed in a motorcycle accident shortly after his and Thomson's definitive papers appeared in 1928.

Davisson and Germer actively pursued the topic of electron diffraction for about three years after 1927, publishing, together and separately, about twenty more papers on the subject; reference 16 gives three of the most important. By the early 1930's, both Davisson and Germer had turned to new fields: Davisson to electron optics (including early television); Germer to high-energy electron diffraction and later still to electrical contacts. Davisson retired from Bell Labs in 1946 and spent the remaining twelve years of his life in Charlottesville, Virginia, summering as usual in Maine. Germer regained his interest in low-energy electron diffraction in 1959–60, at which time he and several co-workers at Bell Labs perfected a technique, eventually referred to as the "post-acceleration" technique,[17] which had been devised in 1934[18] and then abandoned, by Wilhelm Ehrenberg. With this work Germer was able to follow up with great success the study of surfaces, to which he had been attracted in his original work with Davisson; the field of low-energy electron diffraction (LEED) is now widespread and very active. Germer retired from Bell Labs in 1961 and remained active in this "new" field and in his favorite recreation, mountain climbing, until his death in 1971.

In trying to answer the question of "Why Davisson and Germer, and not someone else?" one's thoughts leap to such things as the "luck" of the broken tube in 1925 and the trip to England in

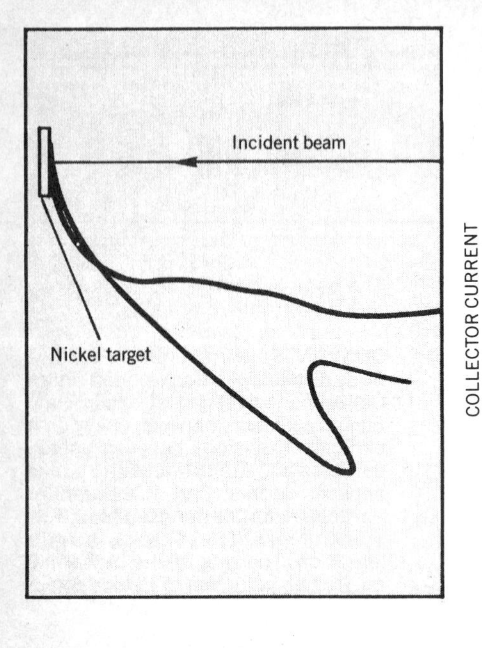

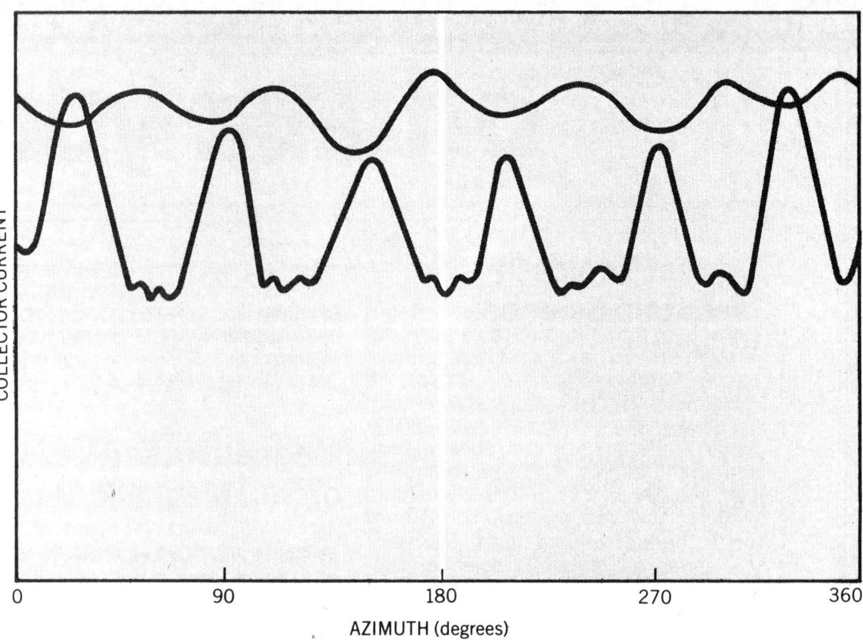

New colatitude and azimuth curves. The black lines show the appearance of the colatitude (left) and azimuth (right) distributions of the scattered electrons when Davisson took the curves to England in 1926. The colored curves are from data taken after 6 January 1927, when the first "quantum bump" was observed. The azimuth curves also confirm the threefold symmetry of the nickel crystal. Figure 7

1926. Davisson and Germer themselves freely admitted the key importance of these events. But to dwell on them exclusively would be a mistake. Neither of these events would even have been remembered had they not been followed by thorough, careful and creative experiment and reflection. Perhaps of equal importance is the habit of attention to technical detail established by Davisson in his student days and extended in the long series of Davisson–Kunsman and earlier Davisson–Germer experiments. Another important factor is the time for pure research provided by Western Electric–Bell Labs, and the technical support in areas such as high vacua and electrical detection techniques available at that industrial laboratory.

All in all, this case history on the discovery of electron diffraction appears to illustrate the complex nature of the world that is physics, the difficulty of singling out any one factor as being responsible for a great discovery, and the importance of establishing and nurturing the ties that bind together the generations of physicists, as well as the physicists of each generation.

References

References to correspondence, personal remarks and other archival material are documented in the author's PhD dissertation, "C.J. Davisson, L. H. Germer, and the Discovery of Electron Diffraction," University of Minnesota, 1973, available from Xerox University Microfilms, 3000 North Zeeb Road, Ann Arbor, Michigan 48106, Order No. 74-10 505.

1. C. J. Davisson, L. H. Germer, Phys. Rev. **30,** 705 (1927).
2. US Reports **283,** 665 (1931).
3. C. J. Davisson, L. H. Germer, Phys. Rev. **15,** 330 (1920).
4. C. J. Davisson, C. H. Kunsman, Science **54,** 523 (1921).
5. M. Born, Nature **119,** 354 (1927).
6. W. Elsasser, Naturwissenschaften **13,** 711 (1925); E. G. Dymond, Nature **118,** 336 (1926).
7. G. P. Thomson, A. Reid, Nature **119,** 890 (1927).
8. C. J. Davisson, L. H. Germer, Nature **119,** 558 (1927).
9. C. J. Davisson, Bell Lab. Record **4,** 257 (1927).
10. C. J. Davisson, L. H. Germer, Phys. Rev. **29,** 908 (1927).
11. C. Eckart, Proc. Nat. Acad. Sci. **13,** 460 (1927); A. L. Patterson, Nature **120,** 46 (1927); F. Zwicky, Proc. Nat. Acad. Sci. **13,** 518 (1927).
12. L'Institut International de Physique Solvay, *Electrons et Protons: Rapports et Discussions du Cinquieme Conseil de Physique,* Gauthier-Villars, Paris (1928), pages 92, 127, 165, 173, 274, 288.
13. N. Bohr, *Atomic Theory and the Description of Nature,* Macmillan, New York (1934), page 56; italics supplied.
14. M. Planck, J. Franklin Inst. **204,** 13 (1927).
15. G. P. Thomson, *The Inspiration of Science,* Oxford U.P., London (1961); reprinted by Doubleday, Garden City, New York (1968), page 163.
16. C. J. Davisson, L. H. Germer, Proc. Nat. Acad. Sci. **14,** 317 (1928); Proc. Nat. Acad. Sci. **14,** 619 (1928); Phys. Rev. **33,** 760 (1929).
17. A. U. MacRae, Science **139,** 379 (1963).
18. W. Ehrenberg, Philosoph. Mag. **18,** 878 (1934). □

Extension of the Low Pressure Range of the Ionization Gauge

ROBERT T. BAYARD AND DANIEL ALPERT
Westinghouse Research Laboratories, and Physics Department, University of Pittsburgh, Pittsburgh, Pennsylvania
April 13, 1950

THE ionization gauge[1] has been generally considered useful for the measurement of gas pressures ranging down to 10^{-8} mm Hg.[2] This low pressure limit is not the lowest pressure attainable with modern vacuum techniques, but has been shown to be a limitation of the conventional ionization gauge itself. An investigation made in this laboratory to determine the cause of this limitation has resulted in a new ionization gauge whose range has been considerably extended. The new gauge has been used to measure pressures of the order of 10^{-10} mm Hg.

The conventional ionization gauge resembles structurally the triode vacuum tube. The grid, however, serves as anode and is operated at a positive potential (100 to 250 v), while the plate is negative (10 to 50 v), both with respect to the cathode. Electrons from the cathode, before being collected at the grid, can create positive ions by collision with gas molecules. The number of ionizing collisions for a given electron current is considered to be proportional to the gas density. The current to the ion collector, or plate, is thus a measure of the pressure.

Dushman[3] has shown that the ion current for a given electron emission is indeed linearly related to the pressure between 3×10^{-6} mm Hg and 10^{-3} mm Hg. On the assumption that this relation continues to hold true, ionization gauges have been used to measure lower pressures. Workers in the field today, however, agree that while currents less than that corresponding to 10^{-8} mm Hg[4] could easily be measured, they are not observed. On the other hand many investigators[5] have had indications that considerably lower pressures than 10^{-8} mm Hg have actually been attained. This suggests the existence of a residual current to the ion collector which is independent of pressure. Nottingham[6] has suggested that this residual current is due to soft x-rays which release photo-electrons from the ion collector, the x-rays being created at the grid by the incidence of the thermionic electrons.

Independent evidence for the existence of a residual current comes from the observed characteristics of the ionization gauge itself. If for a given pressure and electron current, the ion collector current i_c is measured as a function of grid potential v_g, the plot of i_c *vs.* v_g, shown in Fig. 1a, has a characteristic shape for pressures above 10^{-7} mm Hg. The gas ionization curve shown (upper line) is typical of data taken on RCA type 1949 gauges. Note that i_c rises rapidly with v_g up to 200 volts and varies slowly with potential above this value. When the pressure is considerably lowered (ion gauge reading 10^{-8} mm Hg) the curve of i_c *vs.* v_g is radically different as shown (lower line). This residual curve continues to rise with grid potential, the slope on a log-log plot being between 1.5 and 2. In the intermediate range of pressures (10^{-7}–10^{-8} mm Hg) the characteristic corresponds to a superposition of the "gas ionization" curve and the "residual" curve. These results, obtained independently in this laboratory, are in agreement with data published in 1931 by Jaycox and Weinhart[7] and with unpublished findings of W. B. Nottingham.

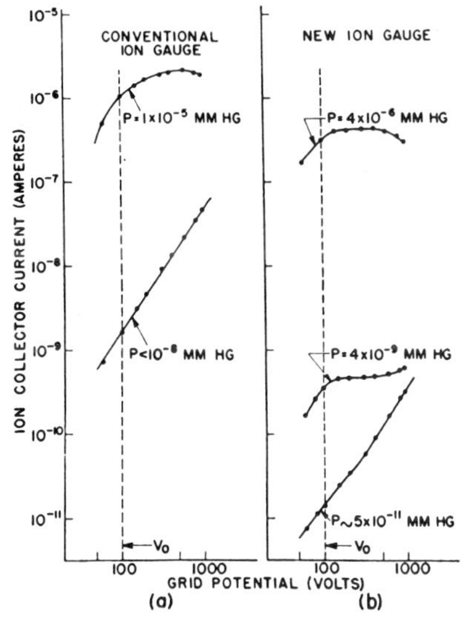

FIG. 1. (a) Typical data for RCA type 1949 gauge. (b) Typical data for the new type gauge. V_0 is the normal operating voltage.

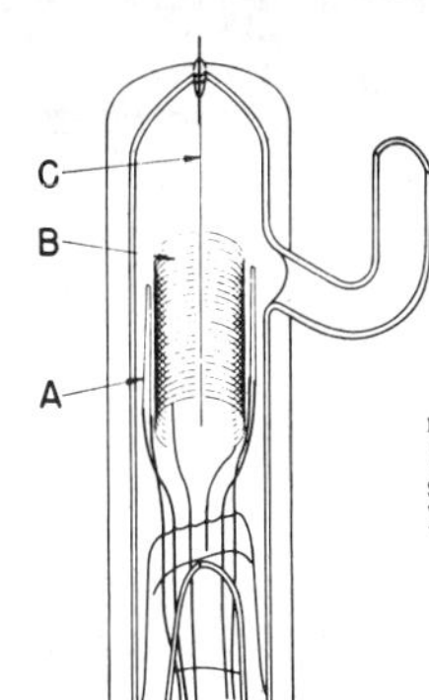

FIG. 2. Cutaway view of the new gauge showing filament *A* (an auxiliary filament is also shown), grid *B*, and ion collector *C*.

With the explicit purpose of examining the hypothesis that this residual current is due to x-rays, a design was sought for an ion gauge which would operate in the usual manner but whose ion collector would intercept only a small fraction of the x-rays produced at the grid. The ionization gauge which evolved from these considerations is shown in Fig. 2. The new gauge has the usual three elements, but is inverted from the conventional arrangement; the filament *A* is outside the cylindrical grid *B*, while the ion collector *C*, consisting of a fine wire, is suspended within the grid. The volume enclosed by the grid is made comparable to that between the grid and plate of the ordinary ionization gauge. The positive potential on the grid forms a barrier to the ions formed inside the grid-enclosed volume so that they are eventually collected at the center wire. The ion collection efficiency is thus comparable to that of the conventional ionization gauge. On the other hand, the geometrical cross section of the ion collector to radiation from the grid is approximately one hundred times smaller than that of the conventional cylindrical collector.

The properties of the new ionization gauge are shown by the curves of Fig. 1b. At higher pressures the characteristics of the new gauge are very similar to those of the RCA gauge, but it is evident that the new tube continues to have typical "gas ionization" characteristics at much lower pressures. At 4×10^{-9} mm Hg the collector current is predominantly due to gas ionization, and it is not until the pressure is less than 10^{-10} mm Hg that the residual current predominates.[8] Note that the slope of the residual curve is the same as that for the RCA gauge. On the other hand the value of the collector current for the new gauge, at a given grid voltage, is one hundred times lower than that for the RCA gauge. This ratio of residual currents for the two types of ion gauges is the same as the ratio of their geometrical cross sections to radiation mentioned above. These general characteristics of the new gauge clearly substantiate the x-ray hypothesis and indicate that ionization gauges can be built to measure even lower pressures.

The sensitivity of the new ionization gauge has been measured by calibration with a McLeod gauge at nitrogen pressures in the neighborhood of 10^{-4} mm Hg and by comparison with a standard ionization gauge at various pressures between 10^{-5} and 10^{-7} mm Hg. As seen from the typical curves in Fig. 1, the sensitivities of the two types of ionization gauges are essentially the same. The new gauge gives indication of pressures at least 100 times lower than the standard ion gauge. It has a minimum of metal surface, is easily outgassed, and is operated on a standard power supply. These characteristics make the gauge a useful tool for studies in the field of high vacuum.

[1] For a complete discussion of ion gauges see for example S. Dushman, *Scientific Foundations of Vacuum Techniques* (John Wiley and Sons, Inc., New York, 1949), p. 332–366.
[2] Pressures mentioned in this letter are equivalent nitrogen pressures.
[3] S. Dushman, Phys. Rev. **17**, 7 (1921).
[4] This current is usually of the order of 10^{-9} ampere for 0.010 ampere electron current and 100–150 v grid potential.
[5] (a) P. Anderson, Phys. Rev. **47**, 958 (1935). (b) W. B. Nottingham, J. App. Phys. **8**, 762 (1937). (c) I. R. Senitsky, Phys. Rev. **78**, 331 (1950). (d) The so-called flash filament method of L. Apker, whereby pressures are estimated from measurements on the rate of accumulation of gas on a clean wire, has been used by several investigators including the author to measure pressures of the order of 10^{-10} mm Hg. (e) The mass spectrometer group in our laboratory has had indications that in a well-baked out spectrometer tube the residual gas pressure is 10^{-9} mm Hg or less.
[6] Reference 1, p. 359.
[7] E. K. Jaycox and H. W. Weinhart, Rev. Sci. Inst. **2**, 401 (1931).
[8] The pressure represented by the lowest curve, as judged by the deviation of the lower portion from a straight line, is estimated to be of the order of 5×10^{-11} mm Hg.

Originally published in Rev. Sci. Instrum. **21**, 571–572 (1950).

ULTRAHIGH VACUUM

In 1950 the limit of man-made vacuum appeared to be a pressure equivalent to 10^{-8} millimeter of mercury. A breakthrough in technique now provides pressures four orders of magnitude lower

by H. A. Steinherz and P. A. Redhead

Originally published in Sci. Amer, 2–13 (March 1962).

The prefixes "super-" and "ultra-" are rather freely used in scientific literature these days. These superlatives may prove embarrassing when it comes time to describe the next round of advances, but they do reflect an explosive rate of progress that has carried many techniques beyond what seemed to be the attainable limits only a short time ago.

The history of ultrahigh vacuum, a good case in point, goes back to 1950. Before that year high-vacuum practitioners had been improving their pumps, valves, seals and other components and had been reaching lower and lower pressures. In the middle 1940's they seemed to have reached a dead end. Using the best pumps and the most exquisite care in the design and operation of the systems, they could reach pressures approaching a hundred-millionth of a torr. (The torr, named for the 17th-century vacuum pioneer Evangelista Torricelli and now the standard unit in vacuum technology, is defined as the pressure necessary to support a column of mercury one millimeter high.) Apparently further refinements availed nothing. The gauges still indicated 10^{-8} torr. It was generally supposed that the pumps must somehow fail at this pressure. (The extremely small pressures with which this article deals are most conveniently expressed in negative powers of 10. The fraction 1/10 is written 10^{-1}, 1/100 as 10^{-2} and so on. Thus 10^{-8} is one hundred-millionth of a torr.)

It is worth pausing a moment to appreciate the achievement that a pressure of 10^{-8} torr represents. It is about one hundred-billionth the pressure of the atmosphere at sea level, which means that only one of every hundred billion air molecules originally present in the vacuum chamber is left after pumping. At a pressure of one atmosphere each molecule travels, on the average, a few millionths of a centimeter before bumping into another molecule. At 10^{-8} torr, if it were not for the walls of the vacuum chamber, a typical molecule would travel almost 500,000 centimeters—some three miles—before encountering another. The pressure exerted by air at one atmosphere on a container (or on the mercury column of a pressure gauge) is the result of 3×10^{23} molecular impacts against each square centimeter of the container walls each second. At 10^{-8} torr the number of impacts is reduced, also by a factor of one hundred billion, to 3.8×10^{12} per square centimeter per second. This is still a lot of impacts, but not enough to hold up a column of mercury even one atom high. Obviously the definition of the torr given above no longer has any operational significance. The torr is nonetheless used to define these extremely low pressures, with the tacit understanding that it can be redefined in terms of a meaningful property such as the rate of impacts against a container wall, or the number of molecules per cubic centimeter (molecular density).

In 1947 Wayne B. Nottingham of the Massachusetts Institute of Technology suggested that the barrier at 10^{-8} torr was an illusion, caused by a failure in measurement rather than in pumping. The only instrument generally available that is capable of measuring pressures below about 10^{-4} torr is the ion gauge. The standard device in use during the 1940's consisted of a hot-wire cathode surrounded by a positively charged grid, which is in turn enclosed in an ion-collecting shell. The whole arrangement is enclosed in an envelope that is connected to the vacuum chamber, so that the gauge is in effect a part of the chamber. Electrons emitted from the central cathode move rapidly toward the grid. In the course of their journey some of them collide with molecules of the gas to be measured, knocking electrons out of its molecules and producing positive ions. All the electrons are eventually collected on the grid. The positive ions move to the negatively charged collector, each one causing a tiny pulse of current to flow in the collector circuit. The number of ions produced depends on the density of the gas (that is, on the number of molecules per cubic centimeter), and so the collector current is an index to the molecular density.

Analyzing the operation of the ion gauge, Nottingham realized that there must be an additional process at work in it. Electrons bombarding the grid produce low-energy X rays. Some of this radiation would strike the ion collector and release electrons from its surface by means of the familiar photoelectric effect. So far as the current meter in the external circuit is concerned, the departure of a negative electron from the collector has exactly the same effect as the arrival of a positive ion. In short, the meter would register some current even if there were no gas whatever in the gauge and therefore no ions. Nottingham's calculations showed that this irreducible photoelectric current corresponded to a pressure of about 10^{-8} torr.

The Bayard-Alpert Gauge

A couple of years later Robert T. Bayard and Daniel Alpert, then at the Westinghouse Research Laboratories, hit on a simple modification of the ion gauge that both proved the correctness of Nottingham's analysis and greatly extended the limits of operation of the instrument. Essentially they switched the positions of the cathode and the ion

LARGE OIL-DIFFUSION PUMPS are located in Space Environments Laboratory of Fairchild Camera and Instrument Corporation's Defense Products Division at Syosset, N.Y. The three pumps, each 32 inches in diameter, are teamed with three others to attain 10^{-8} torr (equivalent to a 300-mile altitude) in a 2,825-cubic-foot chamber in which space vehicles and sensors are to be tested.

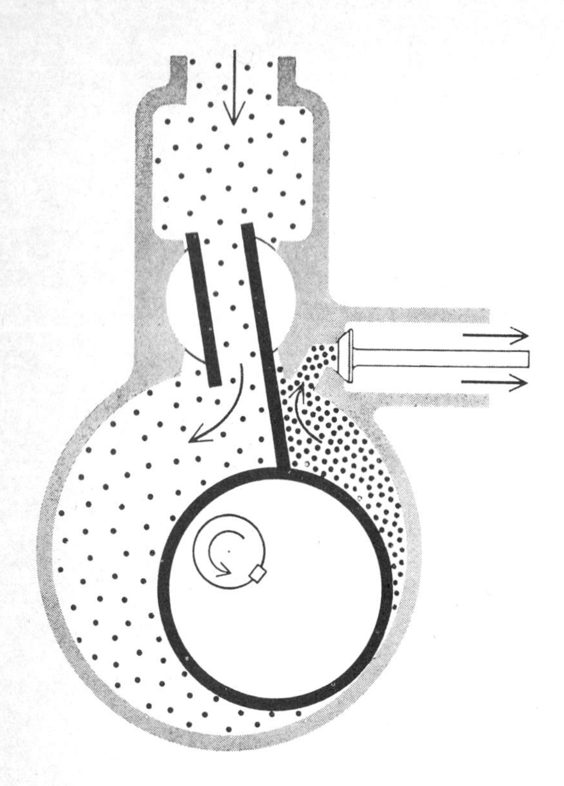

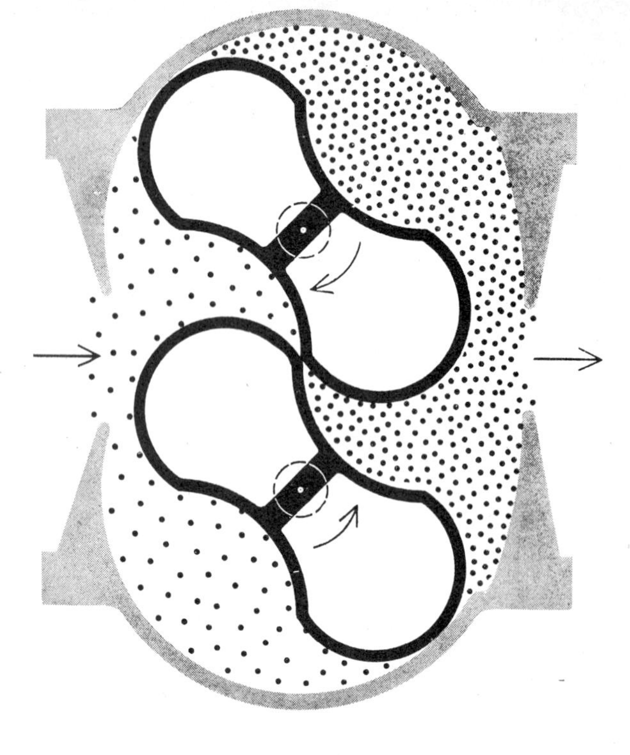

MECHANICAL PUMPS are used in the first stage of evacuation. The rotary type (*left*) depends on the action of an eccentric rotating piston to sweep gas out of a chamber. Blowers (*right*) contain two tightly fitted counterrotating lobes that trap and expel molecules.

collector. In the Bayard-Alpert gauge the cathode consists of a heated wire outside the grid, and the collector is a thin wire running down the axis of the instrument [*see illustration at top left on page* 8] Carrying a negative charge, the collector still picks up most of the ions formed in the gas. But because of its greatly reduced surface area it intercepts far less X radiation than a cylindrical plate does and therefore produces a much smaller photoelectric current. Bayard and Alpert showed that the residual current is equivalent to a pressure of about 10^{-11} torr.

With the new gauge Alpert was able in 1950 to break into the range of ultrahigh vacuum, below 10^{-8} torr. In fact, the instrument played a doubly crucial role. One of the difficulties with ion gauges had always been that they acted to change the very property they were supposed to measure. The reason is that the gas ions driven to the collector are trapped; hence they are removed from the vacuum chamber and the pressure drops. In other words, the ion gauge is also a pump.

Alpert proceeded to capitalize on the drawback. He pumped down a small glass chamber to almost 10^{-8} torr by conventional means, and sealed off the chamber with a newly designed all-metal valve that required no organic sealing compound. (At these pressures conventional sealing compounds give rise to large amounts of gas.) Then he simply let the ion gauge continue to operate. Soon it was registering a pressure of 5×10^{-10} torr. Beyond this point the pressure would not go, although in principle the Bayard-Alpert gauge should have been capable of producing and measuring a pressure 10 times lower. Alpert then built a special type of mass spectrometer (which will be discussed later) to measure the pressures due to the various gases in the vacuum chamber. He found that the pressure limit in his system was set by the diffusion of helium atoms in the air through the glass walls of the chamber.

This work is as clear-cut an example of a scientific first as is ever likely to turn up. Yet, typically, once Alpert showed how to measure pressures in the ultrahigh-vacuum region it became clear that others had entered it before him. The new gauge demonstrated what some investigators had long suspected, that the diffusion pumps used in producing high vacuums could themselves penetrate the "barrier" at 10^{-8} torr. A review of the results of earlier work indicated that pressures not far from 10^{-9} torr had probably been attained as early as 1931.

Nevertheless, the opening of the age of ultrahigh vacuum must be dated from Alpert's remarkable experiments. After they had been published a growing number of workers entered the field, contributing further improvements to equipment and technique. Today a pressure of 10^{-12} torr is attainable, and still lower pressures are clearly in prospect. At 10^{-12} torr the molecular density is down to 33,000 molecules per cubic centimeter and the mean free path of a nitrogen molecule is about 50,000 kilometers, or 30,000 miles, long.

The Uses of Ultrahigh Vacuum

Before discussing the tools and methods of ultrahigh-vacuum technology, it seems appropriate to ask why they were developed. It is likely that sheer curiosity and the urge to explore would have produced them eventually. But there were far more practical lures, in both fundamental and applied research, to attract men and money to the task.

The original impetus toward ultrahigh

vacuum came from the requirements of experimenters studying the physics and chemistry of solid surfaces. Most of the properties of such surfaces can be studied only on surfaces that have been cleaned of adsorbed gases and will stay clean long enough for the appropriate measurements to be made. Metals and semiconductors can be cleaned on an atomic scale by heating them in a vacuum or by spraying their surfaces with a low-energy discharge of an inert gas. How long the pristine surface lasts then depends on the rate at which it is bombarded by adsorbable molecules from the surrounding gas. At a pressure of 10^{-6} torr, considered a good vacuum only a few years ago, a layer of gas one molecule thick will form on a metal surface in about one second. If the surface has been cleaned by heating in a vacuum of 10^{-6} torr, it is usually completely covered before the sample has cooled to its initial temperature, and before many of the desired measurements can be carried out. At 10^{-9} torr the "monolayer time" increases to about 20 minutes, and it becomes possible to observe the clean surface for a reasonable length of time.

A remarkably wide variety of the properties of a material are affected by gases adsorbed on its surface. One obvious example is the force of friction. Indeed, the true friction between metallic surfaces has so far been measured for only a few metals. The standard handbook figures really refer to surfaces lubricated by adsorbed gas layers. Again, the emission of electrons from a solid surface (by thermionic emission, photoelectric emission or other processes) is sensitive to surface contamination, and ultrahigh-vacuum techniques are now being widely used to study the phenomenon. Some electrical properties of semiconductors are also strongly affected by adsorbed gas. Finally, the process of adsorption itself is more amenable to study when it can be observed in slow motion, so to speak, in an ultrahigh-vacuum chamber.

In any experiment involving gases of high purity at low pressures, ultrahigh-vacuum techniques are indispensable. For example, if it is required to maintain a purity of one part per million in a gas sample at a pressure of 10^{-3} torr, the vacuum chamber must be evacuated to less than 10^{-9} torr before the sample is introduced. If the experiment involves an electrical discharge, the difficulties are compounded. When radiation or gas ions from the discharge strike the vacuum chamber, they release contam-

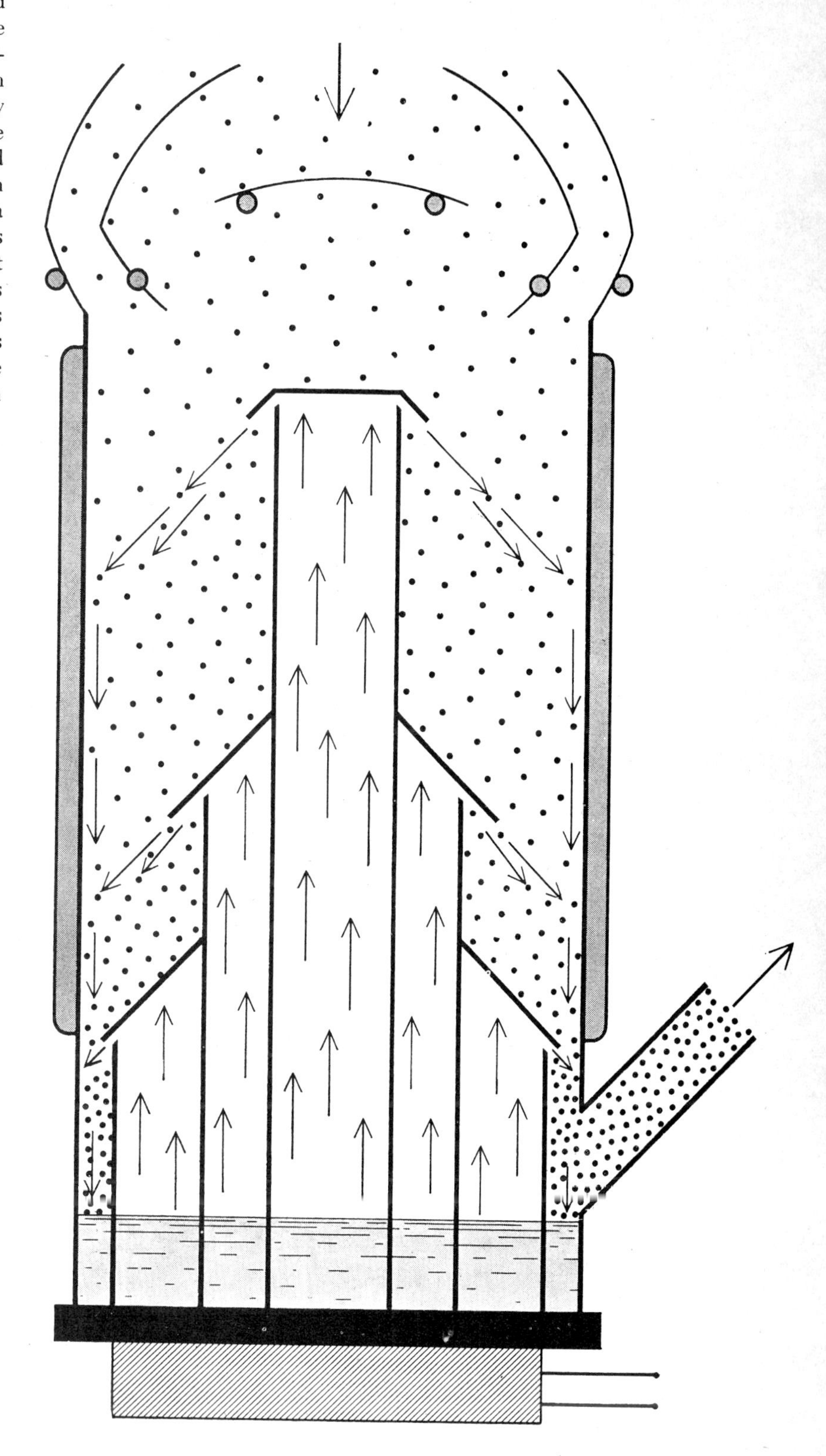

DIFFUSION PUMP is the work horse of high vacuum. A liquid, usually oil or mercury, is vaporized by a heater at the bottom. The vapor rises and is deflected downward (*black arrows*) as a high-speed jet that entrains gas molecules (*colored dots*) diffusing out of the vessel above and carries them out of the pump. Back-diffusion of pumping vapor into the chamber is inhibited by baffles, cooled by coils (*gray circles*), on which the vapor freezes.

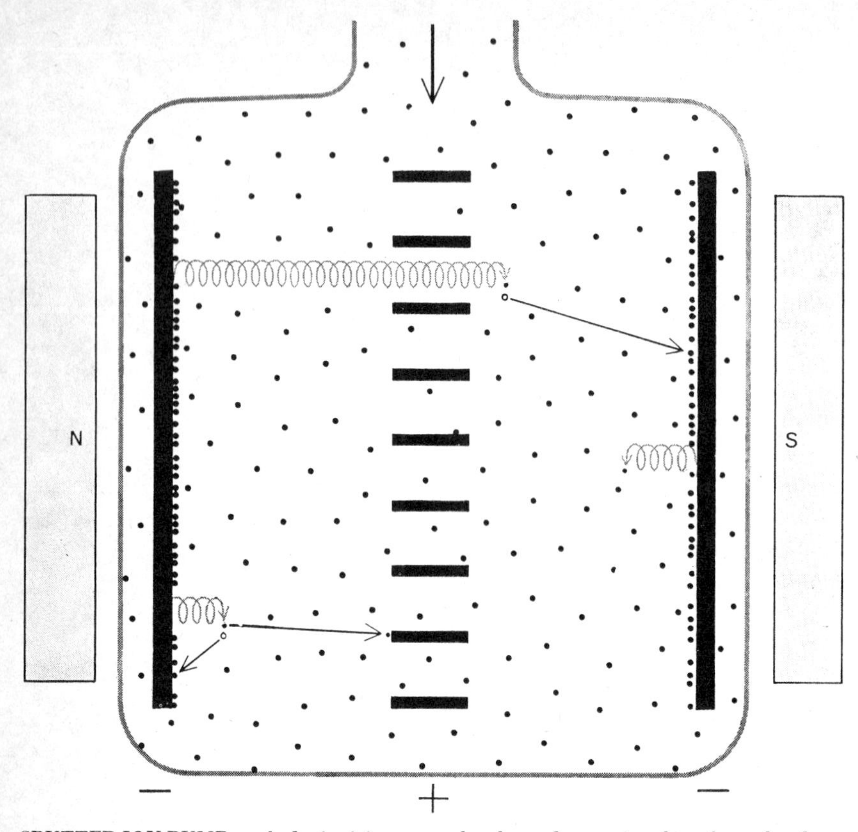

SPUTTER ION PUMP works by ionizing gas molecules and removing them from the chamber (*top*) to be evacuated. Electrons emitted by the cathode plates are accelerated (in spirals because of the magnetic field) by the anode. When they strike gas molecules, they create positive ions (*open colored circles*). These are attracted to the negative collector plates, from which fresh titanium metal is "sputtered" to provide clean pumping surfaces.

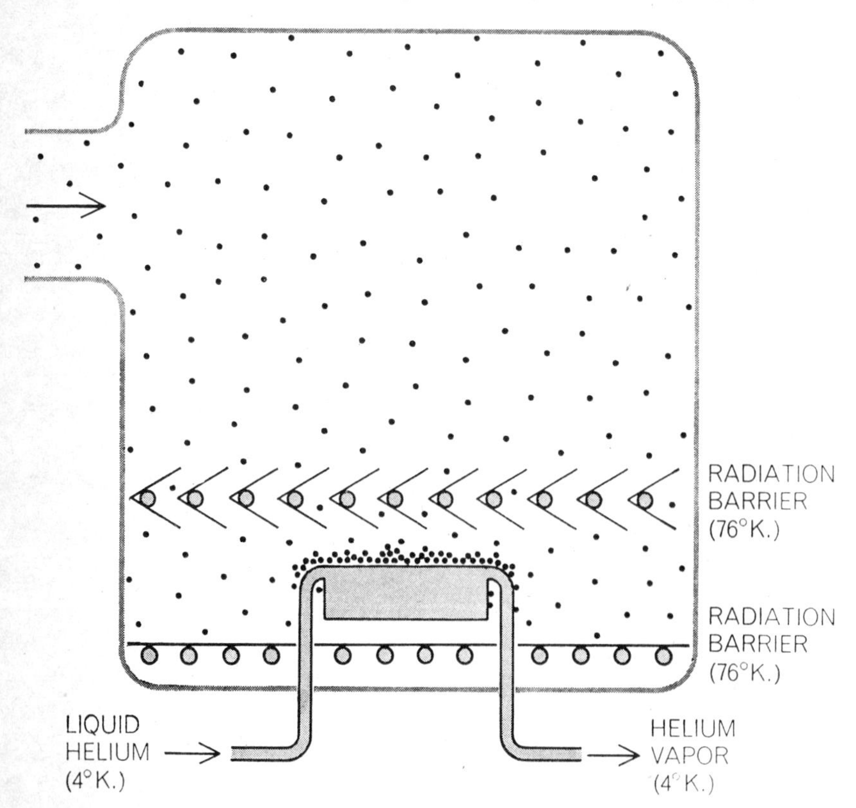

CRYOGENIC PUMP is another device that immobilizes gas molecules and thus removes them from a system. It does so by condensing the molecules on a surface cooled by liquid helium. The V-shaped baffles prevent heat radiation downward but allow molecules to pass.

inating gases from the walls.

Maintaining clean surfaces is essential not only in laboratory studies but also in manufacturing many of the new miniaturized solid-state devices. These are made by depositing extremely thin layers of different materials, one over another on a base plate. The composition and dimensions of the layers must be precisely controlled, which is possible only under ultrahigh-vacuum conditions.

Conventional vacuum-tube electronics furnished one of the major incentives for the development of high-vacuum technology. With a few exceptions normal vacuum techniques are still adequate for tubes. In any vacuum tube, however, residual gases produce a number of undesirable effects, including increased noise, increased grid current and damage to the cathode through bombardment by positive ions. When the standards of performance are exacting, as in low-noise or high-power vacuum tubes, the pressure in the tube envelope must be reduced to the ultrahigh-vacuum range. High-power klystron tubes, for example, require a vacuum of better than 10^{-9} torr.

A new class of electron tube, still under development, obtains its current through the "field emission" of electrons due to high electric fields at a fine, unheated metal point. Tube failure through filament breakage is thereby avoided. To obtain stable emission requires extremely low pressures. W. P. Dyke at the Linfield Research Institute in Oregon has built tubes that maintain stable emission for thousands of hours but only at a pressure of 10^{-12} torr. To maintain this vacuum the tube envelope must be made of a special aluminosilicate glass that has a low permeability to the helium in the atmosphere.

In all the cases mentioned so far the amount of space that must be evacuated is not very large—in the cubic centimeter range. There are two major applications in which the volume requirements are considerably greater.

One of them is in the machines with which physicists seek to solve the problem of how to control the release of power from the thermonuclear reactions of the light isotopes deuterium and tritium. An important obstacle to the achievement of thermonuclear power is the loss of energy by radiation from the hot gas, or plasma. Any contamination of the gas by the atoms of heavier elements increases the loss enormously, since these atoms radiate much more strongly than lighter ones do. When one of the early experimental machines was operated after evacuating it to 10^{-6} torr, the lowest pressure then attainable, the radiation loss was so great that the plasma temperature reached only a tenth of the calculated value. To reduce the loss to acceptable levels requires pressures in the ultrahigh-vacuum range. Recent improvements in technique have made it possible to maintain pressures in the range of 10^{-10} torr, where contamination is no longer a major problem. If practical thermonuclear power is ever achieved, the reactors will have to be very large. From the viewpoint of the vacuum engineer even today's experimental devices are big enough. For example, the evacuated volume in the Model C "stellarator" at Princeton University is about half a cubic meter. Not only must this chamber be pumped down to 10^{-10} torr before the plasma is introduced but also the pressure of the contaminants must be maintained at this value as impurities are released from the walls of the chamber by the action of the high-energy discharge.

The Simulation of Space

The most insistent, as well as the most stringent, demands on ultrahigh-vacuum technology come from the space-exploration program. At 450 miles above the surface of the earth the pressure of the atmosphere is only 10^{-9} torr, and at 1,200 miles it falls to about 10^{-11} torr, approaching the limit now generally attainable in the laboratory. In interplanetary space the pressure is estimated to be on the order of 10^{-16} torr. This corresponds to a density of about four molecules per cubic centimeter.

Under these ultrahigh (and ultraultrahigh) vacuum conditions materials may behave in unexpected ways, and physical and chemical processes may be radically altered. For example, graphite, an excellent lubricant at atmospheric pressure, becomes an abrasive below 10^{-6} torr. Heat transfer, fluid flow, dielectric behavior and other phenomena change drastically as pressure is decreased.

Not all the changes are necessarily harmful. The resistance of metals to certain types of fatigue increases greatly at low pressure. A strip that will break after a few flexings under ordinary conditions can be bent back and forth for hours in an ultrahigh-vacuum chamber. It appears that in an ultrahigh vacuum tiny cracks that form at each bending reweld themselves when the strip is bent the other way. In air at atmospheric pressure the cracks are covered with an oxide coating as soon as they open. The metal cannot join together again, and as a result the cracks grow larger each time the piece is bent.

To test equipment designed for space vehicles, simulation chambers have been built at a number of laboratories. Small or moderate-sized enclosures that can be evacuated to the lowest pressures now attainable suffice for examining fundamental processes and the properties of materials. But to check the performance of complete components requires chambers of enormous size. The biggest ones go up to some 1,000 cubic meters. Even these huge spaces are now maintained at pressures near 10^{-8} torr.

Producing the Ultrahigh Vacuum

Ultrahigh-vacuum workers face two essential problems: (1) to produce and maintain pressures below 10^{-8} torr and (2) to measure the pressure they have achieved. So far as the first problem is concerned the solution in every case represents a balance between the capacity of the pump and the system gas load: the rate at which gas continuously enters the vacuum chamber from the various parts of the system (including the pump).

Each part contributes its share, which the vacuum engineer seeks to minimize. Following Alpert's lead, the valves in ultrahigh-vacuum systems are made of metal and only low-vapor-pressure sealing compounds are used.

In small systems where the experimenter wants to observe what is going on inside or to be able to change his setup easily, the ultrahigh-vacuum chamber and its connecting tubing are still made of glass. The penetration of atmospheric helium through the glass is offset

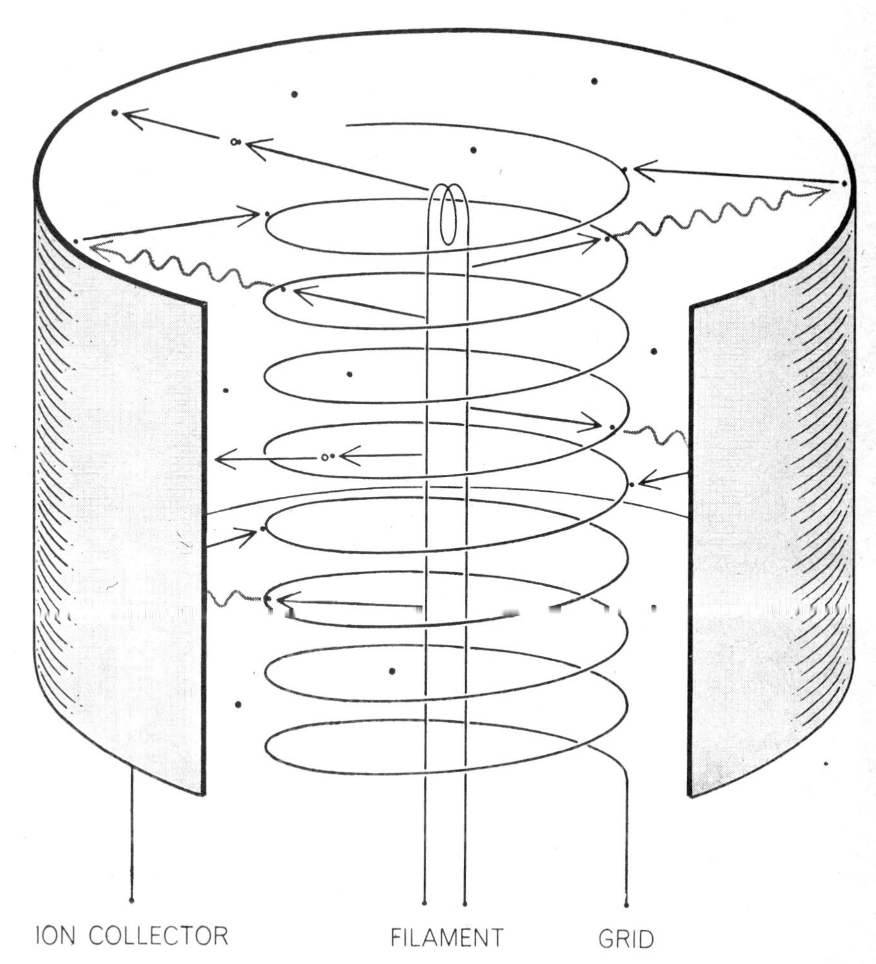

X-RAY PROBLEM made it impossible for early ion gauges to register below 10^{-8} torr. Electrons from the filament created positive ions (*colored circles*) that struck the collector and were counted. But electrons reaching the grid produced X rays (*wavy arrows*). When the X rays struck the large-area collector, they liberated electrons, causing a photoelectric current that could not be distinguished from the current resulting from ion impact.

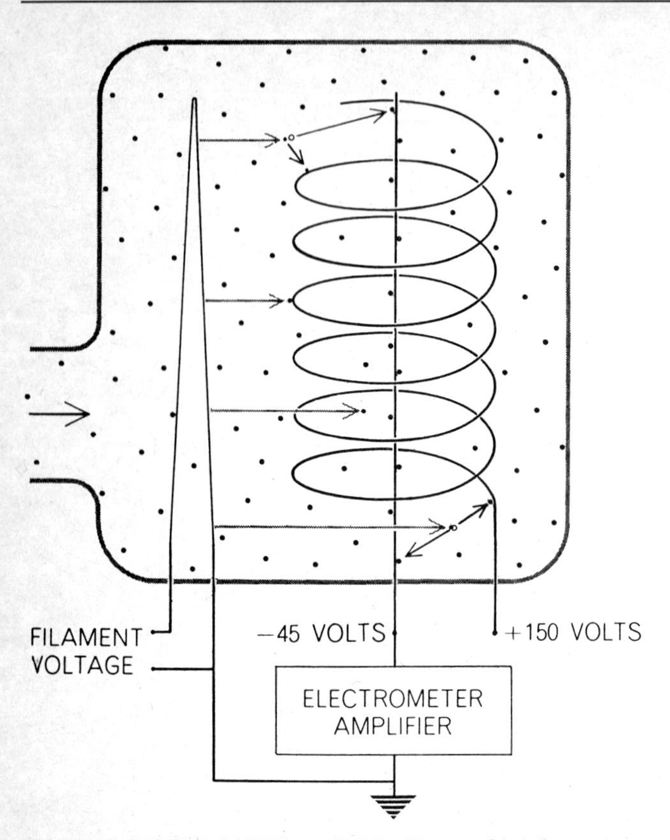

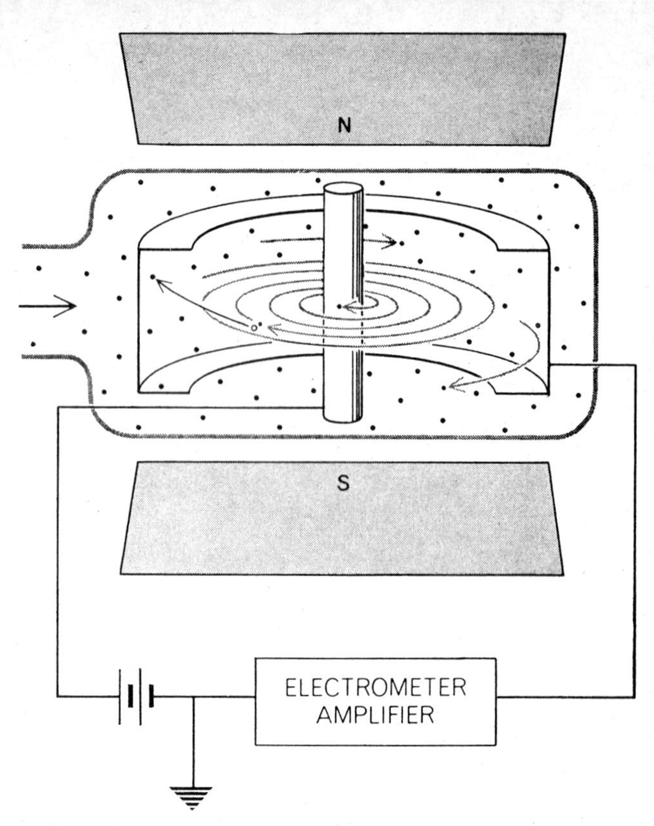

BAYARD-ALPERT GAUGE avoided the X-ray problem by putting the heated-filament cathode outside the grid and making the collector a thin axial wire. The negatively charged collector still gathers positive ions, but because of its small area it intercepts fewer X rays and therefore emits a smaller photoelectric current.

INVERTED MAGNETRON, a cold-cathode gauge, produces electrons by applying a high voltage to unheated electrodes. In such a gauge X rays are proportional to pressure, so no spurious current is produced. Electrons spiraling in toward the central electrode ionize gas molecules, which are collected on the curved cathode.

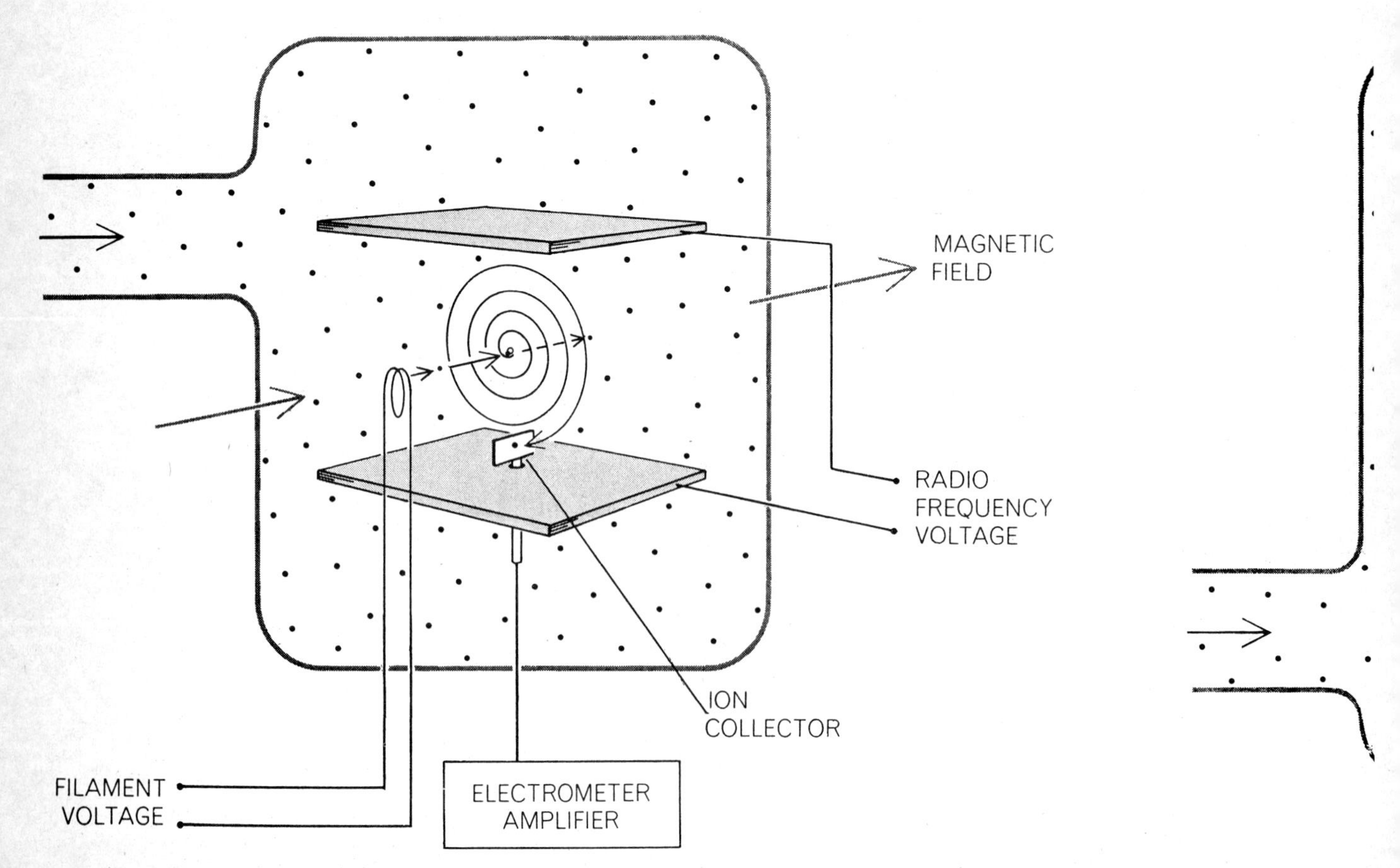

OTHER MASS SPECTROMETERS are diagramed. In the Omegatron (*left*) a radio-frequency voltage is varied to whirl positive ions of different masses into spirals that strike the collector. In the massen filter (*right*) the ion beam is accelerated along the axis of

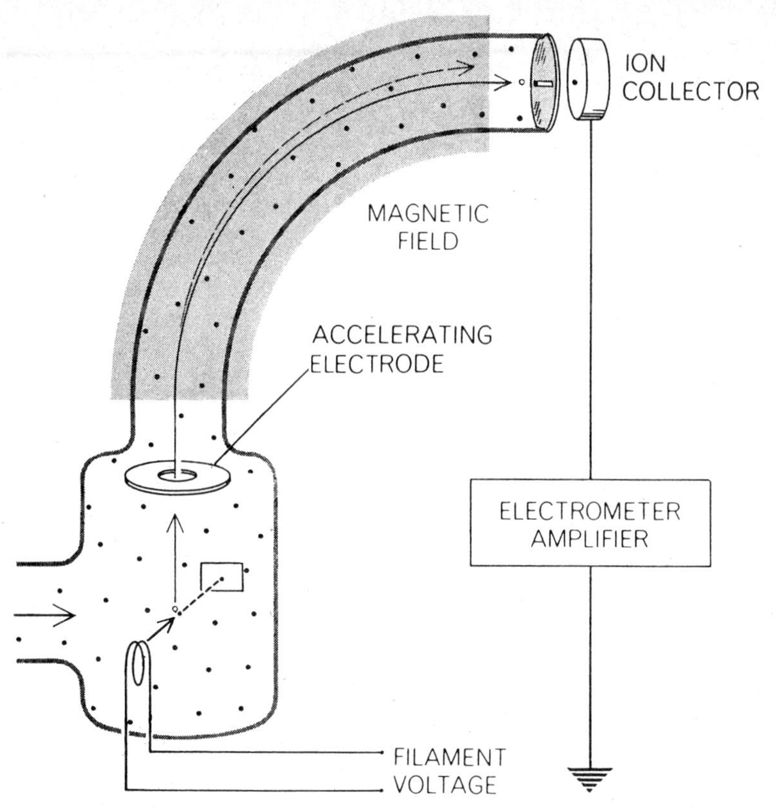

MASS SPECTROMETER can measure the partial pressure of individual gases instead of just total pressure. In this gauge, a modification of a standard spectrometer, gas molecules are ionized by an electron gun (*bottom*) and then accelerated through a magnetic field. By varying the magnetic or electric field one can direct ions of different masses through the slit to strike the collector and so measure the density of one constituent gas at a time.

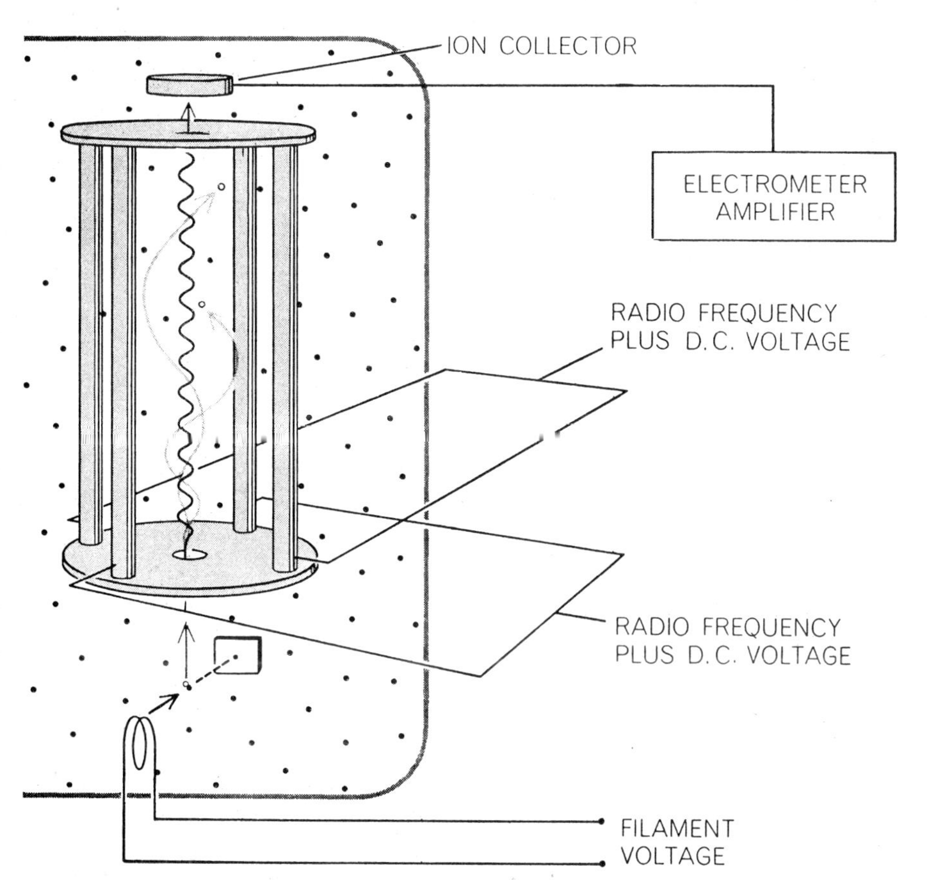

four rods carrying both direct-current and radio-frequency voltages. At each particular voltage ions of only one mass spiral up the axis to the collector; the others hit the rods.

by a pump of sufficiently high capacity. Moreover, as has been mentioned, the composition of the glass itself can be altered to reduce its permeability.

All large ultrahigh-vacuum systems, and nowadays many small ones, are built of metal, usually stainless steel. In the early days (10 years ago) designers had to weld or braze all joints to avoid excessive leakage. This equipment was hard to modify, and even introducing experimental samples into the chamber was awkward. With the recent development of improved metal gaskets and of certain elastomers such as Viton, which vaporizes very slowly even in an ultrahigh vacuum, bolted joints have become practical and metal vacuum systems are much more flexible.

Whatever the system is made of, all its parts must be able to withstand baking for hours at a temperature of several hundred degrees centigrade to drive off gas adsorbed on the surfaces and absorbed within the solid materials. Without this preparation the gas would steadily evolve into the chamber as pumping started and the pressure began to drop. To reach an ultrahigh vacuum would take an inordinately long time.

Even after the most careful preparation, desorption from the walls of the system constitutes a major source of the residual gas that keeps dribbling into an ultrahigh-vacuum system. An idea of the dimensions of the problem is provided by a typical small stainless steel system with walls one millimeter thick and a surface area of 100 square centimeters. With good technique, the rate of evolution can be reduced to about 10^{-10} torr-liter per second. (This is a rate that would raise the pressure of an absolutely empty one-liter chamber to 10^{-10} torr in one second.) Inflow through microscopic leaks in valves and joints would contribute the same amount of gas. Permeation of helium through the walls would add only about a hundredth as much; in metal systems this permeation is negligible. A small laboratory ultrahigh-vacuum pump, with a capacity of 10 liters (a little more than 10 quarts) per second, can maintain this chamber at 2×10^{-11} torr. (The rating of a pump must be referred to the pressure range in which it operates. Working at 2×10^{-11} torr, the pump in the above example can remove in one second the number of gas molecules that a 10-liter chamber contains at 2×10^{-11} torr.)

In principle the same figures should hold for larger systems so long as the pumping rate increases proportionately with the wall surface. Space-simulation chambers often have areas of a million square centimeters or even more. Typically their pumping systems are rated at 100,000 liters per second. With the area multiplied 10,000 times and the pumping speed increased by the same factor, a pressure of 2×10^{-11} torr should still be attainable. In practice, however, unavoidable leaks at the many large gaskets and seals have limited the pressure in the big chambers to about 10^{-8} torr.

The Pumps

The evacuation of any chamber, regardless of size or eventual pressure, begins with mechanical pumping. The pump is almost always the familiar rotary type, in which gas is swept out by an eccentric rotating piston [*see illustration at left on page 4*]. Sometimes, particularly in large systems, the rotary pumps are joined in series with blowers, which remove gas by the action of two counter-rotating lobes [*see illustration at right on page 4*]. Mechanical pumps bring the chamber down to about a thousandth of a torr. Then the ultrahigh-vacuum pumps take over.

Today diffusion pumps are the work horses of high vacuum, as they have been since their invention 40 years ago. Improvements have now extended their usefulness through the whole of the present ultrahigh-vacuum range.

The only moving part in a diffusion pump is a high-speed jet of oil vapor or some other vapor, directed away from the opening of the vacuum chamber [*see illustration on page 5*]. Gas diffusing out of the chamber is entrained in the jet and swept along with it out of the pump. The device is simple and fast and can be made as big or as small as is desired. Its major drawback—the factor that eventually limits the pressure it can reach—is the diffusion of vapor from the pumping fluid back into the vacuum chamber.

In the past few years the limit has been pushed back through a number of technical advances. Pump jets have been redesigned to allow higher boiler pressures and to reduce migration of the fluid into the vacuum chamber. New pump oils are available, consisting of polyphenyl ethers or liquid silicones, that have vapor pressures of less than 10^{-9} torr at room temperature, 10 to 100 times lower than those of the best oils available five years ago.

No matter how low the volatility of the fluid is, some of it is bound to back up in the pump. A good part of the contaminant, however, can be kept out of the vacuum chamber with baffles. The most common arrangement consists of a series of vanes, placed at the entrance to the chamber and cooled to the temperature of liquid nitrogen, on which the oil vapor freezes. Recently "molecular sieve" materials such as zeolite have demonstrated effective baffling action. With the help of zeolite filters operating at room temperature, diffusion-pumped

	ATMOSPHERIC PRESSURE	BEGINNING OF HIGH-VACUUM REGION	BEGINNING OF VERY-HIGH-VACUUM REGION	BEGINNING OF ULTRAHIGH-VACUUM REGION	
PRESSURE (TORR)	760	10^{-3}	10^{-6}	10^{-8}	10^{-9}
NUMBER OF MOLECULES PER SECOND BOMBARDING EACH SQUARE CENTIMETER OF CONTAINER WALLS	3×10^{23}	4×10^{17}	4×10^{14}	4×10^{12}	4×10^{11}
MEAN FREE PATH OF ONE MOLECULE BETWEEN COLLISIONS WITH OTHER MOLECULES (CENTIMETERS)	6.5×10^{-6}	5	5×10^{2}	5×10^{5}	5×10^{6}
NUMBER OF MOLECULES PER CUBIC CENTIMETER	2×10^{19}	3×10^{13}	3×10^{10}	3×10^{8}	3×10^{7}
TYPE OF PUMPS USED	MECHANICAL: ROTARY AND BLOWER		DIFFUSION ——— ION ———		CRYOGENIC ———

CHARACTERISTICS OF VACUUMS are charted, beginning with atmospheric pressure and extending to the currently practical laboratory limit. Values are given approximately, in orders of magnitude. Since "pressure" as such cannot be measured in extremely

systems have been held at 10^{-10} torr for more than 100 days. Refrigerating the zeolite extends its useful life even further and appears to eliminate oil contamination completely.

The decomposition of the pumping fluids has limited oil diffusion pumps to about 10^{-10} torr. Better results have been achieved by replacing oil with mercury. Small mercury diffusion pumps have now reached all the way down to 10^{-12} torr.

In spite of the many advantages of the diffusion pump, it is often replaced in small laboratory systems by the ion pump. Basically this device operates in the same way as the gauge that Alpert used to attain the first certified ultrahigh vacuum: ions, produced by electrons bombarding the gas molecules, are driven by an electric field to a collector, where they are adsorbed. Note that the ion pump works on a quite different principle from that of the diffusion pump. Instead of taking gas entirely out of the system, ion pumps immobilize it within the system. Herein lies one of their weaknesses: low capacity due to the saturation of the collecting surface.

Alpert's gauge had a capacity of only a small fraction of a liter per second. By designing the device primarily for pumping, the speed can be increased to many liters per second [*see top illustration on page 6*]. Containing no foreign pumping fluid, the ion pump contributes far less contamination than the best of the diffusion pumps. It is therefore better suited to operate unattended for long periods.

When the action of an ion pump is analyzed, it turns out to depend on two different processes. Some of the ions reaching the collecting surface are held there by adsorption. Others, however, react chemically with the collector material, forming stable compounds. Chemically active gases are pumped by both mechanisms; inert gases, only by the former. In most ion pumps the collector is made of titanium. The steady hail of ions "sputters" the surface, continuously providing fresh layers of clean titanium metal with which the chemically active gases combine.

All ion pumps suffer to some extent from the same pair of defects. First, they pump chemically active gases much faster than they do inert gases. Second, they re-emit a small fraction of the gas previously pumped. The re-emitted gas acts as an additional source in the vacuum system. Moreover, if one gas is pumped initially and then a second gas is introduced, the pump will remove the second while re-emitting the first. New designs have considerably reduced the size of this effect.

The Cryogenic Pump

Another type of immobilization device now coming into prominence is the cryogenic pump. It removes gas simply by condensation on a very cold surface [*see bottom illustration on page 6*]. Cryogenic pumps can be used in any position, and their speed is not restricted by connecting tubing. These advantages make them particularly valuable for large space-simulation chambers.

When a surface is cooled to the boiling point of helium (4.2 degrees centigrade above absolute zero), all gases are adsorbed, including helium. Pressures of about 10^{-13} torr have already been obtained in small glass systems by partially immersing them in liquid helium. From a study of the adsorption process J. P. Hobson of the National Research Council of Canada has estimated that a half-liter glass sphere evacuated to a pressure of 5×10^{-10} torr by normal methods and then completely immersed in liquid helium should reach about 10^{-30} torr. For better or worse there is no pressure gauge capable of checking this pressure. The calculations indicate, however, that it should be possible, at least in small systems, to produce pressures in the laboratory as low as those thought to exist in interplanetary space.

At the moment, then, pumping has outstripped measurement. But the gauge too has been improved a great deal beyond where Bayard and Alpert left it. Their hot-wire design becomes more effective when the gauge is placed between the poles of a strong magnet. Instead of moving in straight lines the electrons are forced into spirals, which increases their length of travel and therefore their chance of ionizing gas molecules. Changes in the shape and placement of electrodes have further reduced the X-ray current. Nevertheless, the best heated-filament gauges now generally available are only slightly better than the Bayard-Alpert type: their lower limit is about 10^{-10} torr, as opposed to 5×10^{-10}. Recently James M. Lafferty of the General Electric Research Laboratory has built a hot-wire gauge with a magnetic field, capable of measuring pressures as low as 4×10^{-14} torr. Attached to an electron multiplier, its range has been extended even further.

Measurements at 10^{-10} torr and below can now be made with cold-cathode gauges, in which ionizing electrons are produced by applying a high voltage to a pair of unheated electrodes [*see illustration at top right on page 8*]. Lacking a separate grid, the cold-cathode gauge generates no spurious X-ray current. One version of the instrument developed by one of the authors (Redhead) can measure pressures as low as 10^{-12} torr.

The sensitivity of most ultrahigh-vacuum gauges depends on the composition of the gas being measured. In turn, the mixture of molecular species making up the residual gas in an ultrahigh-vacuum system depends largely on the system involved and the pumping methods. For example, a small glass apparatus exhausted with an ion pump contains chiefly helium from diffusion through the walls. On the other hand, hydrogen and hydrocarbons constitute most of the gas in a metal chamber evacuated by an oil diffusion pump. For many purposes it is more important to know the partial pressure of a particular gas or group of gases

LTRAHIGH-VACUUM REGION

0^{-10}	10^{-11}	10^{-12}
$\times 10^{10}$	4×10^{9}	4×10^{8}
$\times 10^{7}$	5×10^{8}	5×10^{9}
$\times 10^{6}$	3×10^{5}	3×10^{4}

high vacuums, density is measured in terms of the other characteristics shown here.

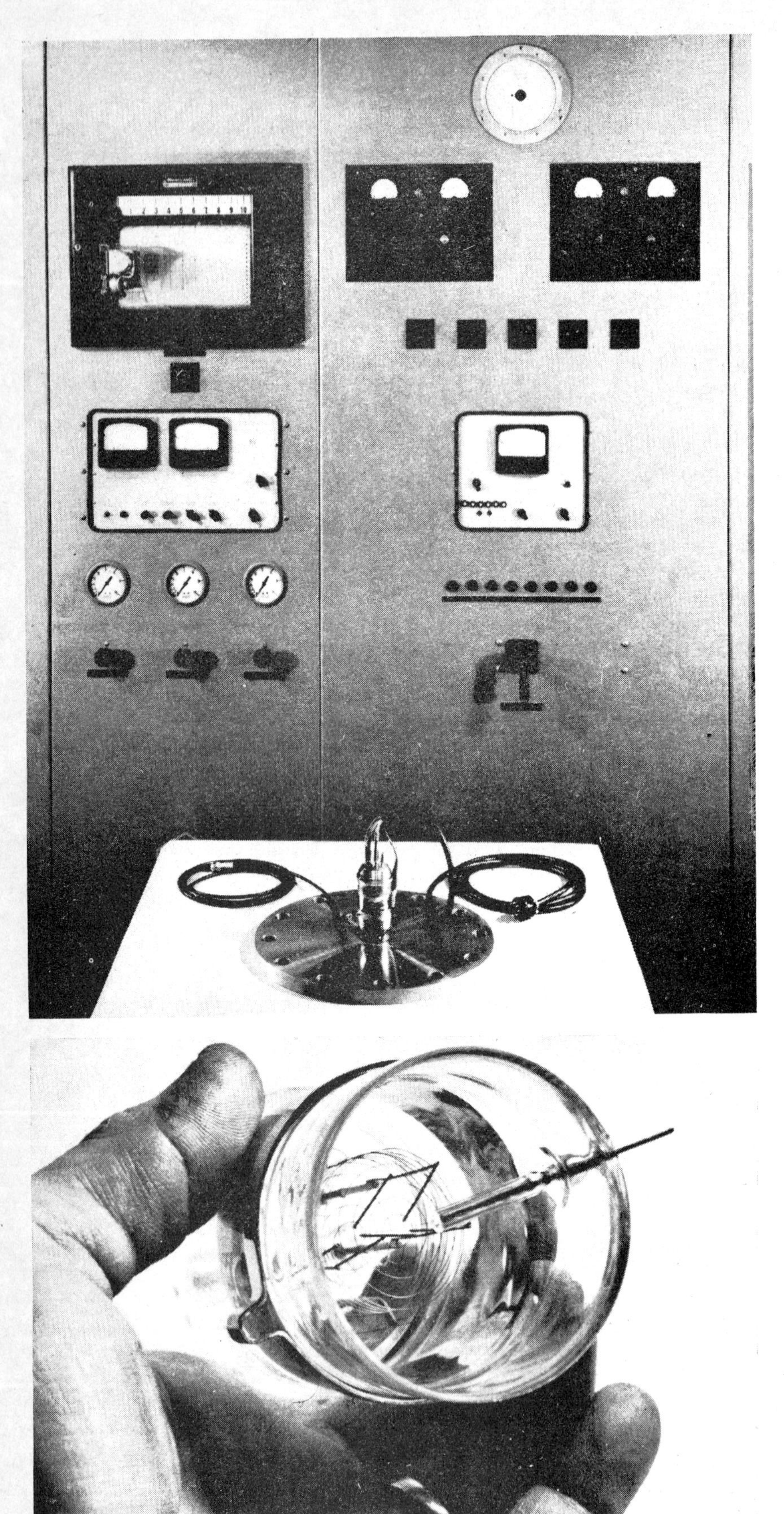

CONTROL PANEL (*top*) for pumps shown on page 3 includes a recorder (*left*) for ultrahigh vacuums and gauges (*right*) for earlier stages and subsidiary pumping systems. On the table in front is the Bayard-Alpert-type gauge to be installed in the chamber. In the close-up of the gauge (*bottom*) the transverse element is the ion collector, the fine spiral wire is the grid and the heavier diagonal wires are the filaments that supply electrons.

than the total pressure. What is needed is a mass spectrometer rather than a total-pressure gauge. Moreover, at very low pressures—below 10^{-12} torr—total-pressure gauges are scarcely less complex than mass spectrometers. Except in a few special cases, total-pressure ionization gauges have little to recommend them in apparatus designed for pressures below 10^{-12} torr.

Mass spectrometers, although they come in a wide variety of forms, all work in fundamentally the same way: the gas to be measured is ionized and the resulting positive ions are made to move at high speed through a magnetic, electric or combined field. Ions with different masses (but the same charge) follow different paths and are thereby distinguished from one another. The current produced by ions that have traced a particular path is an index of the number of gas molecules of a particular mass.

Spectrometers intended for ultrahigh-vacuum work must meet far stricter requirements than are demanded of ordinary analytical instruments. For one thing, they must be extremely sensitive; for another, their operation must not change the composition of the gas in the chamber. This means that they must withstand the preparatory high-temperature baking to which all ultrahigh-vacuum components are subjected.

A number of quite different designs that satisfy the requirements of ultrahigh vacuum are now available. One, developed by W. D. Davis and Thomas A. Vanderslice of the General Electric Research Laboratory, is a modification of a conventional 90-degree sector instrument, in which the ion path is bent through a right angle [*see top illustration on page 9*]. By varying the magnetic field, ions of different mass are directed at the collecting plate, and the corresponding current is measured with the help of an electron multiplier. The instrument can measure partial pressures as low as 10^{-14} torr and can distinguish ions differing in mass by 1 per cent.

Another popular form of spectrometer is the Omegatron, which works like a cyclotron [*see bottom illustration on page 8*]. A radio-frequency field applied at right angles to the magnetic field whirls the ions in curved paths. At each value of the frequency, molecules of a particular mass will follow a smooth spiral to the collector plate, whereas others will be diverted. The Omegatron has a lower limit of about 10^{-11} torr. Because of the shape of the instrument and the strong magnetic field that it requires, the addition of an electron multiplier to extend its range is most difficult.

A device called the massen filter, originally proposed by W. Paul of the University of Bonn in Germany, employs another arrangement of radio-frequency electrodes. It consists of four parallel, equally spaced circular rods carrying both direct-current and radio-frequency voltages [*see bottom illustration on page* 9]. The ion beam is shot down the axis of this structure. For a particular value of the voltages, ions of one mass have a stable path, oscillating closely around the axis and striking the collector plate at the far end. Ions of different masses follow trajectories that make them strike the rods. Changing the voltages changes the mass of the ions collected. With electrodes 25 centimeters long it has been possible to distinguish ions differing in mass by 1 per cent. The addition of an electron multiplier to the massen filter is relatively simple, and the combination has measured partial pressures down to 10^{-15} torr.

For the past few years ultrahigh-vacuum workers have been consolidating their position at about 10^{-12} torr. Today this pressure is obtained and measured consistently, whereas a few years ago it could be reached only by heroic efforts. Now a new cycle is beginning. Present developments in the design of gauges promise to carry them into the range of 10^{-18} torr total pressure. Improvement in pumping can be expected to follow. In fact, as has been indicated, cryogenic techniques may already be capable of producing pressures as low as those in interplanetary space.

ULTRAHIGH-VACUUM SYSTEM made by NRC Equipment Corporation of Newton, Mass., is tested. Oil diffusion pumps (*bottom*) evacuate the chamber on the table to 10^{-9} torr. A furnace (*visible at top right*) swings over the chamber to bake it before experiments.

The Authors

H. A. STEINHERZ and P. A. REDHEAD are respectively manager of the Engineering and Development departments of the NRC Equipment Corporation in Newton, Mass., and a member of the Electron Physics Group of the National Research Council of Canada. Steinherz, who is also an instructor in vacuum technology at Boston University, acquired a B.S. in physics at Yale University in 1947. During the next four years he worked for the Westinghouse Electric Company while studying physics and mathematics at the University of Pittsburgh. From 1951 to 1955 Steinherz was a materials engineer with the General Electric Company. He joined the NRC Equipment Corporation in 1955. Redhead, born in England in 1924, took a B.A. in physics at the University of Cambridge in 1944. He was engaged in research on proximity fuzes and microwave tubes for the British Admiralty until 1947, at which time he went to the National Research Council of Canada.

Bibliography

THE INTERPLAY OF ELECTRONICS AND VACUUM TECHNOLOGY. J. M. Lafferty and T. A. Vanderslice in *Proceedings of the IRE,* Vol. 49, No. 7, pages 1136–1154; July, 1961.

PRODUCTION AND MEASUREMENT OF ULTRAHIGH VACUUM. D. Alpert in *Handbuch der Physik,* Vol. 12, pages 609–663; 1958.

THE PRODUCTION AND MEASUREMENT OF ULTRAHIGH VACUUM (10^{-8}–10^{-13} mm Hg). P. A. Redhead in *Fifth National Symposium on Vacuum Technology Transactions,* pages 148–152; 1959.

SCIENTIFIC FOUNDATIONS OF VACUUM TECHNIQUE. Saul Dushman. Edited by J. M. Lafferty. John Wiley & Sons, Inc., 1962.

VACUUM EQUIPMENT AND TECHNIQUES. Andrew Guthrie and Raymond K. Wakerling. McGraw-Hill Book Company, Inc., 1949.

High-Speed Tube Transportation

In which a system is proposed that would carry passengers from Boston to Washington in 90 minutes. Its vehicles would travel at speeds up to 500 miles an hour through dual evacuated tubes

by L. K. Edwards

Originally published in Sci. Amer., 30–40 (August 1965).

Contemporary technology can orbit an astronaut and send photographs back from Mars, but so far it has not been able to cope with the mundane problem of mass transportation. Proliferating superhighways are clogged as soon as they are built. Air travel is inefficient for short trips, and some air lanes are crowded to the danger point. Railroads move passengers no faster than they did 50 years ago, and they lose money in the process. Is it possible that the techniques and methods of the aerospace industry, and in particular the special approach known as "systems engineering," could be brought to bear on the vexing tasks of moving passengers between the country's closely spaced metropolitan centers and getting commuters to work? This article will describe the evolution, findings and potentialities of one effort to accomplish that end.

The effort began nearly two years ago when, as an engineer in the Space Systems Division of the Lockheed Missiles and Space Company at Sunnyvale, Calif., the writer suggested: Why not develop airplane-speed surface transportation? Lockheed authorized a small group of us to consider the proposition. Our investigation was brought to a focus last fall by the Federal Government's desire to attack the problem of rail transportation in the "Northeast Corridor," the densely urbanized strip along the East Coast from Boston through New York to Washington. After conversations with people in the Northeast Corridor Transportation Project of the Department of Commerce and in the School of Engineering at the Massachusetts Institute of Technology, which is making technical studies for the project, we have evolved two specific proposals. One is a "Corridor" system to provide intercity passenger service, downtown-to-downtown, for several thousand passengers an hour. The other, an "Urban" system, is a commuter network. (Early this year Lockheed decided to concentrate on projects more closely related to its traditional activities; I obtained an option to acquire the company's interest in the transportation project and have since been proceeding on my own responsibility, although with

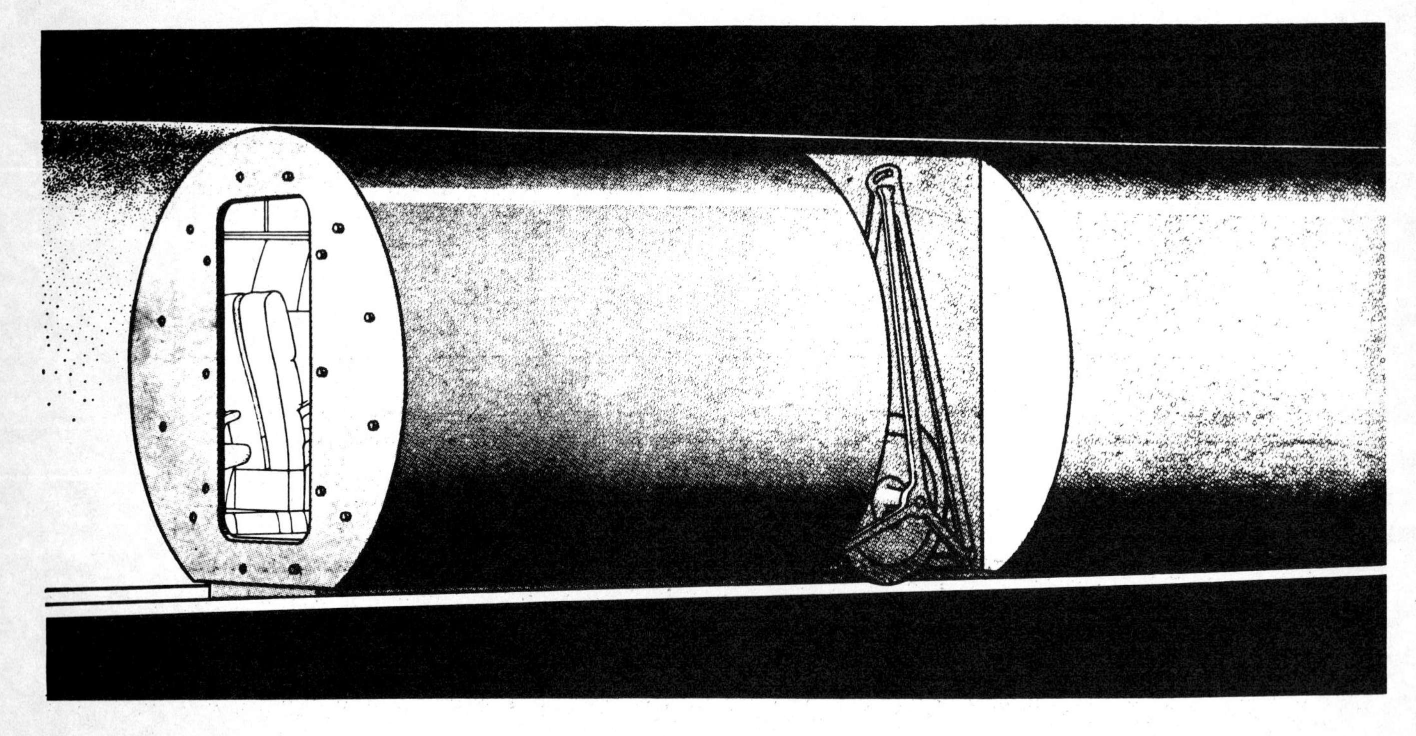

"CORRIDOR" TUBE TRAIN is made up of cylindrical cars 65 feet long running on steel tracks welded to the floor of an underground tube. Each car has four steel wheels fitted with roller bearings. Entrance and exit doors, on opposite sides of the cars, are pres-

some valuable technical support from Lockheed personnel.)

The general elements of the Corridor design emerged rather quickly from a logical sequence based on common sense and a few simple calculations. In retrospect, at least, the logic seems inexorable:

1. To compete with foreseeable improvements in air travel and its associated surface connections, the average speed along the Corridor must be at least 200 miles an hour. Time must be allowed for intermediate stops in addition to those in Boston, New York and Washington, so the cruising speed must be increased accordingly—to something over 400 miles an hour.

2. Such high speeds are unthinkable unless one protects the train from ice and from objects falling or thrown into its path. The vehicle should therefore travel through a continuous enclosure—a tube.

3. Drag forces due to air resistance are substantial at present-day railroad speeds; they increase with the square of the velocity, and they would be prohibitive in the case of aircraft speeds in air at sea-level density. Therefore most of the air should be pumped out of the tube.

4. Given external power plants to exhaust the tube, why not make these the sole source of propulsive power for the train? It is only necessary to admit air at atmospheric pressure behind the train to accelerate it through the evacuated tube; similar pneumatic effects can decelerate the train to a stop.

5. For heavy, continuous traffic a pair of tubes is needed. Convenience of installation and maintenance suggest that they should be installed side by side. They can then be interconnected with controllable valves to improve the efficiency of airflow.

6. At ground level or elevated above it the tubes would be an unsightly nuisance. Moreover, a reasonable degree of straightness would call for large numbers of bridges or tunnels. (The high-speed Tokaido railroad between Tokyo and Osaka in Japan, with a design speed of 150 miles per hour, includes some 40 miles of tunnels—an eighth of its total length.) In the central districts of major cities there is no alternative to tunneling. One should therefore consider putting the tubes in a tunnel all the way from Boston to Washington—a little more than 400 miles.

At this point a practical person might well call a halt, pointing out that anyone with any sense knows that 400 miles of tunnel would be prohibitively expensive. The systems engineer is trained to ask, "Would it?" and to study the pros and cons before either discarding or accepting the idea. On investigation we found that the tunnel, at $4 to $5 million per mile including the dual tubes and valves, may well be less expensive than any other high-speed solution to the Corridor problem. (According to an authoritative estimate it would cost some $6 million per mile to duplicate even the largely conventional Tokaido line in the Northeast Corridor.)

Several factors make the tunnel seem particularly attractive. Because pneumatic propulsion permits tubes of small diameter, the tunnel's cross-section area need be only about 270 square feet, smaller than some of the water tunnels supplying New York, only two-thirds as large as a two-track subway tunnel and only a third as large as a standard railway tunnel. Impressive new machinery has been developed for automatic tunneling through hard rock, and a Corridor tunnel can be dug in bedrock most of the way. In the case of a tunnel, moreover, heavy right-of-way costs and real estate taxes would presumably be replaced by nominal payments for easements. Finally, at depths below 50 feet the temperature does not fall below the freezing point of water; it remains constant all year; there is no weathering. Transportation tunnels typically remain in use for at least a century, whereas most surface-transportation facilities survive only a few decades without substantial renovation.

We therefore decided in favor of the tunnel concept. Mounting the tubes in the tunnel calls for one of the few real innovations we considered necessary for a genuinely high-speed system. The maximum speed attainable in surface transportation is heavily influenced by the smoothness of the surface traversed,

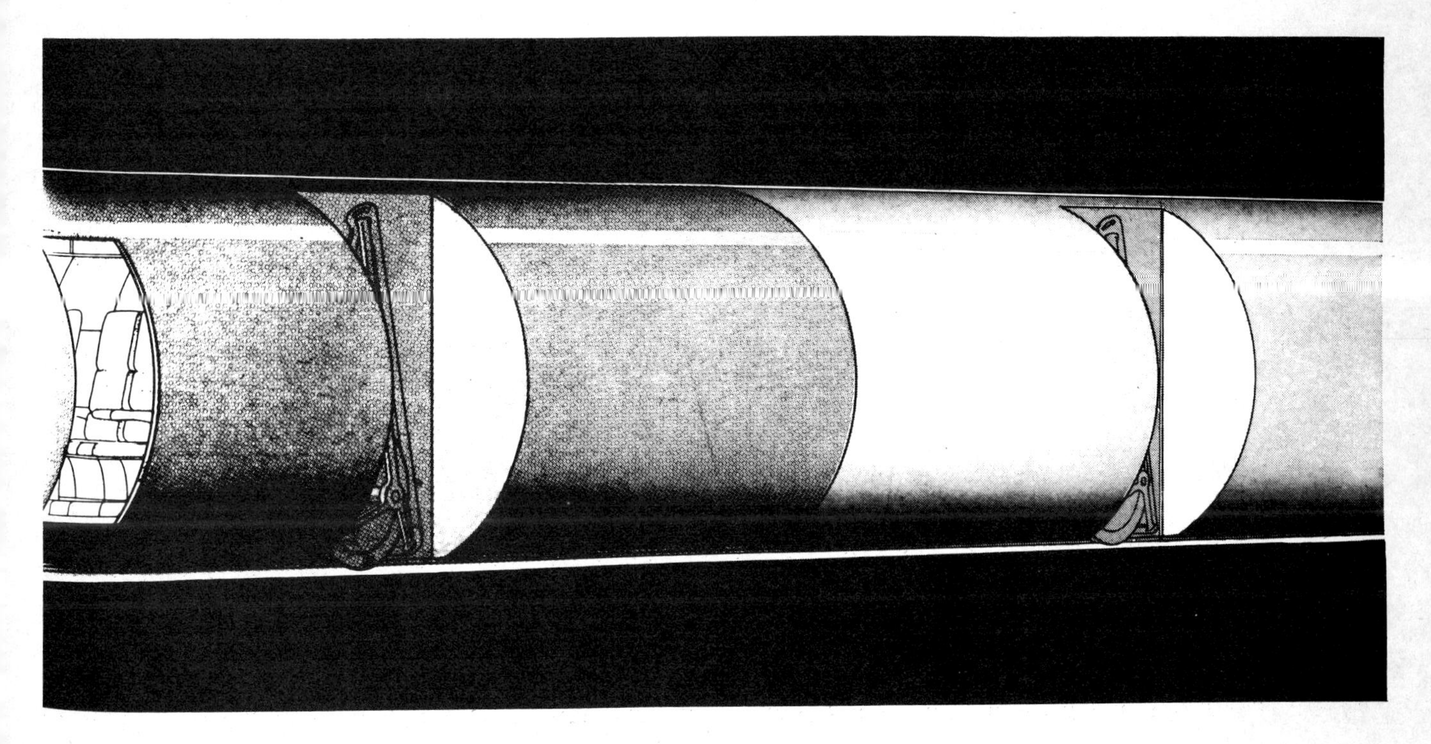

sure-tight sliding panels that retract into the ceiling. The end of an ordinary car rather than of an actual lead car is exposed at the left in order to show the flat faceplate by which the cars are coupled rigidly. Each car will have room for 64 seated passengers.

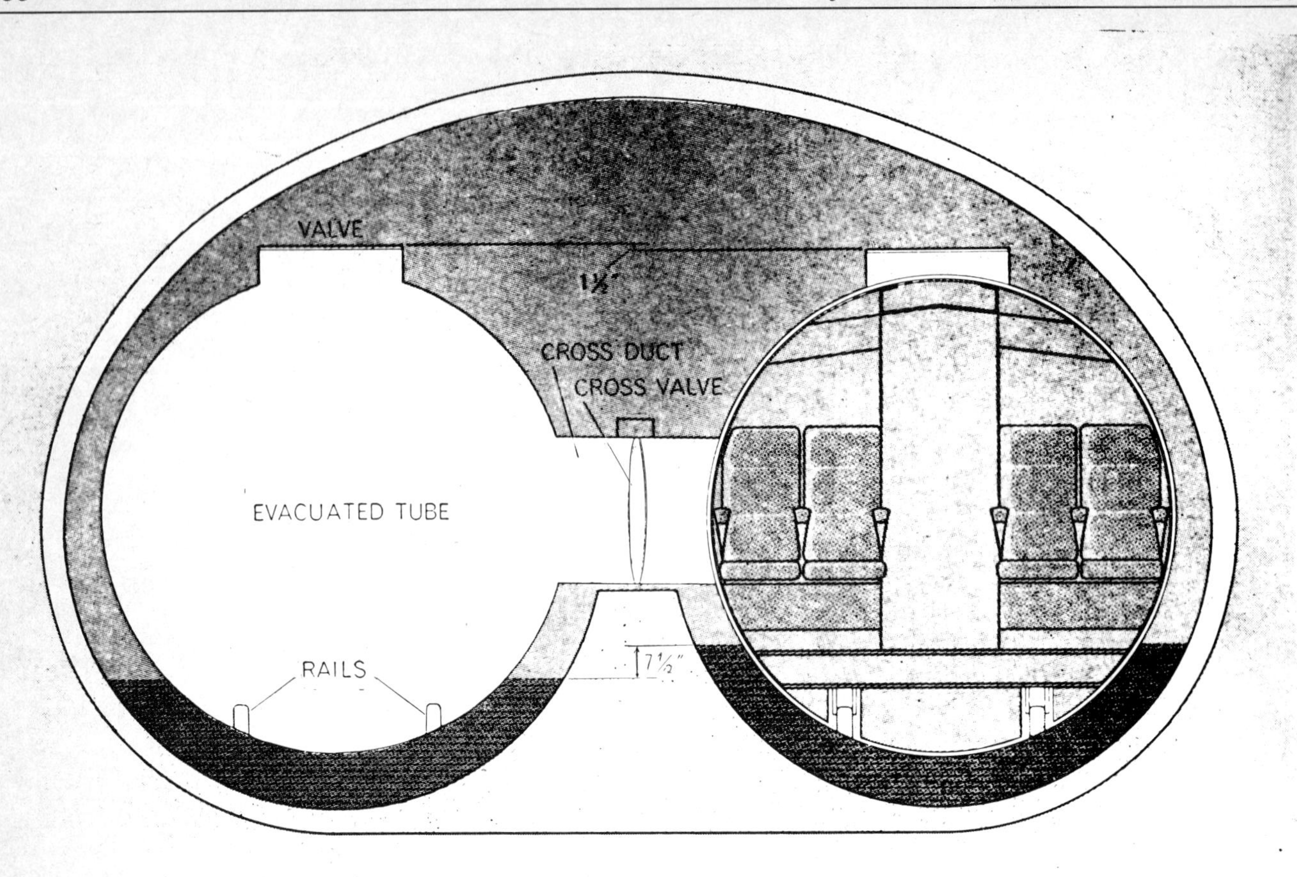

DUAL TUBES are floated in water to ensure a smooth ride. The empty tube will be submerged in the water as shown at left; when a train passes, the tube must displace more water but will only be depressed 1½ inches out of line because the water, being confined in a narrow channel, will itself rise 7½ inches (*right*). The tubes are evacuated by external compressors; the tunnel air is at normal pressure (*color*). Cross valves interconnect the tubes, and other valves, in the top of the tube, open to admit outside air for acceleration.

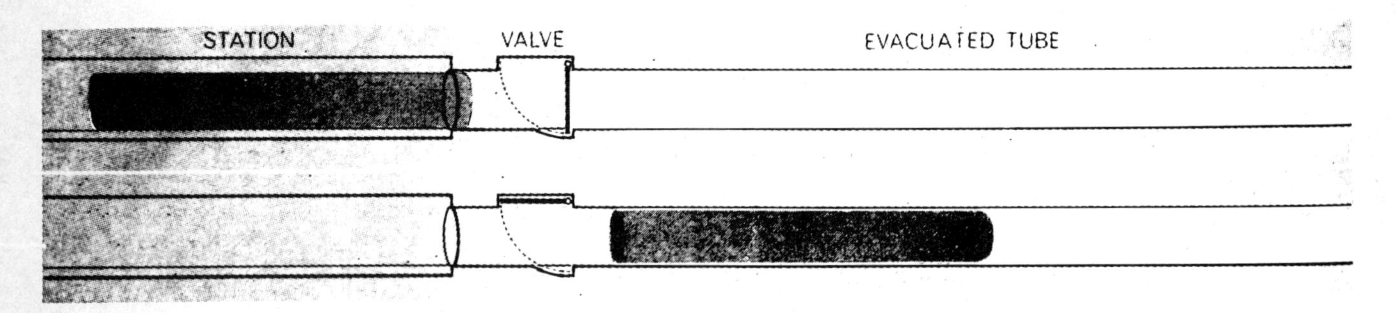

"DYNAMIC AIR LOCK" accomplishes the transition from a standing start to high-speed motion in the evacuated tube. As it waits in the station the train has its nose at the end of the tube, the evacuated portion of which is sealed off by the "start" valve (*top*). The valve swings open (*bottom*) and the train, with a near vacuum in front of it and full pressure behind, is pushed into the tube.

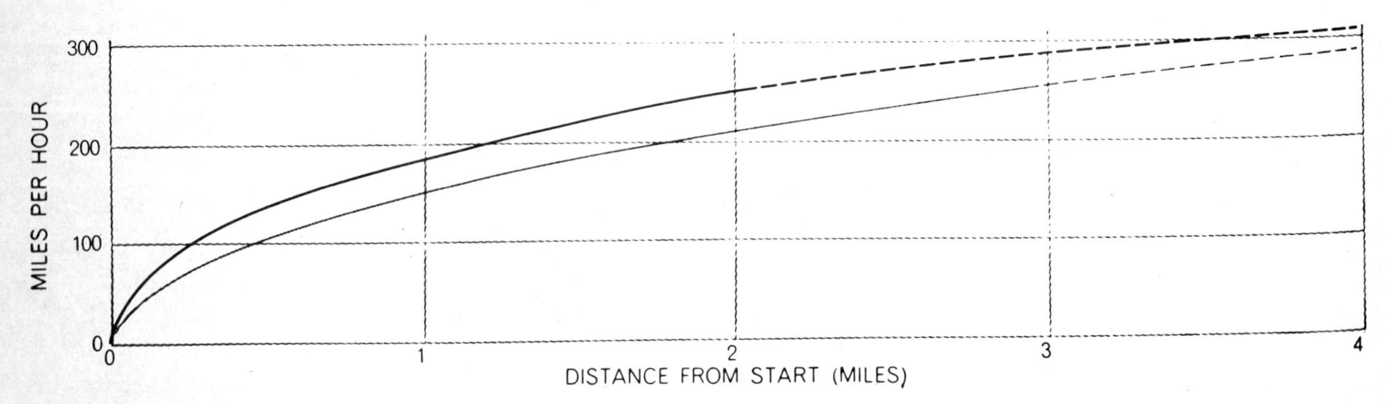

ACCELERATION AND SPEED depend on the train's weight and how much full-pressure air is admitted to the tube. The curves show the acceleration in the first four miles for trains weighing 700,000 pounds (*black*) and a million pounds (*gray*). The unbroken portions of the curves show how many "miles of air" are used for acceleration, or how far the train is when the start valve is closed.

and the smoothness requirement becomes more stringent as speeds increase. The Tokaido railroad's operating speed has been significantly limited, for example, by uneven settling of the roadbed, although it was specially designed. "Smoothness" has two aspects: the initial, or no-load, condition and uniform "compliance," or flexing, under load. The Stephensons learned about this when they first built railroads across England in the 1840's. In spite of expensive and time-consuming efforts to provide a level, solid support under every yard of rail, they found the smoothest ride of all was achieved where the tracks were virtually floated across a bog so soft that a man could hardly stand on it!

In the tube-transportation system we therefore propose to float the tubes in water throughout the level portions of the path. In the no-load condition the tube will be submerged in the water to a depth of about 20 inches [*at left in top illustration on opposite page*]. Regardless of small shifts in the supporting earth, the rigidity of the tube and the self-leveling of the water will keep the alignment of the tube itself undisturbed. When a train comes along, of course, the tube will become heavier and must sink deeper into the water—nine inches deeper in the case of a typical train. The tube will not be seriously deformed by this further immersion, however. Since the water supporting each tube will be constrained in a trough, it will not be able to "seek its own level" under the speeding train; instead the water level will rise 7½ inches and the tube will actually settle only 1½ inches [*at right in top illustration on opposite page*].

This "dynamic flotation" in water can provide the ultimate in uniform compliance under load. It should also have advantages in shock absorption and sound deadening. It does raise a number of questions about the longitudinal water movements and stresses induced in the tube when the train passes and about corrosion, but these problems seem either to be inconsequential or to have ready solutions.

What should the train run on? For half a century patents have been granted on various ways of supporting a vehicle without wheels, and most "advanced" land-transportation concepts incorporate one or another of these schemes. Our investigation confirmed what railway engineers have known all along: with roller bearings a steel-rimmed wheel running on a steel rail is amazingly efficient. It takes only about two pounds of tractive force per ton of railroad train weight to overcome rolling friction—almost regardless of speed. Few airplanes can cruise with less than 130 pounds of thrust per ton of the plane's weight, of which half is chargeable to producing lift. In other words, the wheel-on-rail is more than 30 times more efficient than the airplane in this lift-to-drag relation; helicopters and "ground effect," or hovering, machines tend to be less efficient than airplanes.

There are other advantages to wheels. They avoid the need for an air blower or other active lifting device aboard the vehicle. They position the vehicle in the tube more precisely than other systems can. And it is easy to gear shafts to the wheel axles as a source of auxiliary power for air conditioning and other requirements.

Can wheels run safely at very high speeds? The answer is yes. Specially built "automobiles" have exceeded 500 miles an hour in record runs on wheels with standard roller bearings; the bearings showed no signs of heating or incipient breakdown. An individual bearing on the tube train might suddenly "freeze," but installing the bearings in pairs would allow for that eventuality; moreover, any wheel experiencing such a partial failure could be immediately and automatically lifted off the track for the remainder of the trip. Will the wheel rim fly apart? No, the circumferential stresses are less than those attained in the compressors of some turbojet engines. The wheels can be tested from time to time by "overspeeding" in a test pit, but in general trouble is unlikely—particularly if the wheels are not abused by having to transmit heavy propulsive and braking forces.

Do wheels not require springs? Hardly any, if the roadbed is really smooth. "Consider the billiard ball," an M.I.T. professor once told his class. "It is a hard object rolling on a hard surface. But it does not need springs because the surface is smooth." Aerospace engineers have learned, moreover, that springs and shock absorbers often create more dynamic problems than they solve. (The tube train would, however, have rubber bushings at several points in the wheel mounts to absorb slight shocks and deaden noise.) The lubricants required for wheels will of course have to function in a low-pressure environment, but space technology has solutions available for any problems that arise. The final decision was to use wheels, old-fashioned though they may seem—double-flanged wheels running on continuous steel rails welded to the floor of the tubes.

The passenger car itself is remarkably simple: a 65-foot-long thin-walled cylinder with wells for four wheels [*see illustration on pages 30 and 31*]. The loading on each wheel is well within present railroad practice. The coupling between cars is no more than a strong, flat faceplate with a number of push-button-operated bolts and with no provision at all for flexing. The rigid connection between cars assures pressure integrity throughout the train and keeps objects from falling to the track; it will also avoid the familiar car-to-car swaying motions, which get much worse as the speed increases. Rigid coupling is possible because the track is so straight: the sharpest curve through which the train must turn will have about a five-mile radius, and even this will only be at reduced speeds. With nominal loads at the wheel flanges, the entire train will flex to conform to such gentle curves with much less induced stress than is applied to an airplane wing.

Construction of the cars will be considerably simplified by the nature of the stresses that will be encountered. The greatest loads will result from the maintenance of sea-level pressure inside the car as it moves through a low-pressure environment. These loads produce "hoop tension" in the circular wall, which is ideal. Even the longitudinal stresses turn out to be primarily tension rather than compression, since the train will be accelerated by low pressure in front of it and decelerated by low pressure behind it. The shell can therefore be sheet metal of moderate gauge, with little bracing except at the doors and wheel wells.

At each end of the train we propose to place a special "lead car." This car will have at its forward (or rear) end a pressure enclosure with an emergency exit through it and will include a compartment for the "safety attendant," who replaces the conventional engineer or motorman, a battery-powered low-speed motor for emergency and in-station propulsion, batteries and generators, public telephone booths and perhaps lounge space and dressing rooms.

Entering a clean and quiet station that looks something like the lobby of a large office building, travelers will board the train through gates much like the outer doors of a modern elevator, behind which the openings in the train will be positioned. (Passengers will not see the exterior of the train

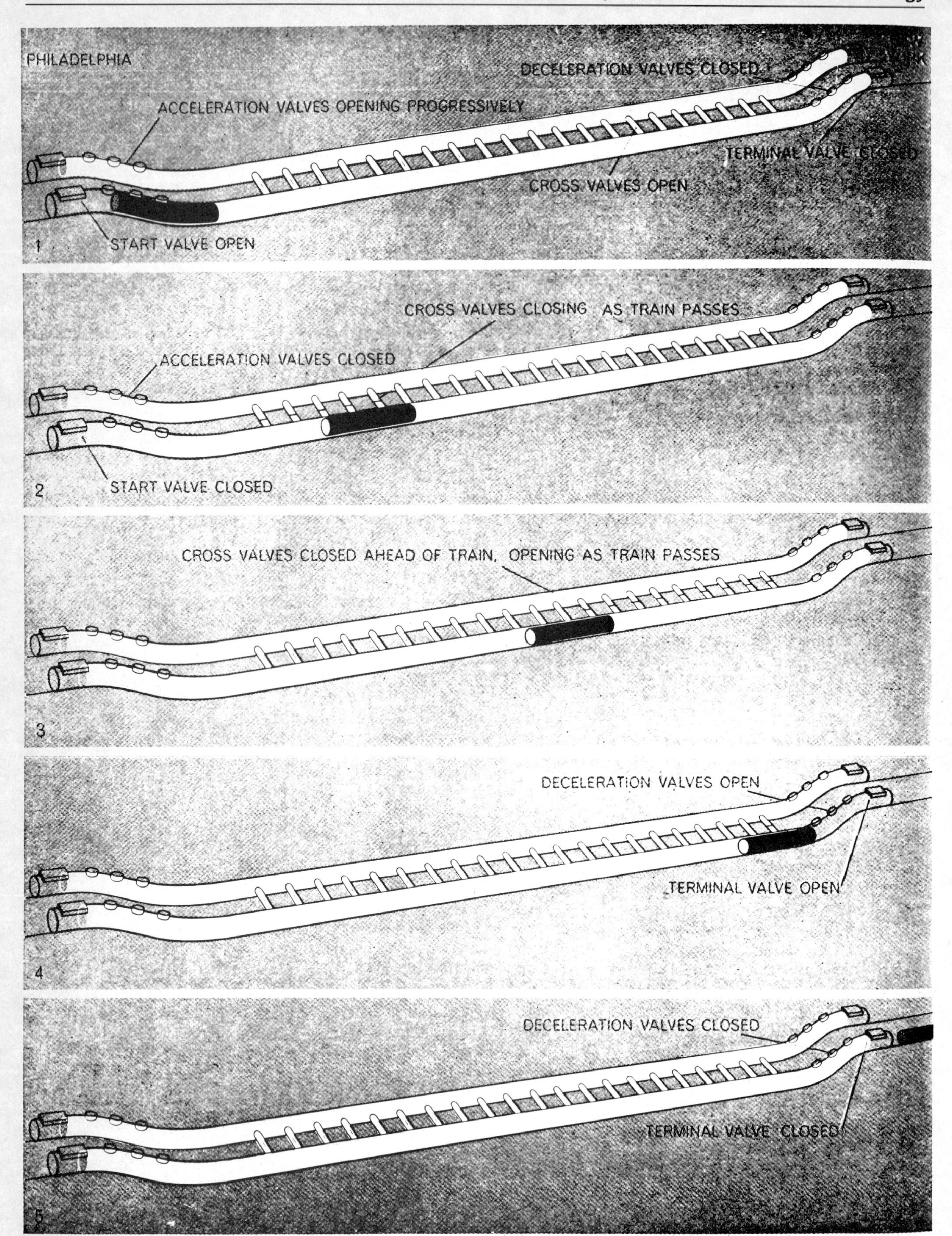

PNEUMATIC PROPULSION would function as shown here on the 85-mile leg from Philadelphia to New York. The train is forced into the tube as shown in the middle illustration on page 32 and is accelerated at maximum rate by the full-pressure air admitted behind it (*1*). After two to three miles the start valve and accelerator valves are closed (*2*); the train continues to accelerate, but at a diminishing rate. The cross ducts close as the train passes in order to contain the air behind it. After a while the train begins to compress the residual air in the tube ahead of it and is decelerated gradually (*3*). When the air ahead of the train reaches atmospheric pressure, the deceleration and terminal valves are opened (*4*). Each of them closes again as the train passes (*5*).

any more than a passenger sees the exterior of an elevator.) The openings in the train will be pressure-tight curved segments that retract into the top of the car, like the doors on some present-day airplanes. The entrance and exit will be on opposite sides.

The car's nine-foot-six-inch inside diameter will allow two-plus-two seating for 64 passengers with spacing comparable to that on luxury airliners. We have tentatively provided luggage space both above and below the seats in the belief that most passengers prefer to keep their luggage with them. There will, of course, be no windows. (There is growing evidence, incidentally, that looking out of windows while traveling on the surface at high speeds would make passengers dizzy, if not actually sick.)

The principle of pneumatic propulsion is obviously not new. Pneumatic delivery tubes were once common in department stores and are still in use in libraries and other institutions. Even pneumatic passenger transportation has a history. One of the oldest such systems on record was built in Ireland during the 1840's. The train was propelled by a piston that extended from the bottom of a car into an externally pumped tube, 15 inches in diameter, laid between the tracks. The train displayed speed, power, efficiency, smoothness, cleanliness and silence far superior to the best steam railroads of the day, and similar installations were soon built in England. All were abandoned within a few years, however, because of one tremendous trifle: a leather flap, which ran the full length of the tube to seal the slot through which the train was connected to the piston, could not withstand the onslaught of the elements and of rats!

New York City's first subway was a pneumatic one, built in 1870 by a proprietor and editor of *Scientific American*, Alfred Ely Beach. In an effort to get a franchise for a subway running the length of Manhattan, he had a one-block tunnel dug under Broadway from Warren Street to Murray Street, using a cylindrical tunneling shield of his own design. A blower propelled the 18-passenger car through the tunnel in one direction and then reversed to "suck" it back. The subway ride was a popular attraction for a year, but Beach never got his franchise; the elevated railroads went up instead.

There is certainly no more attractive way to propel a heavy train at high speed than with the most available medium of all: free air. Sea-level atmospheric pressure applied to the 10,000-square-inch cross section of the train yields a propulsive force of 140,000 pounds, or 70 tons. At only 50 miles an hour this matches the pulling power of five large steam locomotives. Better yet, air can continue to push with this force at speeds as high as 200 miles an hour—at which point the effective power level is 70,000 horsepower. Yet the only power plant is a bank of perhaps four 2,500-horsepower electric motors driving air compressors at one or two points outside the tube. The explanation of this apparent inconsistency is that each train is accelerated for only about two minutes—whereas the external power plants work more or less continuously, storing the energy in the evacuated tube to be drawn on when needed. This is the principle of the pneumatic catapults that launch missiles and torpedoes; a catapult's low-horsepower, long-duty-cycle engine and compressor store a great amount of energy for release in seconds.

In a sense, then, the tube train will be the world's largest catapult, but with a big difference: instead of being almost instantaneous, the acceleration will be gradual. In fact, the rate of both starting and stopping accelerations can be kept below those of jet aircraft and even below those of a compact car. No seat belts will be needed; even standees will not experience accelerations appreciably higher than in present rapid-transit systems. It is the rate of change in acceleration, something called "jerk," that is most critical to comfort, and "jerk" is low in the tube train because the acceleration continues for a comparatively long time.

We already have some experimental data on how a cylindrical object behaves when it is accelerated to high speed in an evacuated tube. For the past five years the Army Research Office in Durham, N.C., in conjunction with Duke University, has been working on a "vac-

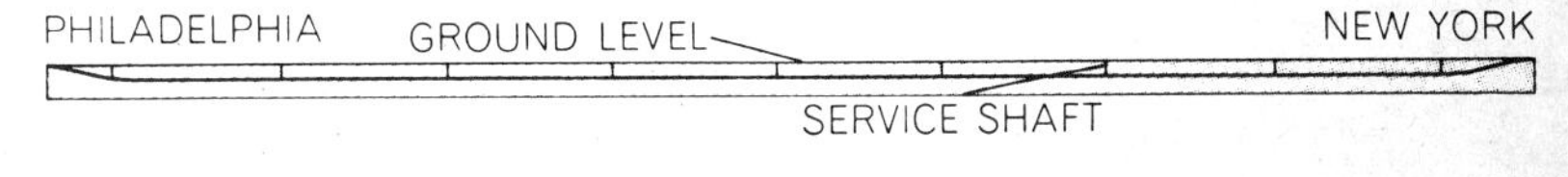

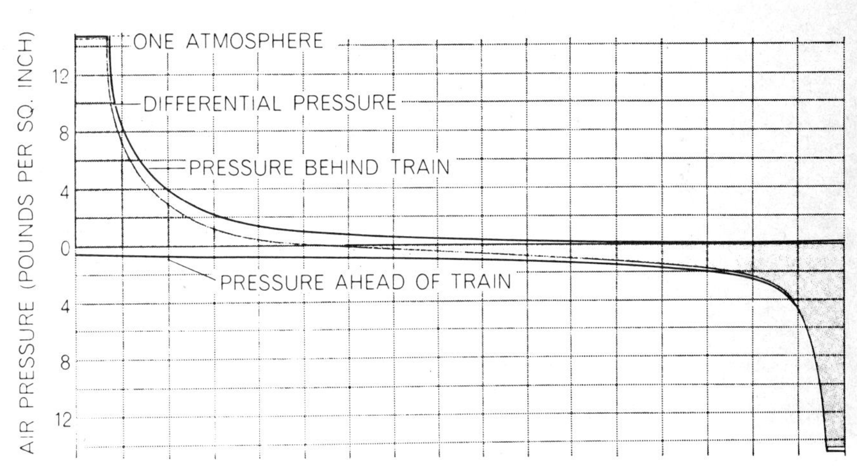

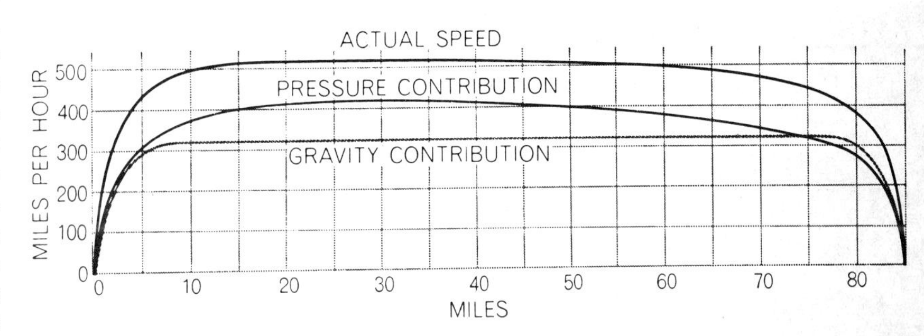

GRAVITY AND VACUUM combine to propel a million-pound train from Philadelphia to New York in 13 minutes. The tube profile has a maximum slope of 15 percent at the ends (*top*). The middle diagram shows how the pressures behind and ahead of the train vary along the route; the difference between the fore and aft pressures accelerates and then decelerates the train. The bottom diagram shows the speed contributions of vacuum and gravity, which add (on an energy basis) to produce a peak speed of about 500 miles an hour.

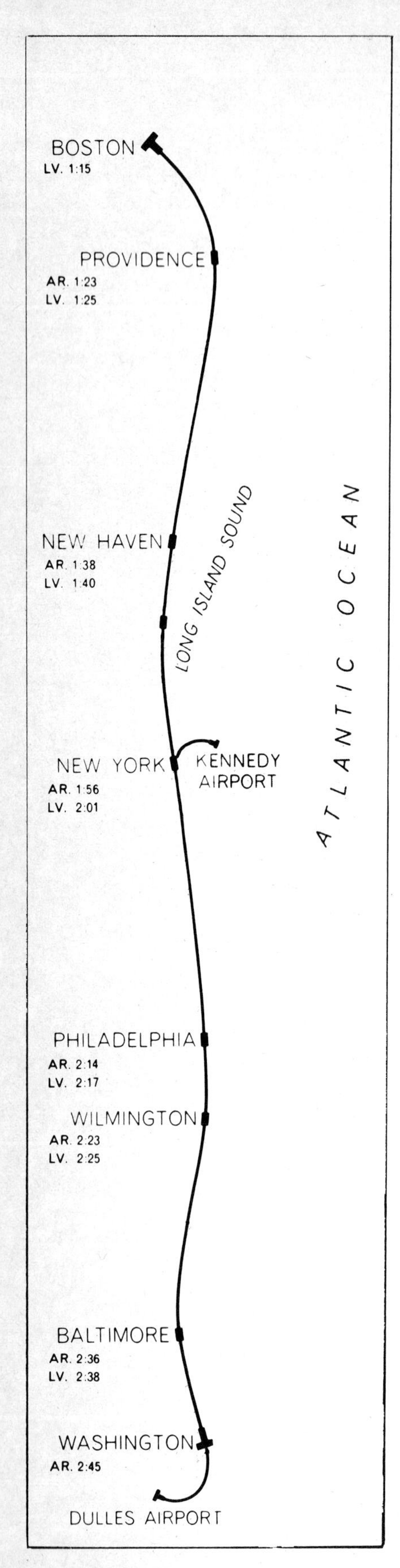

"NORTHEAST CORRIDOR" route, with spurs to airports at New York and Washington, is shown, along with the schedule for the 1:15 out of Boston. Site of one stop in Connecticut remains to be decided.

uum-air missile boost system" designed to get missiles or satellite-launching vehicles moving fast before the first-stage rockets are fired. Tests have been conducted with metal cylinders in tubes ranging from 1¼ inches to six inches in diameter; a six-inch, half-pound hollow projectile, for example, is accelerated to more than 750 miles an hour within the length of a 100-foot tube. The tests have shown that the compressibility of air, which degrades aircraft performance at high speed, actually assists in the operation of such a tube catapult. Another encouraging finding is that there appears to be an inherent centering effect that tends to maintain the projectile's stability and keep it away from the walls of the tube.

Although our tube train is driven by air pressure, no complicated air-lock arrangement is necessary for station stops. While the train is in a station, with its nose at the end of the tube, a large valve closes off the evacuated portion of the tube [*see middle illustration on page 33*]. Passengers leave and enter freely and then the car doors are closed pressure-tight. Now the valve close to the end of the tube is opened; atmospheric air forces the train into the near vacuum in the tube. (The radial clearance between the cars and the tube is to be about an inch; our studies indicate that the power loss through this space should be insignificant, but if experience proves otherwise a brushlike seal could be added at each end of the train.) As the train moves down the tube the valve is kept open for a time to allow continued acceleration at maximum rate. When the proper amount of air has been admitted, depending on the weight of the train, the valve is closed. This traps a slug of air in the tube behind the train; it expands, and in doing so it applies further, although decreasing, force to accelerate the train still more. Within the first four miles the train will have reached a speed of about 300 miles an hour.

The air behind the train becomes increasingly attenuated; the very-low-pressure air in the long tube in front of it is steadily compressed. After a while the pressures are balanced fore and aft; then the decreasing pressure behind the train begins to decelerate it. When the air ahead reaches atmospheric pressure, the valves at the destination end are opened; the train is brought to a stop—and the tube behind it has largely been restored to low pressure [*see illustration on page 34*]. If one ignores the losses due to the circulation of air through the tube and to the trifling amount of rolling friction, the train has had a free trip! In general, the magnitude of the compressor's task—which is to restore the initial near vacuum within the tube—is equal to the amount of these two losses. A normal surface vehicle, on the other hand, consumes energy in proportion to its weight in getting up to speed and then wastes that energy in the form of air turbulence and heat at the brakes before coming to a stop. Aircraft are inherently even more wasteful: a jet airliner consumes billions of foot-pounds of energy in climbing to cruising altitude, all of which is wasted by the descent at the end of the trip.

Pneumatic propulsion, then, is an effective and comparatively efficient means of transportation. When we came to deal with the longer and heavier trains required in the Urban system, however, we found that satisfactory acceleration could not be achieved with atmospheric pressure alone. Fortunately another source of propulsive power is readily at hand—and as "free" as air. It is gravity. We decided to take advantage of the principle of the pendulum. As first-year physics students learn, a pendulum, in the course of a single swing, converts potential energy (the energy stored in the suspended mass of the pendulum) into kinetic energy (the energy of the pendulum's motion) and back again into potential energy. This double conversion is accomplished with a lovely 100 percent efficiency except for the tiny losses resulting from air drag and friction in the suspension. That is why a spring weighing two ounces can swing a four-ounce clock-pendulum weight back and forth for eight days; it would take a 300-pound spring to do the same job if the back-and-forth motion were entirely horizontal and wasted all its energy at the end of each stroke.

It was a simple matter to design a pendulum into our tube-train system. With the decision already made to put the tubes in a tunnel, it would add very little to the cost of the tunnel to slope it downward on both sides of each station. Leaving a station, the train is accelerated by the downgrade, and as it approaches the next station it is decelerated by the upgrade—interchanging potential and kinetic energy with the same high efficiency as a clock pendulum. In the case of the Urban system the pendulum effect can contribute more than half of the total energy required by the tube train. The technique affords other important benefits:

1. It brings most of the tunnel down into deep bedrock, where the cost of tunneling—by blasting or by boring—is reduced and incidental earth shifts are minimized; the rock is more homogeneous in consistency and there is less likelihood of water inflow.

2. The nuisance to property owners decreases with depth, so the cost of easements should be lower.

3. A deep tunnel does not interfere with subways, building foundations, utilities or water wells. The only really deep works of man in the Corridor area, as far as we can determine, are two New York City water conduits. (They were dug as deep as 1,100 feet, incidentally, precisely to save money and ensure long life.)

4. The pendulum ride is uniquely comfortable for the passenger. When a vehicle gains speed by coasting freely down a moderate slope, the passenger feels absolutely no sensation of fore-and-aft acceleration; he can stand up, write and pour water without difficulty. The same thing is true as the vehicle loses speed by coasting uphill at the end of the trip. Tube-train riders will feel some vertical acceleration during brief periods when the path changes slope, but at rates well below the criteria for passenger elevators.

The amount of "free speed" available from the pendulum technique varies with the square root of the depth. At 1,000 feet it would be some 175 miles an hour; at 4,000 feet, nearly 350 miles an hour. We propose to have the Corridor trains cruise at 3,500 feet, deriving 325 miles per hour of speed from gravity.

Is that too deep to go? We do not think so. The 12-mile-long Simplon Tunnel bores 7,000 feet below the surface of the Alps. (It has been giving good service for 60 years, incidentally.) Copper mines go down 5,000 feet and diamond mines 10,000 feet. The earth temperature at 3,500 feet in the Northeast Corridor area is unlikely to exceed 100 degrees Fahrenheit; air conditioning, which would have to be installed in any case, will keep the passengers comfortable. In order to construct the tunnel, vertical shafts 15 feet in diameter would be sunk every 10 miles along the route and fitted with high-speed elevators; these could be left in place to serve as emergency escapes in the unlikely event of a train breakdown within a tube. In the event of an emergency, air would be admitted to the tube from the tunnel, which will be at atmospheric pressure, simply by opening valves along the top of the tube.

In the case of the trip from Philadelphia to New York, the two elementary sources of power, gravity and vacuum, would combine to propel the train 85 miles in 13 minutes, or at an average speed of 390 miles per hour, with a peak speed slightly above 500 miles an hour [*see illustration on page 35*]. A single train will make the run from one end of the Corridor to the other in an hour and a half with seven intermediate stops en route, with New York just 45 minutes from either end. Spurs to Dulles Airport outside Washington and Kennedy International Airport outside New York may be desirable.

In our calculations we have provided for a peak capacity of 9,000 passengers an hour. This would be accomplished by running three trains per hour, each with seats for 1,500 people and liberal rush-hour standing room for another 1,500 who would presumably be willing to stand during such short segments as the Philadelphia–New York run. The capacity of the system could be doubled by running six trains per hour. (The Regional Plan Association has predicted that the rush-hour demand for this system would reach 7,000 passengers per hour in the near future and that the system should be able in time to handle 15,000 per hour.)

A single train will be able to make a round trip in about four hours, thanks to the high speed and the fast "turnaround" (which will not involve turning but merely shunting from one track to another and reversing). The three-trains-per-hour service could therefore be maintained with only 12 trains, plus some spare cars. This would mean a very low aggregate cost for rolling stock, crews and maintenance personnel, although individual employees would be highly trained and could be well paid.

In addition to carrying passengers the tube trains are designed to handle express cargo. This will be done during off-hours, providing ballast for trains that would otherwise accelerate too fast when the passenger load is light. Express cars will have the same length and coupling faceplates as passenger cars but will be open, like hopper cars, to receive standardized, pressure-tight modules preloaded with mail, newspapers and other high-grade cargo; the

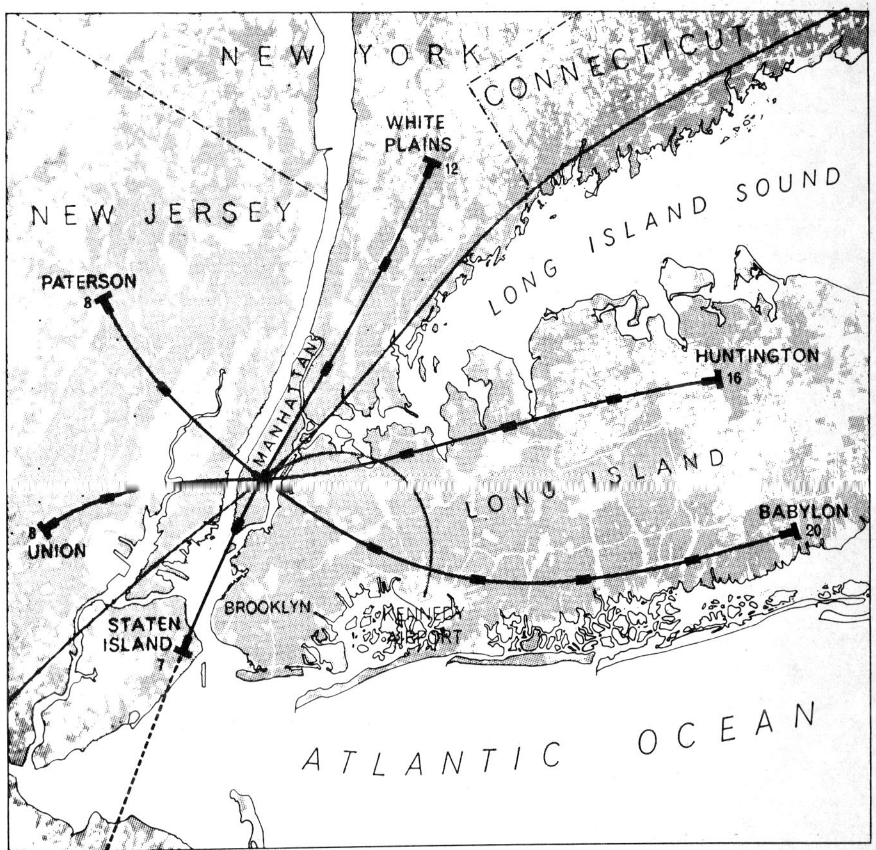

"URBAN" SYSTEM for New York City has six lines radiating from midtown Manhattan. Tentative terminals for the lines and the elapsed time to each from Manhattan allowing for stops en route as indicated, are shown on the map, along with built-up areas (*color*) served by the system. The main line and spur of the Corridor system are shown in gray.

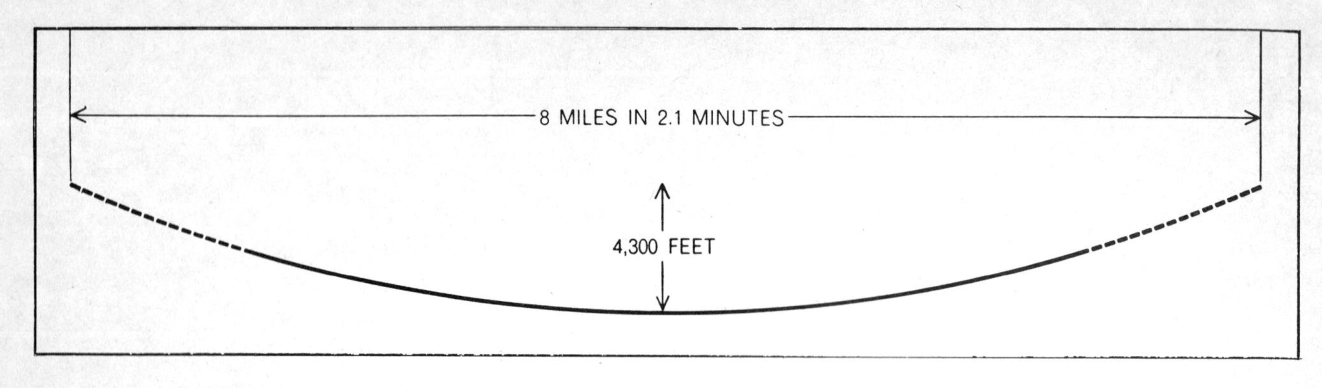

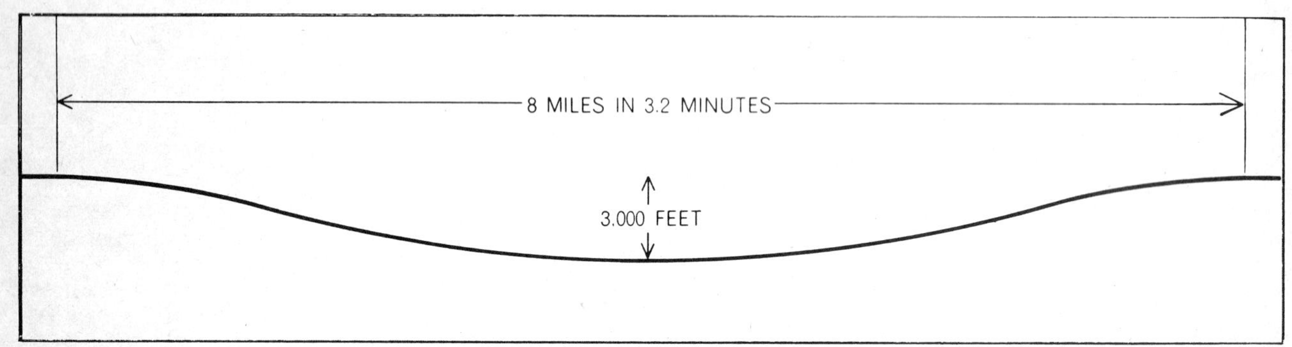

EIGHT-MILE TRIP could be accomplished by a pure passenger pendulum in only 2.1 minutes (*top*). The route, comparable to the arc traced by the end of a 10-mile-long pendulum, would descend 4,300 feet below ground level. The terminal segments of the pendulum route (*broken lines*) would be too steep for a practical system, however. In the proposed Urban system the pendulum path is flattened at each end (*bottom*). The depth of the tunnel becomes 3,000 feet. With pneumatic propulsion, the elapsed time is 3.2 minutes.

modules will be readily transferable to trucks. Any number of these express cars can be added to trains as the passenger load fluctuates.

As we were refining the Corridor system we were urged by several concerned individuals and agencies to see what might be done with a similar approach to help solve the commuter problem, particularly in metropolitan New York. The resulting Urban network is based on the same techniques as the Corridor system. Each Urban train would carry 2,600 passengers seated and 3,400 standing; the size of the seats, the space allowance for standees and the proportion of seats to standees would all be superior to present-day rapid-transit systems. If the train stops about once every eight miles, it can average two miles a minute—including the time spent in stations.

Urban-system lines might radiate in six directions from midtown New York [*see illustration on preceding page*]. Each line would have two tubes, but

TUBE CATAPULT has been tested by the Army Research Office at Durham, N.C., and Duke University. The 100-foot-long tube is evacuated by a compressor. Cylindrical metal projectiles, released at the far end, are forced down the tube by atmospheric pressure, reaching speeds as high as 750 miles an hour. Tests have shown that there is a tendency for the projectiles to remain centered in the tube.

there would be no need for cross-valving. Since the Urban tunnels would slope down and then up again, with no level run as in the Corridor tunnel, floating the tubes in water would not work; instead they can be suspended on springs. With a train every 10 minutes, the capacity of a single tube would be 36,000 passengers per hour; the frequency could be doubled to handle 72,000 per hour.

Each train would be half a mile long. This unusual length has some interesting advantages. At each suburban station there would be a long, narrow parking lot directly above the train, with numerous access roads passing through it. This would decrease local congestion and allow each passenger to park or alight from his bus near the section of the train that would put him closest to his midtown or downtown destination. In the city, passengers would be released into a larger area than that of a conventional station, again relieving congestion.

Most commuters drive or take a bus to their suburban stations. The Urban system would require that they drive a few miles farther. In return, however, they would find good parking, reliable round-the-clock service and a speed that would more than make up for the longer ride to the station and that would be far superior to expressway speed even under the best driving conditions. The system would also lend itself particularly well to feeder-bus service.

As now conceived, the New York Urban commutation system would consist of 140 miles of dual tubes and 19 stations and would require 18 trains. The original cost would be about $1.25 billion. This is about 25 percent more than the amount that, according to one estimate, would be required for thorough modernization of the area's commuter railroads, which are under continual criticism for providing poor service, are losing money on passengers and are trying to get out of the passenger business. The Urban system might well supplant these railroads; it would complement, rather than supplant, the existing subways. A similar Urban system, scaled down somewhat, might make sense for Boston (Route 128 to downtown in five minutes), San Francisco (San José to downtown, with stops about every eight miles, in 22 minutes) and other cities.

"CORRIDOR" TUNNEL might be bored through deep bedrock by hydraulic tunneling machines. This machine-drilled segment of a New York water conduit is 12 feet in diameter.

How close does the speed of the gravity-vacuum system come to the limits of what is practical in surface transportation of passengers? Consider a typical eight-mile Urban segment. Imagine, first of all, a hypothetical vehicle built to travel horizontally with unlimited power. It would accelerate until it would have to decelerate again, accelerating and decelerating at three miles per hour per second, the top rate to which passengers are subjected in present rapid-transit practice. (Actually no existing system has the power to maintain such acceleration beyond 30 to 40 miles an hour.) The speed at midpoint would reach 295 miles an hour and the accelerations would be high and continuous. The elapsed time for the eight miles would be 3.2 minutes, which must be about the theoretical limit for horizontal travel unless everyone is strapped into his seat. The power requirements would be prohibitively high.

Now imagine a "passenger pendulum," a gravity train swooping through a 4,300-foot-deep tunnel with no air resistance [*see top illustration on opposite page*]. It would make the swing in only 2.1 minutes without consuming any power at all, and there would be no sensation of horizontal acceleration. The speed at midpoint would be 360 miles per hour. The steep slopes at each end of the trip are not suitable for mass transportation, however.

Finally, consider the proposed gravity-vacuum system with a fully loaded Urban train (6,000 passengers and a gross weight for the train of 1.6 million pounds). Its path will partially parallel the pendulum's but will level off as it enters the stations, so the maximum depth of the tunnel will be only about 3,000 feet [*see top illustration on opposite page*]

Pneumatic acceleration and deceleration will be added, with one mile of full-pressure air applied at each end. The gravity-vacuum combination will impart an "equivalent horsepower" of 275,000 during acceleration, but the actual horsepower requirement will be no more than 7,700. The acceleration during the first and last miles will be two miles per hour per second, two-thirds as high as the continuous acceleration in the hypothetical horizontal vehicle; during the remainder of the trip the acceleration will be negligible. The elapsed time of 3.2 minutes will nevertheless match that of the horizontal trip, with a speed at midpoint of 335 miles an hour.

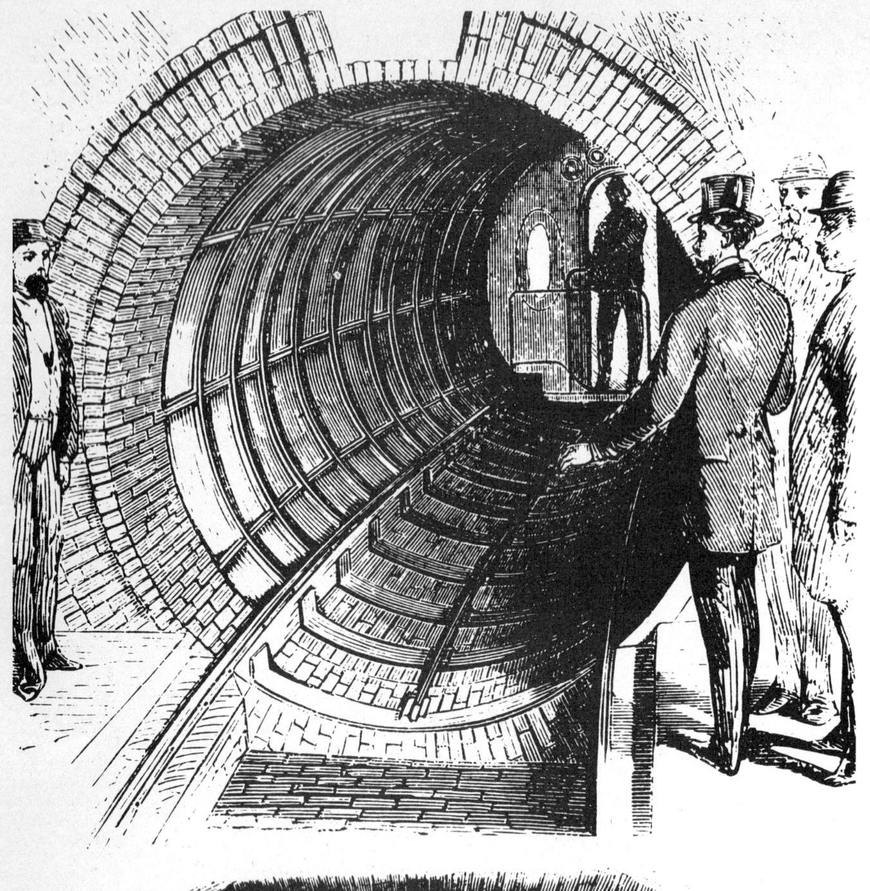

PNEUMATIC TRAIN was constructed under Broadway in New York City in 1870 by Alfred Ely Beach, editor of SCIENTIFIC AMERICAN. These engravings, which illustrated an account in the magazine, show the eight-foot tunnel (*top*) and the "richly upholstered" car (*bottom*).

It seems reasonable to expect, then, that the proposed tube train will cover a moderate distance in an elapsed time that cannot be matched by any other vehicle traveling horizontally and accommodating standing passengers. No advances in propulsion or braking technology can avoid this conclusion, since human comfort is the critical factor.

The system I have proposed seems to offer a number of broad social and economic benefits in addition to convenience for travelers. Its construction would, of course, employ a large number of workers. It would cut down automobile traffic in the heart of a metropolis, thereby reducing air pollution and the intrusion of expressways and garages. (Each tube, with its capacity of 36,000 passengers per hour, would be equivalent to 14 jammed lanes of expressway.) The casualties from grade-crossing, highway and airplane accidents would be reduced. Subsidies to passenger-carrying railroads might be obviated and railroads, free to concentrate on freight, might compete more effectively with trucks. Finally, putting much of the surface transportation underground would beautify the countryside and make it possible to redeem blighted areas of many cities and smaller communities.

How soon could the Corridor or the Urban system be completed? The pace of the project would probably be set not by technological development but by the speed of tunneling; it would normally take perhaps three years from ground-breaking to operational use. With strong leadership, the raising of capital and the necessary legislative processes might take as short a time as one year. If the best "concurrency" techniques of the weapons-system industries were brought to bear, then portions of either system could surely be opened for business on the 100th anniversary of the completion of the first transcontinental railroad. The Golden Spike was driven on May 10, 1869, ushering in a major era of growth in the U.S.

Our forefathers built a great railroad system that is still suitable today for heavy freight, but not for passengers. Will we merely patch up the passenger system and hand it on to our children along with its marginal service, saturated capacity and insatiable appetite for subsidy? Or will we select a new system with a new order of convenience, sound economics and capacity for growth—and do it now, so that we can enjoy it too?